"This book arises from the life-long teaching of a highly regarded educator with topics of essential foundation to readers of interests to signal-processing algorithms. It also contains unique treatments from a respected researcher with an insight one cannot find elsewhere."

K. J. Ray Liu
University of Maryland

"Yao has written an extensive and inclusive book on signal processing, focused on the aspects most relevant to communication and radar systems and based on his teaching and research experience. Beyond being valuable for an advanced course on signal processing, the thoroughness of its mathematical treatment, the inclusion of topics usually not found in textbooks, and the wealth of homework problems make this book an excellent resource for reference and self-study."

Ezio Biglieri
Universitat Pompeu Fabra

Signal Processing Algorithms for Communication and Radar Systems

Based on time-tested course material, this authoritative text examines the key topics, advanced mathematical concepts, and novel analytical tools needed to understand modern communication and radar systems. It covers computational linear algebra theory, VLSI systolic algorithms and designs, practical aspects of chaos theory, and applications in beamforming and array processing, and uses a variety of CDMA codes, as well as acoustic sensing and beamforming algorithms to illustrate key concepts. Classical topics such as spectral analysis are also covered, and each chapter includes a wealth of homework problems. This is an invaluable text for graduate students in electrical and computer engineering, and an essential reference for practitioners in communications and radar engineering.

Kung Yao is a Distinguished Professor Emeritus in the Department of Electrical and Computer Engineering at UCLA, and a Life Fellow of the IEEE. He is the co-author of *Detection and Estimation in Communication and Radar Systems* (Cambridge University Press, 2013).

Signal Processing Algorithms for Communication and Radar Systems

KUNG YAO
University of California, Los Angeles

CAMBRIDGE
UNIVERSITY PRESS

CAMBRIDGE
UNIVERSITY PRESS

University Printing House, Cambridge CB2 8BS, United Kingdom

One Liberty Plaza, 20th Floor, New York, NY 10006, USA

477 Williamstown Road, Port Melbourne, VIC 3207, Australia

314–321, 3rd Floor, Plot 3, Splendor Forum, Jasola District Centre, New Delhi – 110025, India

79 Anson Road, #06–04/06, Singapore 079906

Cambridge University Press is part of the University of Cambridge.

It furthers the University's mission by disseminating knowledge in the pursuit of education, learning, and research at the highest international levels of excellence.

www.cambridge.org
Information on this title: www.cambridge.org/9781108423908
DOI: 10.1017/9781108539159

First published 2019

Printed in the United Kingdom by TJ International Ltd. Padstow Cornwall

A catalogue record for this publication is available from the British Library.

Library of Congress Cataloging-in-Publication Data
Names: Yao, Kung, author.
Title: Signal processing algorithms for communication and radar systems / Kung Yao, University of California, Los Angeles.
Description: Cambridge, United Kingdom ; New York, NY, USA : Cambridge University Press, 2019. | Includes bibliographical references and index.
Identifiers: LCCN 2018039216 | ISBN 9781108423908 (hardback : alk. paper)
Subjects: LCSH: Signal processing. | Algorithms.
Classification: LCC TK5102.9 .Y36525 2019 | DDC 621.382/23–dc23
LC record available at https://lccn.loc.gov/2018039216

ISBN 978-1-108-42390-8 Hardback

My wife
Mary Yu Yao

My children and their spouses
Dr David C. Yao, Dr Molly F. Walsh Yao, Dr David Tien, Erica C. Yao Tien, Roger C. Yao, and Charleen W. Ang Yao

My grandchildren
Isabel Z. Tien, Philip Z. Tien, Vivian J. Yao, Alex Z. Tien, Sullivan J. Yao, Madeline J. Yao, and Andrew J. Yao

Contents

Preface

The materials in this book have been used as the basis for two graduate courses taught in the Signals and Systems Track of the Electrical and Computer Engineering (formerly known as Electrical Engineering) Department of UCLA for many years. Students on these courses were studying communications, telecommunications, signal processing, control, and optimizations fields. Students majoring in electromagnetics and antenna also often took these two courses, and students working in aerospace industries, radar, avionics, and high-speed real-time information processing systems also were interested in these courses.

While each chapter of the book is self-contained and has no prerequisites, it is assumed all the students had already attended courses in linear systems, probability, and elementary random processes (including treatments on wide-sense stationarity and Gaussian processes). It is also assumed that the students are familiar with the use of Matlab for performing computations and simulations. Most of the students in these two courses had first taken the graduate course on "Detection and Estimation for Communication and Radar Systems," using the textbook by K. Yao, F. Lorenzelli, and C.E. Chen, published by Cambridge University Press, 2013.

Each chapter contains a detailed list of references and a set of homework problems. The solutions of odd-numbered problems are available from the Cambridge University Press website. It is the belief of the author that it is not possible to fully understand the materials in this book without doing some of the homework problems.

The author is appreciative of the contributions over many years by researchers covering all the topics in this book. He is also appreciative of the contributions by Dr Flavio Lorenzelli for some of materials in Chapters 5, 6, and 7 generated by him when he taught a graduate class at UCLA covering some of the topics in this book.

1 Applications of Spectral Analysis

1.1 Introduction

Our aim is to apply qualitatively applications of spectral analysis to various signal processing problems in applied science, communication, control, and avionic/aerospace systems. In this way, we will be motivated to consider the problems of determining the complexity and implementation of spectral analysis by means of parametric and non-parametric modeling.

1.2 Modulated Frequency-Shifted Keyed (M-FSK) Modulation in Communication Systems

First, consider a simple communication system illustrating the use of frequency-shifted keyed (FSK) modulated signals. Consider the earliest well-known Bell System 103 type full-duplex modem used for low data rate (300 Baud rate or less) transmission between a computer terminal and a central computer developed in the 1960s. This problem models a two-way binary data communication system. That is, upon encoding the ASCII data from the terminal or computer (either 64 or 128 hypothesis signals) as 6 bits or 7 bits data (plus parity, start, and stop bits), the basic transmission system is modeled as a repeated binary channel.

The binary data of a "one" (called a *mark*) and a "zero" (called a *space*), from old telegraph notation, can be modeled as two sinusoids (also called "tones") of amplitudes A and frequencies f_1 and f_0 of duration T. That is, the received data are of the form

$$x(t) = \begin{cases} A \sin 2\pi f_0 t + n(t), & \text{“space”} \\ A \sin 2\pi f_1 t + n(t), & \text{“mark”} \end{cases} \quad 0 \leq t \leq T, \tag{1.1}$$

where $n(t)$ represents the noise on the telephone line. In fact, since the transmission is full-duplex, there are two sets of FSK signals of the form given in (1.1). One set represents that of the originate modem and the other represents that of the receiver modem. Indeed, the spectral contents of the telephone line may look like that of Fig. 1.1. The detector at either modem must declare the presence of one of the two transmitted tones. A suboptimum non-coherent receiver based on spectral domain filtering is less complex than an optimum coherent receiver. In high SNR conditions encountered in typical telephone lines, a non-coherent receiver is quite adequate.

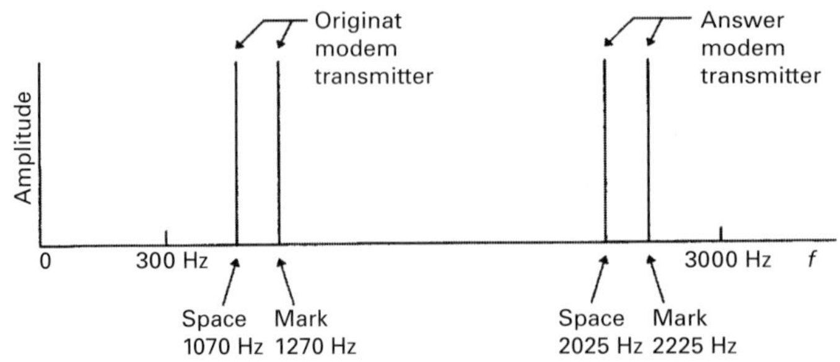

Figure 1.1 Full-duplex low data rate modem spectra

On the other hand, a general M-ary waveform is modeled by

$$x(t) = \begin{cases} A \sin 2\pi f_0 t + n(t) \\ A \sin 2\pi f_1 t + n(t) \\ \vdots \\ A \sin 2\pi f_{M-1} t + n(t) \end{cases} \qquad 0 \le t \le T. \tag{1.2}$$

The M-ary FSK (M-FSK) waveform is a practical form of modulation permitting higher data rate transmission. Since the detection can be done non-coherently, its receiver structure (consisting of a bank of M bandpass filters) is simpler and it may be more robust to various system degradations and interferences compared to a coherent phase-shifted keyed (PSK) system. Of course, compared to an M-PSK system, an M-FSK system is less efficient from the SNR point of view.

1.3 CW Doppler Radar

Let the transmitted signal $s(t)$ be given by

$$s(t) = A \sin 2\pi f_0 t, \qquad -\infty < t < \infty.$$

Let the first-order approximation of the *range* be given by $R \approx R_0 + \dot{R}(t - t_0)$. The delay time is expressed by

$$\Delta t = \frac{2R}{c},$$

where c is the velocity of propagation. The received signal is given by

$$\begin{aligned} s_R(t) &= A_R \sin 2\pi f_0 (t - \Delta t) \\ &= A_R \sin\left(2\pi f_0 t - 4\pi f_0 \frac{R}{c}\right) \\ &= A_R \sin\left(2\pi f_0 t - 4\pi f_0 \dot{R}\frac{t}{c} - 4\pi f_0 \frac{(R_0 - \dot{R} t_0)}{c}\right) \\ &= A_R \sin\left(2\pi f_0 t + 2\pi f_{dt} + \theta_0\right), \end{aligned}$$

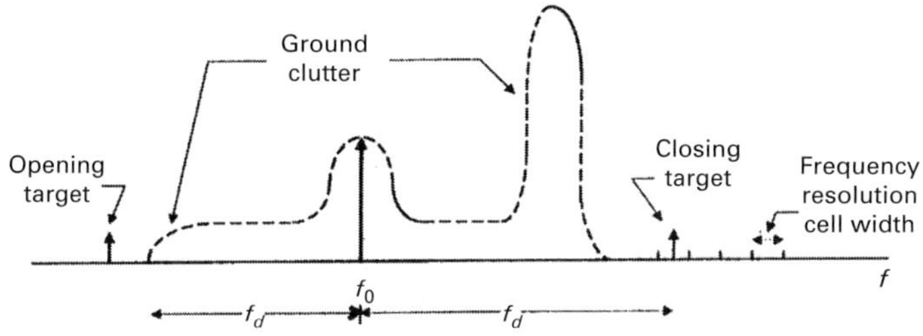

Figure 1.2 Spectral domain of an airborne Doppler radar receiver

where A_R is the received amplitude, $f_0 \equiv c/\lambda$, θ_0 is a fixed phase offset, and the "Doppler frequency" shift is defined by $f_d \equiv -2\dot{R}/\lambda$.

For a transmitted signal of frequency f_0, the received frequency is given by $(f_0 + f_d)$. When the target is *closing* (i.e., moving toward the transmitter), the range decreases and $\dot{R} < 0$, and thus the Doppler frequency f_d is positive. When the target is *opening* (i.e., moving away from the transmitter), the range increases and $\dot{R} > 0$, thus producing a negative Doppler frequency. Fig. 1.2 shows the spectral content of the receiver of an airborne Doppler radar. The ground reflection returns from all directions to produce the large band of "clutters" centered about the transmitted frequency f_0. The narrow width clutters centered about f_0 come from the ground patch essentially directly below the aircraft. The large clutter returns at the right edge of the band are due to the clutter returns in the mainlobe of the radar antenna. A fast-closing target may clear the clutter region and appear above the upper edge of the clutter region. Similarly, a fast-opening target may clear the clutter region and appear below the lower edge of the clutter region. Targets that appear in the clutter region may be much more difficult to detect than those in the clear regions. The size of the frequency resolution cell determines the resolution of the velocity of the target. The resolution cell is determined by the parameters of the data and the specific spectral analysis technique used in the radar receiver.

1.4 Speech Processing

Modern speech processing uses spectral analysis in many ways. Speech waveforms are non-stationary and at best can be modeled as quasi-stationary over some short duration of 5–20 ms. Speech can be considered as voiced (e.g., vowels such as /a/, /i/, etc.) or unvoiced (e.g., /sh/, etc.), or both. Time and frequency examples of a voiced waveform are given in Fig. 1.3 and those of an unvoiced waveform in Fig. 1.4. As can be seen, voiced waveform are quasi-periodic in the time-domain and thus harmonically related in the frequency domain. The fine structure of the spectrum is due to the vibrating vocal cord and the envelope structure (formant) is due to the modulation of the source with the pharynx and the mouth cavity. On the other hand, unvoiced waveforms are more random and spectrally broadbanded.

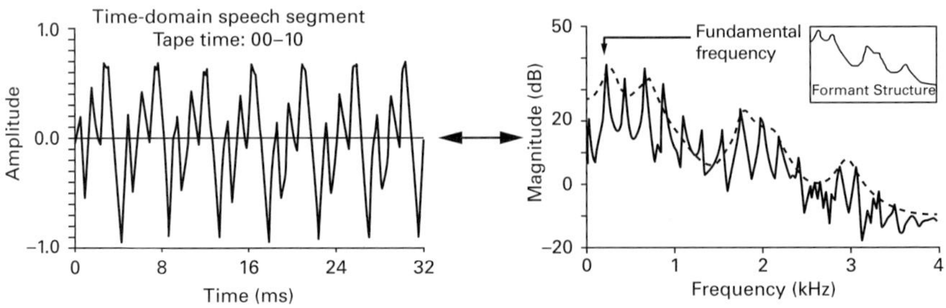

Figure 1.3 Time and spectral domains of a voiced waveform (IEEE Permission, Spanias, Proceedings of the IEEE Oct. 1994)

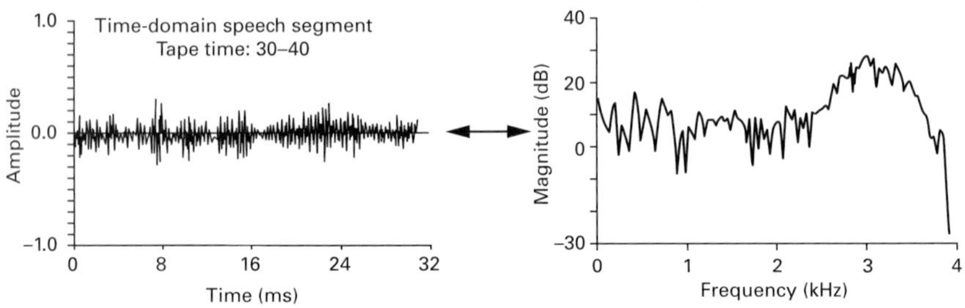

Figure 1.4 Time and spectral domains of an unvoiced waveform (IEEE Permission, Spanias, Proceedings of the IEEE Oct. 1994)

Various spectral analysis and synthesis techniques have been used to characterize speech waveforms. Autoregressive (AR) parameters based on an all-pole system for the modeling of a vocal tract based on the linear prediction (LP) technique was proposed in the 1970s. Since then, spectral techniques based on short-time Fourier transform (STFT) have been proposed. In the last four years, motivation for robust low-rate speech coding has led to the use of vector quantization (VQ) of LP coding (LPC). Quite sophisticated spectral analysis–synthesis techniques are used in modern code excited linear prediction (CELP) coding. In addition, spectral analysis techniques (e.g., STFT and wavelet techniques) have been proposed for voice recognition and speaker identification.

1.5 Other Applications

From classical spectroscopy, it is well known that different heated bodies emit different spectral radiations, and thus can be used for identification purposes. In radio astronomy, line and continuous spectra in the radio and x-ray band have been used for characterization of supernova stars. In atmospheric and oceanographic sciences, spectral contents of wind and ocean waves are used to characterize storm conditions. In earth resource satellites, different spectral optical and infrared bands yield information for earth resource explorations. In mechanical systems, spectral information characterizes different modes of vibrations. Prediction of bearing failure of induction motors based on stator current spectra lines has been used with success. In avionic systems, spectral analysis of radar reflected jet engine modulated (JEM) rotor blades has been used to identify the aircraft via its engine types. In medical sciences, spectral analysis of electroencephalograms has characterized schizophrenia, Parkinson disease, Huntington disease, and other neurologically caused illness. Spectral analysis and synthesis techniques have been used for many years in applied science and engineering. It is clear modern spectral analysis techniques will be able to solve even more sophisticated engineering and scientific problems in the future.

1.6 Conclusion

In Section 1.2, we first introduced the binary FSK communication system that appeared in the 1960s and then considered the M-ary FSK communication system, where the detection problem can be considered as a spectral analysis problem. In Section 1.3, the detection of a Doppler radar return waveform can be considered to be a spectral analysis problem [2]. In Section 1.4, frequency aspects of the human speech waveform were introduced [3]. In Section 1.5, various aspects of spectral analysis in physical problems were treated [4].

1.7 References

Some details on the M-FSK coherent and non-coherent modulation problems can be found in [1]. Various aspects of a Doppler radar system can be found in [2]. Details on the discussion of speech processing can be found in [3]. Various aspects of spectral analysis in physical problems are treated in [4].

[1] M. Schwartz, W.R. Bennett, and S. Stein, *Communication Systems and Techniques*, McGraw-Hill, 1966.

[2] G.W. Stimson, *Introduction to Airborne Radar*, Hughes Aircraft Company, 1983.

[3] A.S. Spanias, "Speech Coding: A Tutorial Review," *Proceedings of the IEEE*, 1994, pp. 1541–1582.

[4] D.B. Percival and A.T. Walden, *Spectral Analysis for Physical Applications*, Cambridge University Press, 1993.

1.8 Exercises

1. Read the historical development of the Bell 101 and 203 modulations in Wikipedia.

2. If we just use the M-FSK modulation as considered in (1.2), the possible phase discontinuity when shifting from one frequency to another frequency can cause a large spectral sidelobe outside of the desired frequency band. Continuous-phase FSK has been introduced to mitigate this problem. See *Digital Communications*, 3rd edn., by J.G. Proakis, McGraw-Hill, 1995, Section 4.3.3, p. 190.
3. Read the history of Doppler radar in Wikipedia.
4. Read about the use of spectral analysis in speech processing in Wikipedia.

2 Discrete Fourier Transform, Fast Fourier Transform, and Convolution

2.1 Discrete Fourier Transform and the Classical Fourier Transform

From the classical Fourier transform theory, if we have an integrable continuous-time signal $x(t)$ available on the real line, its Fourier transform can be obtained from

$$X(f) = \int_{-\infty}^{\infty} x(t)e^{-i2\pi ft}dt, \tag{2.1}$$

for any real-valued frequency f. Similarly, if an integrable continuous-frequency signal $X(f)$ is available on the real line, its inverse transform can be obtained from

$$x(t) = \int_{-\infty}^{\infty} X(f)e^{i2\pi ft}df \tag{2.2}$$

for any real-valued time t. Furthermore, upon substituting $X(f)$ of (2.1) into (2.2), we have

$$\int_{-\infty}^{\infty} \left\{ \int_{-\infty}^{\infty} x(\tau)e^{-i2\pi f\tau}d\tau \right\} e^{i2\pi ft}df = x(t) \tag{2.3}$$

at points of continuity of $x(\cdot)$. Similarly, upon substituting $x(t)$ of (2.2) into (2.1), we have

$$\int_{-\infty}^{\infty} \left\{ \int_{-\infty}^{\infty} X(u)e^{i2\pi ut}du \right\} e^{-i2\pi ft}dt = X(f) \tag{2.4}$$

at points of continuity of $X(\cdot)$. Thus, the transform pair forms unique invertible operations.

In most cases, we are not interested in obtaining the entire spectrum of the continuous-time waveform. Instead, we are interested in the power level at some designated frequencies. Therefore we want to restrict the above transformations to discrete-time and discrete-frequency variables. In order to use digital computers, CCD devices, etc. to perform the processing, we want to find an equivalent (if possible) or some approximate finite summations in place of the integrals in (2.1) and (2.2). Intuitively, we want to take N samples in time at

$$t_n = n\Delta t, \qquad n = 0, 1, \ldots, N-1, \tag{2.5}$$

where $\triangle t = T$ is some fixed sampling time duration somehow related to the frequency content of the signal. We also take N samples in the frequency domain at

$$f_k = k\triangle f, \quad k = 0, 1, \ldots, N-1, \tag{2.6}$$

where $\triangle f = 1/NT$ represents the frequency resolution of N data available over an NT duration. Then replacing the integral in (2.1) and (2.2) by the summation, we have

$$\hat{X}\left(\frac{k}{NT}\right) = \sum_{n=0}^{N-1} x(nT)e^{-i2\pi(k/NT)(nT)}\triangle t, \quad k = 0, \ldots, N-1, \tag{2.7}$$

and

$$\hat{x}(nT) = \sum_{k=0}^{N-1} X\left(\frac{k}{NT}\right)e^{i2\pi(k/NT)(nT)}\triangle f, \quad n = 0, \ldots, N-1. \tag{2.8}$$

The crucial question is whether $\{\hat{X}(k/NT),\ k = 0, \ldots, N-1\}$ and $\{\hat{x}(nT),\ n = 0, \ldots, N-1\}$ form a valid transform pair. If we substitute $\hat{X}$ in (2.7) into the right-hand side of (2.8) as X, does $\hat{x}(nT)$ equal $x(nT)$? Similarly, if $\{\hat{x}(nT), n = 0, \ldots, N-1\}$ of (2.8) is substituted into the right-hand side of (2.7) as x, does $\hat{X}(k/NT)$ equal $X(k/NT)$? In other words, do the discretized versions of (2.1) and (2.2) given by (2.7) and (2.8) satisfy the unique invertible operations in (2.3) and (2.4)?

As we shall see later, the normalized discrete Fourier transform (DFT) of (2.7) and the normalized inverse discrete Fourier transform (IDFT) of (2.8) are not generally equivalent but form approximations to the corresponding original continuous Fourier transforms. Specifically, consider the application of the Poisson summation formula to (2.2) in converting this integral into a summation. Let $h(t)$ be an equi-spaced train of delta functions in the time-domain with period T_1 given by

$$h(t) = \sum_{m=-\infty}^{\infty} \delta(t + mT_1), \quad -\infty < t < \infty. \tag{2.9}$$

Then the Fourier transform of $h(t)$ is

$$H(f) = \frac{1}{T_1}\sum_{k=-\infty}^{\infty} \delta(f - k/T_1), \quad -\infty < f < \infty, \tag{2.10}$$

which is also an equi-spaced train of delta functions in the frequency domain. By the convolution property,

$$\int_{-\infty}^{\infty} x(\tau)h(t-\tau)d\tau = \int_{-\infty}^{\infty} X(f)H(f)e^{i2\pi ft}df, \quad -\infty < t < \infty. \tag{2.11}$$

Upon substituting $h(t)$ into the left-hand side of (2.11) and $H(f)$ into the right-hand side of (2.11),

$$\tilde{x}(t) \stackrel{\Delta}{=} \sum_{m=-\infty}^{\infty} x(t + mT_1) = \frac{1}{T_1}\sum_{k=-\infty}^{\infty} X(k/T_1)e^{i2\pi kt/T_1}, \quad -\infty < t < \infty, \tag{2.12}$$

is the Poisson summation formula for $x(t)$. Since $\tilde{x}(t)$ defined in (2.12) is a periodic function with period T_1, $\tilde{x}(t)$ has the Fourier series coefficient

$$\frac{1}{T_1}X\left(\frac{k}{T_1}\right) = \frac{1}{T_1}\int_{-T_1/2}^{T_1/2} \tilde{x}(t)e^{-i2\pi kt/T_1}dt, \qquad -\infty < k < \infty. \tag{2.13}$$

Because $\tilde{x}(t)$ is periodic, we need only to consider $\tilde{x}(t)$ within $[-T_1/2, T_1/2)$. Let

$$t = nT, \qquad n = 0, 1, \ldots, N-1 \tag{2.14}$$

for $\tilde{x}(t)$. Setting the full period in (2.9)–(2.13) as T_1, we have

$$T_1 = NT. \tag{2.15}$$

Then (2.12) becomes

$$\tilde{x}(nT) = \frac{1}{NT}\sum_{k=-\infty}^{\infty} X\left(\frac{k}{NT}\right)e^{i2\pi k(nT)/NT}. \tag{2.16}$$

Since any integers, $-\infty < k < \infty$, can be expressed as

$$k = j + rN \qquad \text{for} \qquad j = 0, 1, \ldots, N-1, \; -\infty < r < \infty. \tag{2.17}$$

Using (2.17) in (2.16),

$$\tilde{x}(nT) = \frac{1}{NT}\sum_{j=0}^{N-1}\left\{\sum_{r=-\infty}^{\infty} X\left(\frac{j+rN}{NT}\right)\right\}e^{i2\pi nj/N}e^{i2\pi nr}, \; n = 0, 1, \ldots, N-1. \tag{2.18}$$

In (2.17), since n and r are integers, $e^{i2\pi nr} = 1$. Denote

$$\tilde{X}\left(\frac{j}{NT}\right) \triangleq \sum_{-\infty}^{\infty} X\left(\frac{j}{NT}+\frac{r}{T}\right), \qquad j = 0, 1, \ldots, N-1. \tag{2.19}$$

Thus, the expression in (2.18) can be written as

$$\tilde{x}(nT) = \frac{1}{NT}\sum_{j=0}^{N-1}\tilde{X}\left(\frac{j}{NT}\right)e^{i2\pi nj/N}, \qquad n = 0, 1, \ldots, N-1. \tag{2.20}$$

Comparing (2.20) to the intuitively conceived discretized transform expression of (2.8), we see that they are identical other than the normalization factor $(1/NT)$. However, (2.20) relates the N frequency-domain values of $\tilde{X}(\cdot)$ to the N time-domain sampled values of $\tilde{x}(\cdot)$. While the expression of (2.20) is exact, $\tilde{X}(\cdot)$ is not equal to $X(\cdot)$ but is related by (2.19), and $\tilde{x}(\cdot)$ is not equal to $x(\cdot)$ but is related by (2.12).

The result of (2.20) shows that if the signal $x(t)$ satisfies

$$x(nT) = \sum_{m=-\infty}^{\infty} x(nT + mNT), \qquad n = 0, 1, \ldots, N-1, \tag{2.21}$$

$$X\left(\frac{j}{NT}\right) = \sum_{r=-\infty}^{\infty} X\left(\frac{j}{NT}+\frac{r}{T}\right), \qquad j = 0, 1, \ldots, N-1, \tag{2.22}$$

Table 2.1 Comparison between $x(nT)$ and $\tilde{x}(nT)$

$1 = x(0)$	$\tilde{x}(0) = x(0) + x(3) + x(6) + \cdots = 1.05$
$0.607 = x(0.5)$	$\tilde{x}(0.5) = x(0.5) + x(3.5) + x(6.5) + \cdots = 0.638$
$0.368 = x(1)$	$\tilde{x}(1) = x(1) + x(4) + x(7) + \cdots = 0.387$
$0.223 = x(1.5)$	$\tilde{x}(1.5) = x(1.5) + x(4.5) + x(7.5) + \cdots = 0.235$
$0.135 = x(2)$	$\tilde{x}(2) = x(2) + x(5) + x(8) + \cdots = 0.142$
$0.082 = x(2.5)$	$\tilde{x}(2.5) = x(2.5) + x(5.5) + x(8.5) + \cdots = 0.086$

then the transform pair of (2.7) and (2.8) is valid. A sufficient condition for (2.22) to be valid is that $x(t)$ is bandlimited to $[-1/2T, 1/2T)$. A sufficient condition for (2.21) to be valid is that $x(t)$ is time limited to $[0, NT)$. Mathematically, a bandlimited continuous-time signal cannot be time limited and similarly a continuous-time limited signal cannot be bandlimited. In practice, by taking the sampling duration T sufficiently small and N sufficiently large, the requirements of (2.21) and (2.22) can be satisfied arbitrarily well. Then the normalized DFT and IDFT are reasonable approaches to evaluate the Fourier and inverse Fourier transformations of continuous-time and continuous-frequency signals.

Example 2.1 Consider $x(t) = U(t)e^{-t}, -\infty < t < \infty$. Its Fourier transform is given by $X(f) = 1/(1 + i2\pi f), -\infty < f < \infty$. Table 2.1 lists the left-hand side and right-hand side of (2.21) for $T = 0.5$ and $N = 6$. There is about 5% error for each $\tilde{x}(nT)$ as compared to $x(nT), n = 0, \ldots, 5$.

2.2 Sampling

For a continuous-time waveform which is both time limited and frequency limited, the DFT given by (2.7) provides the exact computation of the Fourier transform. Since it is impossible to have both time and frequency limited waveforms, we would like to know how small a T should be chosen and how large an N needs to be to approximate the continuous-time Fourier transform given a continuous-time signal. Obviously, the sampling rate depends on the frequency content of the signal.

Let $X(f), -\infty < f < \infty$, be the Fourier transform of the continuous-time signal $x(t), -\infty < t < \infty$. Then

$$X(f) = \int_{-\infty}^{\infty} x(t)e^{-i2\pi ft}dt, \tag{2.23}$$

and

$$x(t) = \int_{-\infty}^{\infty} X(f)e^{i2\pi ft}df. \tag{2.24}$$

Suppose we sample $x(t)$ every T seconds, then we denote the sampled sequence by

$$x[n] = x(nT), \qquad -\infty < n < \infty. \tag{2.25}$$

Denote the Z-transform of $\{x[n], -\infty < n < \infty\}$ by

$$\tilde{X}(z) = \sum_{n=-\infty}^{\infty} x[n]z^{-n}$$

and

$$x[n] = \frac{1}{2\pi i}\oint_C \tilde{X}(z)z^{n-1}dz.$$

If C is taken along the unit circle given by $z = e^{i2\pi fT}$, then

$$\tilde{X}(e^{i2\pi fT}) = \sum_{n=-\infty}^{\infty} x[n]e^{-in2\pi fT} \tag{2.26}$$

and

$$x[n] = T\int_{-1/2T}^{1/2T} \tilde{X}(e^{i2\pi fT})e^{i2\pi nfT}df. \tag{2.27}$$

Substituting (2.24) in (2.25),

$$\begin{aligned} x[n] &= \int_{-\infty}^{\infty} X(f)e^{i2\pi nfT}df \\ &= \sum_{m=-\infty}^{\infty}\int_{(2m-1)/2T}^{(2m+1)/2T} X(f)e^{i2\pi nfT}df \\ &= T\int_{-1/2T}^{1/2T}\left\{\frac{1}{T}\sum_{m=-\infty}^{\infty} X\left(f+\frac{m}{T}\right)\right\}e^{i2\pi nfT}df. \end{aligned} \tag{2.28}$$

By comparing (2.28) with (2.27), since the trigonometric system $\{e^{i2\pi nfT}, -\infty < n < \infty\}$ is complete in the $\mathcal{L}_2$ sense,

$$\tilde{X}(e^{i2\pi fT}) = \frac{1}{T}\sum_{m=-\infty}^{\infty} X\left(f+\frac{m}{T}\right), \qquad -\frac{1}{2T} \le f < \frac{1}{2T}. \tag{2.29}$$

This implies that $\tilde{X}(e^{i2\pi fT})$ is periodic and the spectrum of the sampled discrete-time signal consists of the folded version of the continuous-time signal. If $X(f)$ is bandlimited to $[-f_o, f_o)$, then by setting $f_o = 1/2T$, (2.29) becomes

$$\tilde{X}(e^{i2\pi fT}) = \frac{1}{T}X(f), \qquad -\frac{1}{2T} \le f < \frac{1}{2T},$$

and the spectrum of the sampled discrete-time signal is the same (other than a $1/T$ scaling factor) as that of the spectrum of the continuous-time signal. This shows that for a strictly bandlimited signal $X(f)$ of bandwidth $[-f_o, f_o)$, by using the Nyquist sampling rate of $1/T = 2f_o$, the sampled data $\{x[n], -\infty < n < \infty\}$ contain the same spectral information as the original continuous-time signal. If $X(f)$ is not bandlimited to $[-f_o, f_o)$, but we still use sampling rate of $1/T = 2f_o$, then the folding

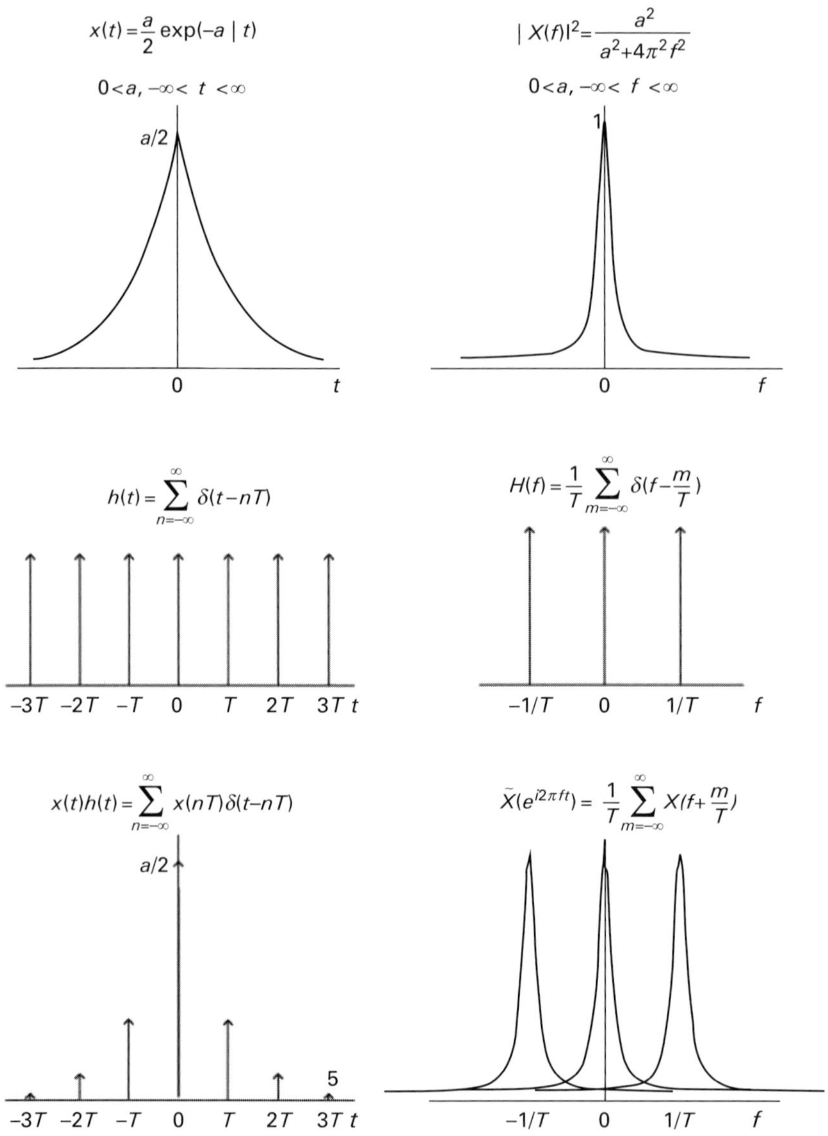

Figure 2.1 Pictorial representation of the sampling theorem

effect commonly called aliasing occurs in (2.29). This situation can best be explained by visualizing Fig. 2.1. When the signal is sampled correctly, the spectrum $X(f)$ can be extracted from $\tilde{X}(f)$ by passing $\tilde{X}(f)$ through an ideal low-pass filter. However, if the spectral contents of $X(f)$ outside of $[-1/2T, 1/2T)$ are non-zero but small, then the ill effects of aliasing are minimal and the sampling operation has introduced only minimal distortion.

2.3 Discrete Fourier Transform

Let $\mathbf{x} = [x[0], x[1], \ldots, x[N-1]]^T$ be a sequence of N complex-valued data. Its spectral contents are obtained by taking the Z-transform of $\mathbf{x}$

$$\tilde{X}(z) = \sum_{n=-\infty}^{\infty} x[n] z^{-n}. \tag{2.30}$$

If we restrict our interests to N frequencies of $\theta_k = 2\pi k/N, k = 0, 1, \ldots, N-1$, then for any given k we can take $z = e^{i\theta_k}$ and (2.30) becomes

$$X[k] = \tilde{X}(e^{i\theta_k}) = \sum_{n=0}^{N-1} x[n] e^{-i2\pi kn/N} = \sum_{n=0}^{N-1} x[n] \omega_N^{nk}, \quad k = 0, 1, \ldots, N-1, \tag{2.31}$$

where $\omega_N = e^{-i2\pi/N}$. From here on, the subscript N of ω_N will be omitted whenever possible. The transformation of $\mathbf{x} = [x[0], x[1], \ldots, x[N-1]]^T$ to $\mathbf{X} = [X[0], X[1], \ldots, X[N-1]]^T$ defined by (2.31) is called the discrete Fourier transform (DFT). We note (2.31) is identical to (2.7) except for a scale factor. The summation form of the DFT in (2.31) can be expressed in the matrix–vector multiplication form of the DFT as given by

$$\mathbf{X} \triangleq \mathcal{F}(\mathbf{x}) = \mathbf{A}\mathbf{x}, \tag{2.32}$$

where

$$\mathbf{A} = \begin{pmatrix} \omega^0 & \omega^0 & \ldots & \omega^0 \\ \omega^0 & \omega^1 & \ldots & \omega^{N-1} \\ \vdots & \vdots & & \vdots \\ \omega^0 & \omega^{N-1} & \ldots & \omega^{(N-1)^2} \end{pmatrix}. \tag{2.33}$$

From (2.31) or (2.32), we observe that DFT is a linear transformation and possesses much of the transform properties of the Fourier transformation of signals on the entire real line. In particular, the inverse discrete Fourier transform (IDFT) exists and is defined in Lemma 2.1.

Lemma 2.1 *Let* $\mathbf{X}$ *be the DFT of* $\mathbf{x}$. *Then the IDFT of* $\mathbf{X}$ *is defined by*

$$x[n] = \frac{1}{N} \sum_{k=0}^{N-1} X[k] e^{i2\pi nk/N} = \frac{1}{N} \sum_{k=0}^{N-1} X[k] \omega^{-nk}, \quad n = 0, 1, \ldots, N-1, \tag{2.34}$$

or

$$\mathbf{x} = \mathcal{F}^{-1}(\mathbf{X}) = \mathbf{A}^{-1}\mathbf{X}, \tag{2.35}$$

where

$$\mathbf{A}^{-1} = \frac{1}{N} \begin{pmatrix} \omega^0 & \omega^0 & \ldots & \omega^0 \\ \omega^0 & \omega^{-1} & \ldots & \omega^{-(N-1)} \\ \vdots & \vdots & & \vdots \\ \omega^0 & \omega^{-(N-1)} & \ldots & \omega^{-(N-1)^2} \end{pmatrix}. \tag{2.36}$$

By direct evaluation, the DFT(or IDFT) given by (2.31) or (2.32) (equivalently (2.34) or (2.35)) requires N^2 complex multiplications and $(N^2 - N)$ complex additions. However, using some clever manipulations of the input vector, it is possible to use at most $((N/2)\log_2 N)$ complex multiplications. This improved algorithm for performing DFT is called the fast Fourier transform (FFT).

2.4 Fast Fourier Transform

This section considers a simplified derivation of the FFT algorithm when N has a power of 2.

Lemma 2.2 *The number of complex multiplications needed in the FFT algorithm to perform (2.32) is at most equal to* $((N/2)\log_2 N)$.

Proof: Let

$$x_o[n] = x[2n+1] \quad for \quad n = 0, 1, \ldots, \frac{N}{2} - 1$$

and

$$x_e[n] = x[2n] \quad for \quad n = 0, 1, \ldots, \frac{N}{2} - 1.$$

Then (2.31) can be expressed as

$$\begin{aligned} X[k] &= \sum_{n=0, n=odd}^{N-1} x[n]\omega_N^{nk} + \sum_{n=0, n=even}^{N-1} x[n]\omega_N^{nk} \\ &= \sum_{n=0}^{\frac{N}{2}-1} x_o[n]\omega_N^{k(2n+1)} + \sum_{n=0}^{\frac{N}{2}-1} x_e[n]\omega_N^{2nk}, \quad k = 0, 1, \ldots, N-1. \end{aligned} \tag{2.37}$$

But $\omega_N^2 = \omega_{N/2}$. Thus, (2.37) becomes

$$X[k] = \omega_N^k \sum_{n=0}^{\frac{N}{2}-1} x_o[n]\omega_{N/2}^{kn} + \sum_{n=0}^{\frac{N}{2}-1} x_e[n]\omega_{N/2}^{kn}, \quad k = 0, 1, \ldots, N-1. \tag{2.38}$$

However, for kn to be the index of a $N/2$-point DFT, kn must be smaller than or equal to $((N/2) - 1)^2$. Denote $X_o[k]$ and $X_e[k]$ to be the $(N/2)$-points of $[x_o[0], x_o[1], \ldots, x_o[(N/2)-1]]^T$ and $[x_e[0], x_e[1], \ldots, x_e[(N/2)-1]]^T$, $k = 0, 1, \ldots, (N/2)-1$, respectively. Since n is less than $N/2$,

$$X[k] = \omega_N^k X_o[k] + X_e[k], \quad k = 0, 1, \ldots, \frac{N}{2} - 1.$$

Let $k = \ell + N/2$ where $\ell = 0, 1, \ldots, (N/2) - 1$. Then for $k = N/2, \ldots, N-1$,

$$\begin{aligned} X[k] &= \omega_N^{(\ell+\frac{N}{2})} \sum_{n=0}^{\frac{N}{2}-1} x_o[n]\omega_{N/2}^{n(\ell+\frac{N}{2})} + \sum_{n=0}^{\frac{N}{2}-1} x_e[n]\omega_{N/2}^{n(\ell+\frac{N}{2})} \\ &= -\omega_N^{\ell} \sum_{n=0}^{\frac{N}{2}-1} x_o[n]\omega_{N/2}^{n\ell} + \sum_{n=0}^{\frac{N}{2}-1} x_e[n]\omega_{N/2}^{n\ell} \\ &= -\omega_N^{\ell} X_o[\ell] + X_e[\ell] \\ &= -\omega_N^{(k-\frac{N}{2})} X_o\left[k - \frac{N}{2}\right] + X_e\left[k - \frac{N}{2}\right]. \end{aligned}$$

Therefore

$$X[k] = \begin{cases} \omega_N^k X_o[k] + X_e[k], & 0 \le k \le \frac{N}{2} - 1, \\ -\omega_N^{(k-\frac{N}{2})} X_o[k - \frac{N}{2}] + X_e[k - \frac{N}{2}], & \frac{N}{2} \le k \le N - 1. \end{cases} \tag{2.39}$$

We have decomposed an N-point DFT into two $N/2$-point DFTs. From (2.39), if $X_o[k], k = 0, 1, (N/2) - 1$, and $X_e[k], k = 0, 1, \ldots, (N/2) - 1$, are available, only $N/2$ multiplications are required to compute $X[k], k = 0, 1, \ldots, N - 1$. Since (2.39) is recursive, we can apply (2.39) repeatedly. After $\log_2 N$ iterations, the final DFT involves only two terms without multiplications. Thus, we need $(N/2)\log_2 N$ complex multiplications to perform the DFT. In practice, since there are no multiplications involved with ± 1 and $\pm i$, the actual number of multiplications will be slightly less than $(N/2)\log_2 N$. Fig. 2.2 shows the construction of an eight-point DFT from two four-point DFTs that are in turn constructed from four two-point DFTs.

We also state without proof the following lemma.

Lemma 2.3 *The number of complex multiplications required in the IFFT algorithm to perform (2.35) is at most equal to* $(N/2)\log_2 N$.

2.4.1 Decimation-In-Time (DIT) and Decimation-In-Frequency (DIF)

There are two ways to implement the FFT, namely the decimation-in-time (DIT) and the decimation-in-frequency (DIF). Fig. 2.2 represents the DIT implementation since at each stage the input sequence (which is in time units) is divided into smaller sequences (by the odd and even indices) for subsequent processing. The input data indices are shuffled and can be found by using the reverse of the binary representation of the index bits.

In the DIF form of FFT, the output data (which are in the frequency domain) are divided into smaller sections. Fig. 2.3 shows a DIF eight-point FFT. Note that the butterfly in the DIF is not the same as that in the DIT form.

In both the DIT and DIF forms of FFT, at each stage there are $N/2$ butterflies that perform the computations. Within each butterfly, the inputs and outputs can share the same memory locations except the computation of ω_N^k. The evaluation of ω_N^k can use one of the following ways.

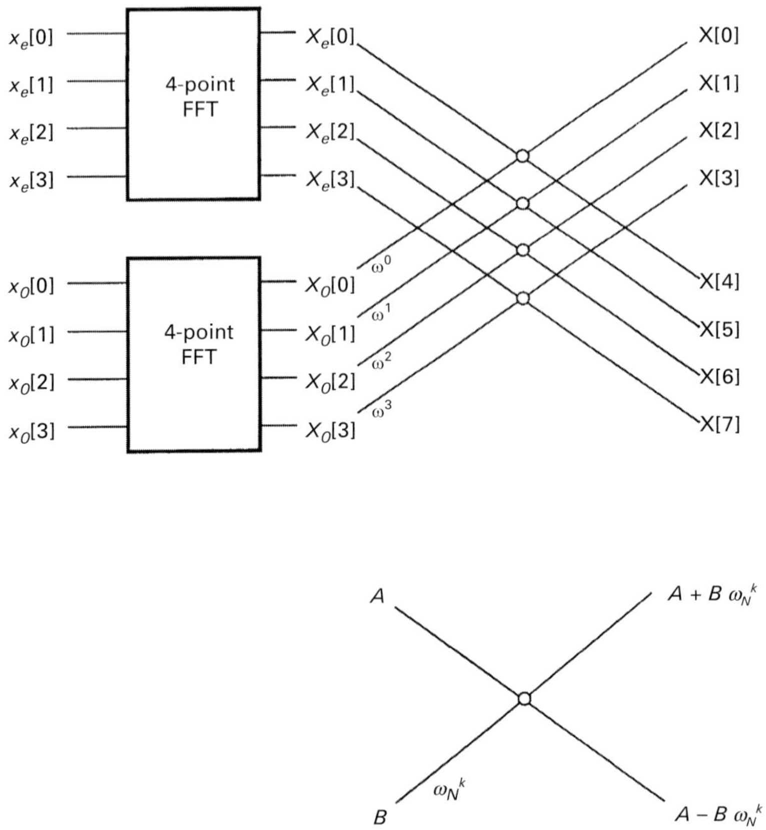

Figure 2.2 Construction of an eight-point DFFT

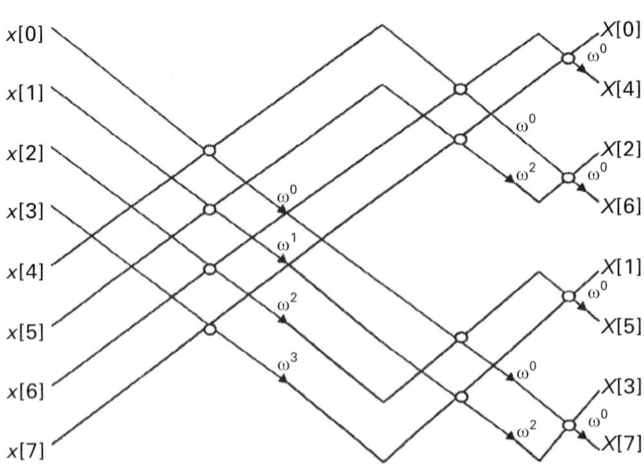

Figure 2.3 An eight-point DIF form FFT

1. Store $\{\omega_N^k\}$ in read only memory for all values of N and k.
2. Use $\omega_N^k = \cos(2\pi k/N) - i\sin(2\pi k/N)$ in real time.
3. Use the recursion formula

$$\omega_N^k = \omega_N^{k-\ell} \cdot \omega_N^\ell,$$

with $0 \le \ell \le k$ and the initial condition $\omega_N^0 = 1$.

2.5 Cyclic Convolution

In various system applications, FFT is not only used for efficient DFT spectral analysis purpose, but can be used efficiently to perform convolutions. While the linear convolutional property of the classical continuous-time Fourier transform is no longer applicable to DFT, the following cyclic convolutional property is available by using DFT/FFT and IDFT/IFFT.

For a given sequence of data, $\mathbf{x} = [x[0], x[1], \ldots, x[N-1]]^T$, its periodic extension can be defined by $x_p[n] = x[n \bmod N]$ for all integral n. Denote $\mathbf{x}_p = [x_p[0], x_p[1], \ldots, x_p[N-1]]^T$. Similarly, for a given $\mathbf{h} = [h[0], h[1], \ldots, h[N-1]]^T$, its periodic extension can be defined by $h_p[n] = h[n \bmod N]$ for all integral n. Denote $\mathbf{h}_p = [h_p[0], h_p[1], \ldots, h_p[N-1]]^T$. Then the cyclic convolution $\mathbf{y}_p = [y_p[0], y_p[1], \ldots, y_p[N-1]]^T$ of $\mathbf{x_p}$ and $\mathbf{h_p}$ is defined by

$$y_p[k] \triangleq \sum_{m=0}^{N-1} x_p[m] h_p[k-m], \qquad k = 0, 1, \ldots, N-1. \tag{2.40}$$

The cyclic convolution can also be expressed in terms of the DFT of $\mathbf{x}$ and $\mathbf{h}$ as follows.

Lemma 2.4 *Let* $\mathbf{X} = [X[0], X[1], \ldots, X[N-1]]^T$ *denote the DFT of* $\mathbf{x}$ *and* $\mathbf{H} = [H[0], H[1], \ldots, H[N-1]]^T$ *denote the DFT of* $\mathbf{h}$. *Then the DFT of* $\mathbf{y}_p$ *denoted by* $\mathbf{Y}_p$ *is given by*

$$\mathbf{Y}_p = \mathcal{F}(\mathbf{y}_p) = \mathbf{X} \cdot \mathbf{H}, \tag{2.41}$$

or

$$\mathbf{y}_p = \mathcal{F}^{-1}(\mathbf{X} \cdot \mathbf{H}), \tag{2.42}$$

where $\mathbf{X} \cdot \mathbf{H}$ *denotes component-by-component multiplication of each vector.*

Proof: (2.42) follows from the IDFT of (2.41). The kth component of $\mathbf{y}_p$ is then given by

$$\begin{aligned} y_p(k) &= \frac{1}{N} \sum_{n=0}^{N-1} e^{i2\pi kn/N} \left\{ \sum_{m=0}^{N-1} x(m) e^{-i2\pi mn/N} \right\} \left\{ \sum_{\ell=0}^{N-1} h(\ell) e^{-i2\pi \ell n/N} \right\} \\ &= \frac{1}{N} \sum_{m=0}^{N-1} \sum_{\ell=0}^{N-1} x(m) h(\ell) \left\{ \sum_{n=0}^{N-1} e^{i2\pi(k-m-\ell)n/N} \right\}. \end{aligned} \tag{2.43}$$

The bracketed term in (2.43) is equal to

$$\sum_{n=0}^{N-1} e^{i2\pi(k-m-\ell)n/N} = \begin{cases} N, & k-m-\ell \equiv 0(\text{mod}N), \\ 0, & \text{otherwise.} \end{cases} \tag{2.44}$$

In (2.44), $a \equiv b(\text{mod}N)$ means $a = b + nN$ for any integer n. Combining (2.43) and (2.44) yields (2.40).

2.6 Conclusion

Section 2.1 showed how the classical Fourier transform can be approximated by the DFT. Section 2.2 considered sampling in the time-domain and the aliasing distortion for non-bandlimited analog signals. Section 2.3 defined the DFT. By exploiting various symmetries in the complex exponentials, the fast version of the DFT, named the FFT was obtained. The decimation-in-time and the decimation-in-frequency versions of FFT were also shown. Section 2.5 presented the concept of cyclic convolution and its possible efficient evaluation using the FFT algorithm.

2.7 References

In Sections 2.1, 2.2, and 2.3, the relationship between the DFT and classical Fourier transform were presented. More details on this issue can be found in [1] and [2]. Section 2.4 introduced the FFT which is credited to Colley and Tukey [3]. Earlier, a German physicist Runge [4] presented an algorithm similar to the FFT in 1903, but was not known to signal processing people. The works of other people on earlier results similar to the FFT algorithm are discussed in [5]. Both the decimation-in-time and the decimation-in-frequency versions of FFT can be found in many digital signal processing books, including that of [2]. Section 2.5 showed that cyclic convolution can be computed efficiently using the FFT algorithm [2].

[1] E.O. Brigham, *The Fast Fourier Transform and Its Applications*, Prentice-Hall, 1988.
[2] A.V. Openheim and R.W. Schafer, *Digital Signal Processing*, Prentice-Hall, 1975.
[3] J.W. Colley and J.W. Tukey, "An Algorithm–for the Machine Calculation of Complex Fourier Series," *Math. Computation*, 1965, pp. 297-301.
[4] C. Runge, "Uber die Zerlegung Empirisch Periodischer Funktionen in Sinnus-wellen," *Z.A. Math. Physik*, 1903, pp. 443–456.
[5] C.W. Colley, P.A.W. Lewis, and P.D. Welch, "Historical Notes on the FFT," *IEEE Trans. Audio Electroacoustics*, 1967, pp. 76–79.

2.8 Exercises

1. Derive the Fourier transform $F(f)$ of the following time-domain signals $f(t)$ by explicit integrations (show all the steps in your derivations).
   a. Consider $f(t) = I_{[-a,a]}(t)$, where $I_{[a,b]}(t)$ is the indicator function defined by $I_{[a,b]}(t) = 1,\ a \leq t \leq b;\ = 0, t \notin [a,b]$. Hint: Just do the integration.
   b. Consider $f(t) = (1 - |t/a|)I_{[-a,a]}(t),\ 0 < a$. Hint: Do the integrations in two parts (first for the negative t values and then for the positive t values). Note: $(1/2)(1 - \cos(2\pi f a)) = \sin^2(\pi f a)$.
   c. Consider $f(t) = \exp(-a|t|),\ -\infty < t < \infty,\ 0 < a$. Hint: Do the integrations in two parts (first for the negative t values and then for the positive t values).
   d. Consider $f(t) = \exp(-at^2),\ -\infty < t < \infty,\ 0 < a$. Hint: Complete the square in the exponent $t^2 + (b/a)t = (t + (b/2a))^2 - (b/2a)^2$. Then use the fact the integral of any Gaussian probability density function (pdf) is one.
2. In order to understand how well the discrete DFT and IDFT approximate the continuous-time FT and IFT, we want to consider the time-domain error

$$e_t(n, N, T) = \tilde{x}(nT) - x(nT),\ n = -(N/2), \ldots, (N/2) - 1, \tag{2.45}$$

$$\tilde{x}(nT) = \sum_{m=-\infty}^{\infty} x(nT + mNT), \tag{2.46}$$

and the frequency-domain error

$$e_f(j, N, T) = \tilde{X}(j/(NT)) - X(j/(NT)),\ j = -(N/2), \ldots, (N/2) - 1, \tag{2.47}$$

$$\tilde{X}(j/(NT)) = \sum_{r=-\infty}^{\infty} X\left(\frac{j}{NT} + \frac{r}{T}\right). \tag{2.48}$$

By using (2.46) in (2.45) and (2.48) in (2.47), the two errors can be expressed as

$$e_t(n, N, T) = \sum_{\substack{m=-\infty \\ m\neq 0}}^{\infty} x(nT + mNT),\ n = -(N/2), \ldots, (N/2) - 1, \tag{2.49}$$

$$e_f(j, N, T) = \sum_{\substack{r=-\infty \\ r\neq 0}}^{\infty} X(\frac{j}{NT} + \frac{r}{T}),\ j = -(N/2), \ldots, (N/2) - 1. \tag{2.50}$$

In order not to consider each value of n in (2.49) and j in (2.50), define the averaged time-domain error and averaged frequency-domain error by

$$E_t = (1/N) \sum_{n=-N/2}^{N/2-1} e_t(n, N, T), \tag{2.51}$$

$$E_f = (1/N) \sum_{j=-N/2}^{N/2-1} e_f(j,N,T). \tag{2.52}$$

Now, consider $x(t) = e^{-0.5t^2}, \; -\infty < t < \infty$.

a. Use the Fourier transform of $x(t)$ to find $X(f) = \int_{-\infty}^{\infty} e^{-0.5t^2} e^{-i2\pi ft} dt$ explicitly. Hint: Consider its characteristic function.

b. Pick some N to show some interesting aspects of E_t, E_f, and $E_{av} = (E_t + E_f)/2$, by plotting their values versus $T = 0.1, \ldots, 1.5$.

c. Pick some T to show some interesting aspects of E_t, E_f, and E_{av}, by plotting their values versus $N = 4, \ldots, 20$.

Make any general comments on the time-bandwidth tradeoffs for this particular $x(t)$ function.

3. Let $x = [x[0], \ldots, x[3]] = [2,3,4,5]$ and $h = [h[0], \ldots, h[3]] = [3,4,5,6]$.

a. Find the cyclic convolution of x and h.

b. Find the linear convolution $y[k] = \Sigma_{m=0}^{3} x[m]h[k-m]$, $k = 0, \ldots, 6$, with $h[m] = 0$, for $m < 0$ and $m > 3$.

c. Pad 3 zeros to the end of x and h and find its cyclic convolution.

d. Let $x_0 = [x,0,0,0]$ and $h_0 = [h,0,0,0]$. Let $X_0 = \text{DFT}\{x_0\}$ and $H_0 = \text{DFT}\{h_0\}$. Find the $\text{IDFT}\{X_0 \cdot H_0\}$.

e. Find the complexity (i.e., number of multiplications and number of additions) of linear convolution for two N-point complex-valued vectors by direct convolution. Find the complexity of linear convolution using the FFT/IFFT method for two $\lceil 2N-1 \rceil$-point vectors (i.e., denote $\lceil x \rceil$ to be the smallest power of 2, greater or equal to x).

4.

$$X(f) = \begin{cases} 1, & -0.5 \le f < 0.5, \\ 0, & \text{elsewhere} \end{cases}$$

show $x[n] = x(n) = \delta_{n,o}, \; -\infty < n < \infty$, where $x(n) = \int_{-\infty}^{\infty} X(f) e^{-i2\pi fn} df$. Furthermore, if $X(f) \neq 0, 0.5 < f$, then in general $x[n] \neq x(n)$, $-\infty < n < \infty$. However, for any $0 \le r \le 1$, and

$$X(f) = \begin{cases} 1, & 0 \le |f| \le \frac{1-r}{2}, \\ 0.5\left[1 + \cos\left(\frac{\pi}{r}\left(|f| - \frac{1-r}{2}\right)\right)\right], & \frac{1-r}{2} \le |f| \le \frac{1+r}{2} \\ 0, & \frac{1+r}{2} \le |f|, \end{cases}$$

show that we still have $x[n] = x(n) = \delta_{n,o}$, $-\infty < n < \infty$. Hint: Plot this $X(f)$, $-\infty < f < \infty$. What special property does it have?

3 Spectral Analysis via Continuous/Discrete Fourier Transformation

3.1 Frequency Resolution of Continuous- Discrete-Time Data

One of the most basic applications of DFT/FFT in communication and radar systems is the detection of the presence or absence of a sinusoid (or multiple sinusoids) and the estimation of its (their) frequency(ies). In this section, we assume a noise-free environment.

3.1.1 Continuous-Time Signal

First, consider a continuous-time single complex sinusoid $x(t)$ of frequency f_1 and amplitude A_1 observed over an infinite observation duration

$$x(t) = A_1 e^{i2\pi f_1 t}, \qquad -\infty < t < \infty. \tag{3.1}$$

The Fourier transform of (3.1) is called the spectrum of $x(t)$ and is given by

$$X(f) = A_1 \int_{-\infty}^{\infty} e^{-i2\pi(f-f_1)t} dt = A_1 \delta(f - f_1), \qquad -\infty < f < \infty. \tag{3.2}$$

We declare $x(t)$ has a single frequency component at f_1 if $|X(f)| > \gamma$ at only the single frequency $f = f_1$ where γ is the threshold of the detection process. For this sinusoid in the absence of noise, we can set γ to be zero.

Example 3.1 Consider a real-valued sinusoid of frequency f_1 and amplitude A_1 given by

$$x(t) = A_1 \cos(2\pi f_1 t) = (A_1/2)[e^{i2\pi f_1 t} + e^{-i2\pi f_1 t}], \; f_1 > 0, \; -\infty < t < \infty.$$

Then its spectrum is given by

$$X(f) = (A_1/2)[\delta(f - f_1) + \delta(f + f_1)], \; -\infty < f < \infty.$$

Similarly, if

$$x(t) = A_1 \sin(2\pi f_1 t) = (A_1/2i)[e^{i2\pi f_1 t} - e^{-i2\pi f_1 t}], \; f_1 > 0, \; -\infty < t < \infty,$$

then its spectrum is given by

$$X(f) = (A_1/2i)[\delta(f - f_1) - \delta(f + f_1)], \; -\infty < f < \infty.$$

However, for both real-valued sinusoids, their spectrum magnitudes are identical and given by

$$|X(f)| = (|A_1|/2)[\delta(f - f_1) + \delta(f + f_1)], \quad -\infty < f < \infty.$$

Next, consider two complex sinusoids specified over an infinite observation duration

$$x(t) = A_1 e^{i2\pi f_1 t} + A_2 e^{i2\pi f_2 t}, \qquad -\infty < t < \infty. \tag{3.3}$$

The Fourier transform of (3.3) is

$$X(f) = A_1\delta(f - f_1) + A_2\delta(f - f_2), \qquad -\infty < f < \infty. \tag{3.4}$$

We can estimate the two frequencies f_1 and f_2 regardless of how close f_1 and f_2 are and how small $|A_1|$ and $|A_2|$ are as long as both of them exceed the threshold level γ.

Now, consider the above two problems observed only over the duration $[0, T)$. For a single complex sinusoid,

$$x(t) = A_1 e^{i2\pi f_1 t}, \qquad 0 \le t < T, \tag{3.5}$$

and its spectrum is given by

$$X(f) = A_1 e^{-i\pi(f-f_1)T} \frac{\sin \pi(f - f_1)T}{\pi(f - f_1)}, \qquad -\infty < f < \infty. \tag{3.6}$$

Fig. 3.1 shows the spectrum magnitude $|X(f)|$ of $x(t)$. Instead of the delta function given by (3.2), $X(f)$ now has a mainlobe from $-1/T$ to $1/T$ and many sidelobes. Furthermore, $X(f)$ has an infinite number of nulls at $(f_1 + \frac{m}{T}), m = \pm 1, \pm 2, \ldots$ In order to determine f_1 as the peak of $|X(f)|$, the threshold level must be set above the peak of the highest sidelobes. In Fig. 3.1, γ must be set to be greater than 0.21 of the central peak.

Let us consider the case of two complex sinusoids observed over $[0, T)$,

$$x(t) = A_1 e^{i2\pi f_1 t} + A_2 e^{i2\pi f_2 t}, \qquad 0 \le t < T. \tag{3.7}$$

Then

$$X(f) = A_1 e^{-i\pi(f-f_1)T} \frac{\sin \pi(f - f_1)T}{\pi(f - f_1)} + A_2 e^{-i\pi(f-f_2)T} \frac{\sin \pi(f - f_2)T}{\pi(f - f_2)}, \quad -\infty < f < \infty. \tag{3.8}$$

By using the notation of

$$\mathrm{sinc}(x) = \frac{\sin(\pi x)}{\pi x}, \tag{3.9}$$

the spectrum magnitude squared of (3.7) becomes

$$\begin{aligned} |X(f)|^2 &= |A_1|^2 T^2 \mathrm{sinc}^2((f - f_1)T) + |A_2|^2 T^2 \mathrm{sinc}^2((f - f_2)T) \\ &\quad + 2T^2 \mathrm{sinc}((f - f_1)T)\mathrm{sinc}((f - f_2)T)\mathrm{Re}\{A_1 \bar{A}_2 e^{-i\pi T(f_2 - f_1)}\}, -\infty < f < \infty. \end{aligned} \tag{3.10}$$

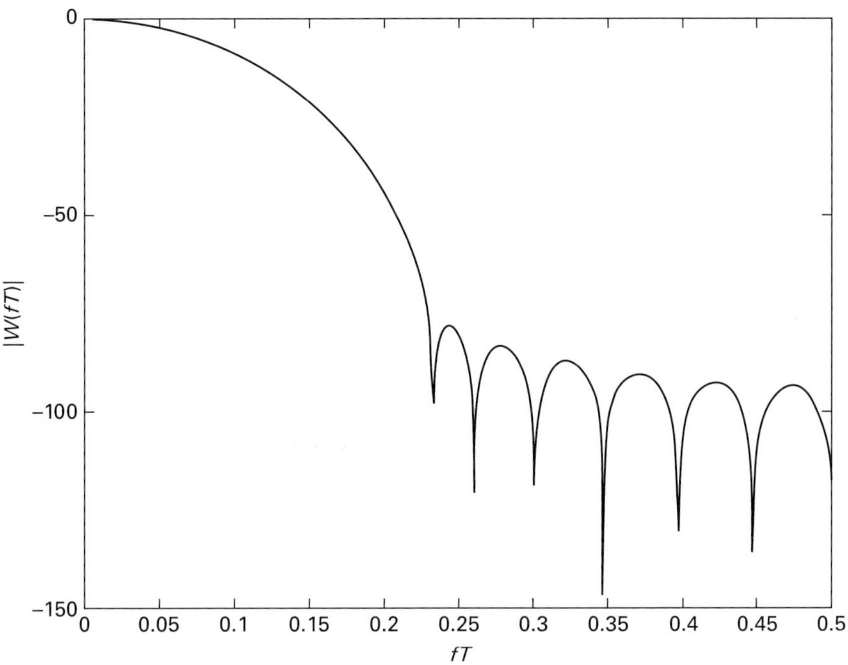

Figure 3.1 Spectrum magnitude of a single complex sinusoid observed over $[0, T)$

Example 3.2 Let $A_1 = A_2 e^{i\phi}$ and $f_2 - f_1 = 1/T$ in (3.10). Depending on the relative phase ϕ, $|X(f)|$ can have either one or two peaks. Consider two extreme cases of ϕ. Then,

$$|X(f)| = \begin{cases} |A_1|T \ |\text{sinc}((f - f_1)T) - \text{sinc}((f - f_2)T)|, & \phi = 0 \\ |A_1|T \ |\text{sinc}((f - f_1)T) + \text{sinc}((f - f_2)T)|, & \phi = \pi \end{cases}, \quad -\infty < f < \infty. \tag{3.11}$$

From (3.11), if $(f_2 - f_1) = 1/T$, we may not be able to determine the existence of the two sinusoids. Fig. 3.2 shows that $|X(f)|$ will have a single large and wide mainlobe for $\phi = \pi$. Specifically, $|X(f)|$ will have a peak at $f = (f_1 + f_2)/2$ with $|X((f_1 + f_2)/2)| = 4|A_1|T/\pi$. This peak is $(4\pi)^{1/2}$ larger than the peak of a single sinusoid in (3.6). The mainlobe has a width of $3/T$ versus that of $2/T$ in (3.6).

Example 3.3 Suppose we reconsider the above example with relative phase $\phi = \pi$ but increase $f_2 - f_1$ to $2/T$. There will be two mainlobes as shown in Fig. 3.3 near the two frequencies f_1 and f_2.

It is also clear from Fig. 3.3 that even if $(f_2 - f_1) > 2/T$, if $|A_2|$ is smaller than the magnitude of the sidelobe of the first complex sinusoid, the second complex sinusoid will not be detected if the threshold is set above $|A_2|$.

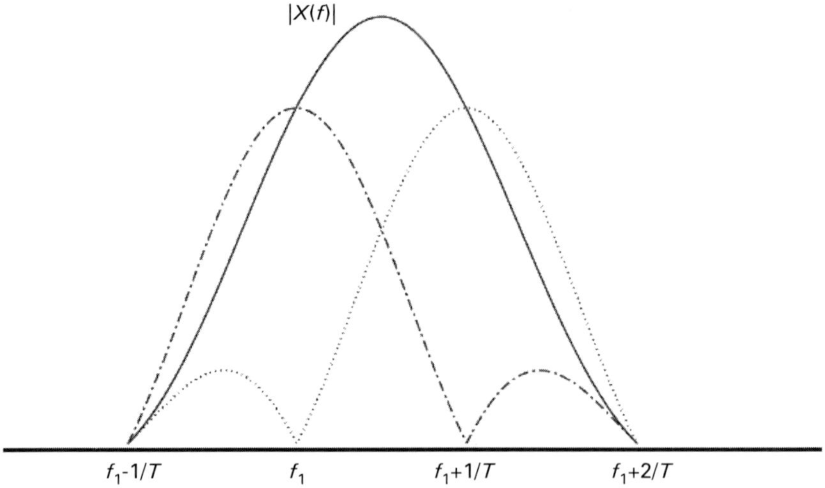

Figure 3.2 Spectrum magnitude of two complex sinusoids of equal amplitudes observed over $[0, T)$ with relative phase $\phi = \pi$ and $f_2 - f_1 = 1/T$

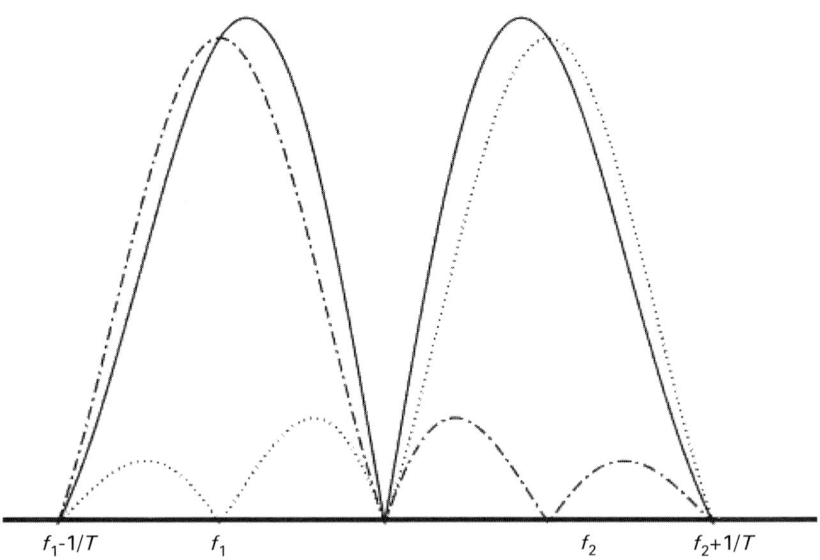

Figure 3.3 Spectrum magnitude of two complex sinusoids of equal amplitudes observed over $[0, T)$ with relative phase $\phi = \pi$ and $f_2 - f_1 = 2/T$

3.1.2 Discrete-Time Data

Consider a sampled single complex sinusoid of frequency f_1 and sampling interval T given by

$$x[n] = A_1 e^{i2\pi f_1 Tn}, \qquad 0 \leq f_1 T < 1, \qquad n = 0, 1, \ldots, N-1. \tag{3.12}$$

The DFT of (3.12) is

$$X[k] = A_1 e^{i\pi(f_1T - \frac{k}{N})(N-1)} \frac{\sin[\pi(f_1T - \frac{k}{N})N]}{\sin[\pi(f_1T - \frac{k}{N})]}, \qquad k = 0, 1, \ldots, N-1. \tag{3.13}$$

Comparing (3.13) with (3.6), we observe that $|X[k]|$ is similar to that of $|X(f)|$ with N replacing T. While the values of $|X(f)|$ are observable for $-\infty < f < \infty$, $|X[k]|$ are available only for $k = 0, 1, \ldots, N-1$. The nulls of $|X[k]|$ are located at

$$\frac{k}{N} = f_1T + \frac{m}{N}, \qquad m = \pm 1, \pm 2, \ldots$$

Example 3.4 Let $N = 8$, $f_1T = 1/8$, and $A_1 = 1$. Then

$$|X[k]| = \begin{cases} 8, & k = 1, \\ 0, & \text{otherwise.} \end{cases}$$

In general, if $f_1T = k_1/N$, $0 \le k_1 \le N-1$, then

$$|X[k]| = |A_1| N \delta_{k,k_1}, \qquad 0 \le k \le N-1. \tag{3.14}$$

Example 3.5 Let $N = 8$, $A_1 = 1$, but let $f_1T = 3/16$. Then $|X[k]|$ has two peaks at $k = 1$ and $k = 2$ as shown in Fig. 3.4.

It is clear from these two examples that if we know there is only one sinusoid, and we observed two peaks at $k = 1$ and $k = 2$ with $|X[1]| = |X[2]|$, then we can conclude $f_1T = 1.5/8 = 3/16$.

Consider two complex sinusoids given by

$$x[n] = A_1 e^{i2\pi f_1 T n} + A_2 e^{i2\pi f_2 T n}, \qquad n = 0, 1, \ldots, N-1. \tag{3.15}$$

The DFT of (3.15) is

$$\begin{aligned} X[k] = {} & A_1 e^{i\pi(f_1T - \frac{k}{N})(N-1)} \frac{\sin[\pi(f_1T - \frac{k}{N})N]}{\sin[\pi(f_1T - \frac{k}{N})]} \\ & + A_2 e^{i\pi(f_2T - \frac{k}{N})(N-1)} \frac{\sin[\pi(f_2T - \frac{k}{N})N]}{\sin[\pi(f_2T - \frac{k}{N})]}, \qquad k = 0, 1, \ldots, N-1, \end{aligned}$$

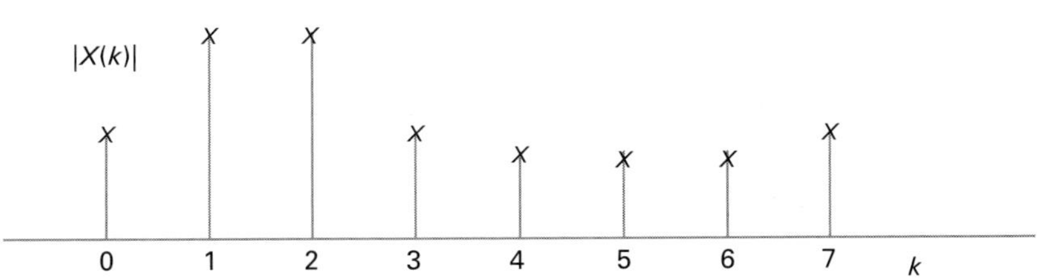

Figure 3.4 8-Point DFT with $f_1T = \frac{3}{16}$

and its magnitude squared is given by

$$|X[k]|^2 = |A_1|^2 \left\{ \frac{\sin[\pi(f_1T - \frac{k}{N})N]}{\sin[\pi(f_1T - \frac{k}{N})]} \right\}^2 + |A_2|^2 \left\{ \frac{\sin[\pi(f_2T - \frac{k}{N})N]}{\sin[\pi(f_2T - \frac{k}{N})]} \right\}^2$$
$$+ 2 \left\{ \frac{\sin[\pi(f_1T - \frac{k}{N})N]}{\sin[\pi(f_1T - \frac{k}{N})]} \right\} \left\{ \frac{\sin[\pi(f_2T - \frac{k}{N})N]}{\sin[\pi(f_2T - \frac{k}{N})]} \right\}$$
$$\times \operatorname{Re}\{A_1 \bar{A}_2 e^{i\pi(N-1)(f_2T - f_1T)}\}, \qquad k = 0, 1, \ldots, N-1. \quad (3.16)$$

Example 3.6 Consider two sampled complex sinusoids with a sufficiently large N such that $N \approx (N-1)$, A_1 and A_2 are real and equal, and $(f_2 - f_1)T = 2/(N-1) \approx 2/N$. Then (3.16) becomes

$$|X[k]| = |A_1| \left| \frac{\sin[\pi(f_1T - \frac{k}{N})N]}{\sin[\pi(f_1T - \frac{k}{N})]} + \frac{\sin[\pi(f_2T - \frac{k}{N})N]}{\sin[\pi(f_2T - \frac{k}{N})]} \right|.$$

Furthermore, let

$$f_1T = \frac{k_1 - 0.5}{N}, \qquad f_2T = \frac{k_1 + 1.5}{N}.$$

Then using $\sin(x) \approx x$ for small x, we have

$$|X[k_1+1]| = |A_1| \left| \frac{N}{1.5\pi} + \frac{N}{0.5\pi} \right| = |X[k_1]| > |X[k_1 - 1]| = |X[k_1 + 2]|.$$

This shows the two peaks can be observed. In general, we need $|f_2 - f_1|T \geq 2/N$ to be able to resolve two complex sinusoids of frequencies f_1 and f_2.

3.2 Windows

3.2.1 Continuous-Time Signal

Let $\{x(t), -\infty < t < \infty\}$ be a continuous-time signal of interest, but which can be observed only over a finite time duration of $[0, T)$. Denote a rectangular window of length T by

$$w(t) = \begin{cases} 1, & 0 \leq t < T, \\ 0, & \text{elsewhere.} \end{cases} \quad (3.17)$$

Then the windowed waveform $y(t)$ is given by

$$y(t) = w(t)x(t), \quad -\infty < t < \infty, \quad (3.18)$$

and its Fourier transform $Y(f)$ is given by

$$Y(f) = \int_{-\infty}^{\infty} W(f - f')X(f')df', \qquad -\infty < f < \infty, \quad (3.19)$$

where $W(f)$ is the Fourier transform of $w(t)$ and $X(f)$ is the Fourier transform of $x(t)$. Ideally, we want the windowed waveform $y(t)$ to be similar to the original signal $x(t)$. In the frequency domain, $Y(f)$ is identical to $X(f)$, if $W(f) = \delta(f)$, $-\infty < f < \infty$, in

(3.19). However, the inverse Fourier transform of $W(f) = \delta(f), -\infty < f < \infty$, yields $w(t) = 1, -\infty < t < \infty$, which is a unity valued window of infinite time duration. The crucial issue is what finite duration window $w(t)$ over $[0, T)$ and zero elsewhere is such that its Fourier transform $W(f)$ has a large narrow mainlobe with low sidelobe peaks. In other words, we seek finite duration windows with their $W(f)$ approximating the delta function in the frequency domain. We note, due to the convolutional nature of (3.19) in the frequency domain, the widths of all the peaks of the original signal $X(f)$ are broadened. A narrower mainlobe in $W(f)$ implies there is less broadening and thus the frequency resolution capability of the original signal $X(f)$ is more preserved. Similarly, lower sidelobes in $W(f)$ of (3.19) imply that the low amplitude signal values of $X(f)$ will be less disturbed in $Y(f)$ due to the presence of other possible strong adjacent frequency interfering waveforms.

Now, consider a folklore lemma in Fourier transformation that relates, in a qualitative manner, the smoothness of $w(t)$ in the time-domain to the decrease of the sidelobes of $W(f)$ in the frequency domain.

Lemma 3.1 *Let the smoothness of a time-domain function $w(t), -\infty < t < \infty$, be characterized as inversely related to the magnitude of its nth-order derivative $\partial^n w(t)/\partial t^n$ for some positive integer n. Then the smoothness of $w(t)$ is associated with the rapidness of the decrease of $W(f)$ in the frequency domain of $-\infty < f < \infty$.*

By taking the nth partial derivative of

$$w(t) = \int_{-\infty}^{\infty} e^{i2\pi ft} W(f)\, df, \qquad -\infty < t < \infty,$$

we obtain

$$\frac{\partial^n w(t)}{\partial t^n} = (i2\pi)^n \int_{-\infty}^{\infty} f^n W(f) e^{i2\pi ft}\, df, \qquad -\infty < t < \infty.$$

In order for $|\partial^n w(t)/\partial t^n|$ to be small for any positive integral n, $W(f)$ must decay rapidly.

Example 3.7 Consider a rectangular window

$$w(t) = \begin{cases} 1, & 0 \le t < T, \\ 0, & \text{elsewhere}, \end{cases}$$

which is not continuous at 0 and T, while a triangular window

$$w(t) = \begin{cases} 2\frac{t}{T}, & 0 \le t \le \frac{T}{2}, \\ 2(1 - \frac{t}{T}), & \frac{T}{2} \le t < T, \\ 0, & \text{elsewhere}, \end{cases}$$

is a continuous function on $(-\infty, \infty)$, but not differentiable at $0, T/2$, and T. Thus, the triangular window is "smoother" than the rectangular window. Indeed, the sidelobe peak of a rectangular window is -13dB below that of its mainlobe peak, while that of a triangular window is -27dB. Other windows with additional smoother properties in the time-domain indeed yield even lower sidelobe peaks.

3.2.2 Discrete-Time Data

Let $x[n] = x(nT), n = 0, 1, \ldots, N-1$, be an input discrete-time signal sequence obtained from the sampling of a continuous-time signal $x(t)$ with a sampling interval of T. If $w[n], n = 0, 1, \ldots, N-1$, denotes the window weights, then the output $y[n], n = 0, 1, \ldots, N-1$, of the windowed sequence is given by

$$y[n] = w[n]x[n], \qquad n = 0, 1, \ldots, N-1. \tag{3.20}$$

The frequency response $Y(e^{i2\pi fT}), 0 \le fT < 1$, of $y[n], n = 0, 1, \ldots, N-1$, is given by

$$Y(e^{i2\pi fT}) = \int_0^1 X(e^{i2\pi(fT-\sigma)})W(e^{i2\pi\sigma})d\sigma, \qquad 0 \le fT < 1, \tag{3.21}$$

where

$$X(e^{i2\pi fT}) = \sum_{n=0}^{N-1} x[n]e^{-i2\pi nfT}, \qquad 0 \le fT < 1, \tag{3.22}$$

and

$$W(e^{i2\pi fT}) = \sum_{n=0}^{N-1} w[n]e^{-i2\pi nfT}, \qquad 0 \le fT < 1. \tag{3.23}$$

Upon substituting (3.22) and (3.23) into (3.21), we obtain

$$Y(e^{i2\pi fT}) = \sum_{n=0}^{N-1} x[n]w[n]e^{-i2\pi nfT} = \sum_{n=0}^{N-1} y[n]e^{-i2\pi nfT}, \qquad 0 \le fT < 1. \tag{3.24}$$

For DFT and FFT applications, we want to consider only discrete normalized frequencies of

$$fT = \frac{k}{N}, \qquad k = 0, 1, \ldots, N-1, \tag{3.25}$$

in (3.22)–(3.24). Then we denote

$$X[k] = \sum_{n=0}^{N-1} x[n]e^{-i2\pi nk/N}, \qquad k = 0, \ldots, N-1. \tag{3.26}$$

$$W[k] = \sum_{n=0}^{N-1} w[n]e^{-i2\pi nk/N}, \qquad k = 0, \ldots, N-1. \tag{3.27}$$

$$Y[k] = \sum_{n=0}^{N-1} x[n]w[n]e^{-i2\pi nk/N} = \sum_{n=0}^{N-1} y[n]e^{-i2\pi nk/N}, \qquad k = 0, \ldots, N-1. \tag{3.28}$$

Symmetry Conditions

For a continuous-time waveform, if $\{w(t), -\infty < t < \infty\}$ is real-valued, then the magnitude of its Fourier transform, $|W(f)|$, is an even function

$$|W(f)| = |W(-f)|, \qquad 0 \le f < \infty. \tag{3.29a}$$

Similarly, if $\{w[n], n = 0, \ldots, N-1\}$ is a real-valued sequence, the magnitude of its DFT, $|W[k]|$, is symmetric about its mid-point. That is, for

$$N \text{ even:} \quad |W[\tfrac{N}{2} - k]| = |W[\tfrac{N}{2} + k]|, \quad k = 1, \ldots, \tfrac{N}{2} - 1, \tag{3.29b}$$

$$N \text{ odd:} \quad |W[\tfrac{N+1-2k}{2}]| = |W[\tfrac{N-1+2k}{2}]|, \quad k = 1, \ldots, \tfrac{N-1}{2}. \tag{3.29c}$$

Now, in addition to the real-valuedness of $w[n]$, suppose we also impose a particular form of symmetry condition on $w[n]$ about its mid-point in the periodic extension sense. Then its DFT, $W[k]$, is also real-valued and symmetric about its mid-point in the periodic extension sense.

Lemma 3.2 *Consider a real-valued window sequence $\{w[n], n = 0, \ldots, N-1\}$ of length N with periodic extension symmetric conditions given by*

$$N \text{ even:} \quad w[\tfrac{N}{2} - m] = w[\tfrac{N}{2} + m], \quad m = 1, \ldots, \tfrac{N}{2} - 1, \tag{3.30a}$$

$$N \text{ odd:} \quad w[\tfrac{N+1-2m}{2}] = w[\tfrac{N-1+2m}{2}], \quad m = 1, \ldots, \tfrac{N-1}{2}. \tag{3.30b}$$

Then its DFT values are also real-valued and satisfy the periodic extension symmetric conditions of

$$N \text{ even:} \quad W[\tfrac{N}{2} - k] = W[\tfrac{N}{2} + k], \quad k = 1, \ldots, \tfrac{N}{2} - 1, \tag{3.31a}$$

$$N \text{ odd:} \quad W[\tfrac{N+1-2k}{2}] = W[\tfrac{N-1+2k}{2}], \quad k = 1, \ldots, \tfrac{N-1}{2}. \tag{3.31b}$$

Proof: For an even N, from (3.27),

$$W[k] = W_R[k] + i\,W_I[k] = \sum_{n=0}^{N-1} w[n]\big[\cos[\tfrac{2\pi nk}{N}] - i\,\sin[\tfrac{2\pi nk}{N}]\big], \quad k = 0, \ldots, N-1. \tag{3.32}$$

The imaginary part of $W[k]$ in (3.32) is then given by

$$W_I[k] = -w[0]\cdot 0 - w[\tfrac{N}{2}]\cdot 0 - \sum_{m=1}^{\frac{N}{2}-1} w[\tfrac{N}{2}+m]\big[\sin[\tfrac{2\pi k}{N}(\tfrac{N}{2}+m)] + \sin[\tfrac{2\pi k}{N}(\tfrac{N}{2}-m)]\big], \tag{3.33}$$

where the symmetry conditions of (3.30a) about $n = \frac{N}{2}$ are used. Upon expansion, the bracketed term in (3.33) is zero for all $m = 1, \ldots, \frac{N}{2} - 1$. Thus, $W_I[k] = 0$, and $W[k] = W_R[k]$ in (3.32) can be expressed as

$$\begin{aligned} W_R[k] &= w[0] + w[\tfrac{N}{2}]\cos[\pi k] + \sum_{m=1}^{\frac{N}{2}-1} w[\tfrac{N}{2}+m]\big[\cos[\tfrac{2\pi k}{N}(\tfrac{N}{2}+m)] + \cos[\tfrac{2\pi k}{N}(\tfrac{N}{2}-m)]\big] \\ &= w[0] + (-1)^k \left[w[\tfrac{N}{2}] + 2\sum_{m=1}^{\frac{N}{2}-1} w[\tfrac{N}{2}+m]\cos[\tfrac{2\pi km}{N}] \right]. \end{aligned} \tag{3.34}$$

To show the symmetry of $W_R[k]$ about $N/2$, consider $k = (N/2) \pm k'$, $k' = 1, \ldots, (N/2) - 1$. Upon using

$$\cos\left[\frac{2\pi km}{N}\right] = \cos[\pi m]\cos\left[\frac{2\pi mk'}{N}\right], \qquad k' = 1, \ldots, \frac{N}{2} - 1$$

in (3.34), we obtain,

$$W_R[\tfrac{N}{2} \pm k'] = w[0] + (-1)^{\frac{N}{2}}(-1)^{\pm k'}\left[w[\tfrac{N}{2}] + 2\sum_{m=1}^{\frac{N}{2}-1}(-1)^M W[\frac{N}{2} + n]\cos[\tfrac{2\pi mk'}{N}]\right],$$
$$k' = 1, \ldots, \frac{N}{2} - 1, \tag{3.35}$$

which is equivalent to (3.31a). For an odd N, the symmetry conditions of (3.30a) for $w[n]$ are used to obtain the real-valuedness and symmetry conditions of $W[k]$ in (3.31b) in an identical manner as above. We note in (3.30), $w[0]$ has no symmetric pair in the original window sequence but indeed equals itself upon a periodic extension.

Example 3.8 Consider a 4-point triangular window sequence with $w[0] = 1/3$, $w[1] = 2/3$, $w[2] = 1$, and $w[3] = 2/3$ satisfying (3.30a) is given in the top left position of Fig. 3.5. Then direct DFT shows $W[0] = 8/3$, $W[1] = -2/3$, $W[2] = 0$, and $W[3] = -2/3$ are real-valued and satisfying (3.31a) is given in the bottom left position of Fig. 3.5. Similarly, consider a 5-point triangular window sequence with $w[0] = 0.1667$, $w[1] = 0.5$, $w[2] = 0.8333$, $w[3] = 0.8333$, and $w[4] = 0.5$ satisfying (3.30b) is given in the top right position of Fig. 3.5. Then $W[0] = 2.8333$, $W[1] = -0.8727$, $W[2] = -0.1273$, $W[3] = -0.1273$, and $W[4] = -0.8727$ are real-valued and satisfying (3.31) is given in the bottom right position of Fig. 3.5.

Example 3.9 Consider a 4-point triangular window sequence with $w[0] = 1/4$, $w[1] = 3/4$, $w[2] = 3/4$, and $w[3] = 1/4$. We note this sequence is "fully symmetric" within the sequence but does not satisfy (3.30a). Then $W[0] = 2$, $W[1] = -(1 + i)/2$, $W[2] = 0$, and $W[3] = (-1 + i)/2$ are not real-valued and do not satisfy (3.31b). Since $w[n]$ is real-valued, the symmetry conditions on $|W[k]|$ of (3.29c) are satisfied.

Discrete-Time Windows

Rectangular Window

Let

$$w[n] = \begin{cases} 1, & n = 0, 1, \ldots, N-1, \\ 0, & \text{otherwise.} \end{cases} \tag{3.36}$$

Then the window in the frequency domain on the interval $[0, 1)$ from (3.22) is given by

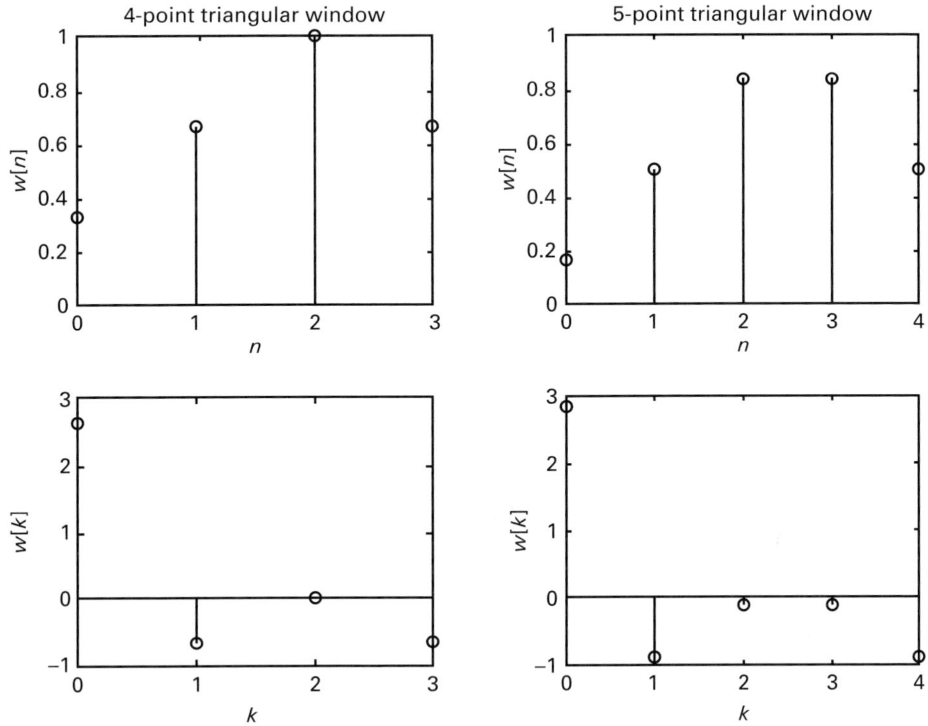

Figure 3.5 4-point and 5-point real-valued triangular windows with periodic extension symmetric conditions in time- and frequency-domains

$$W(e^{i2\pi fT}) = \sum_{n=0}^{N-1} e^{-i2\pi nfT} = e^{i\pi fT(1-N)} \frac{\sin(\pi NfT)}{\sin(\pi fT)}, \qquad 0 \leq fT < 1. \tag{3.37}$$

The DFT of the window from (3.27) is given by

$$W[k] = \sum_{n=0}^{N-1} e^{-i2\pi nk/N} = \begin{cases} N, & k = 0 \\ 0, & k = 1, \ldots, N-1. \end{cases} \tag{3.38}$$

Example 3.10 Consider an 8-point rectangular window defined by (3.36). From (3.38), $W[0] = 8$, $W[k] = 0$, $k = 1, \ldots, 7$. These 8 values also equal $W(e^{i2\pi fT})_{fT=0} = 8$, $W(e^{i2\pi fT})_{fT=k/8} = 0$, $k = 1, \ldots, 7$, of (3.37). In order to determine the frequency response of the window, we need to evaluate values of $W(e^{i2\pi fT})$ in addition to $f = k/8$, $k = 1, \ldots, 7$. One simple approach is to use an N_1-point FFT of the window weights padded with $N_1 - N$ zeros, where N_1 is greater (possibly much greater) than N. A plot of the magnitude of this FFT with $N_1 = 64$ is given in Fig. 3.6. This figure shows the fine frequency-domain behavior of $W(e^{i2\pi fT})$ not available from (3.38) using an N-point DFT on (3.36).

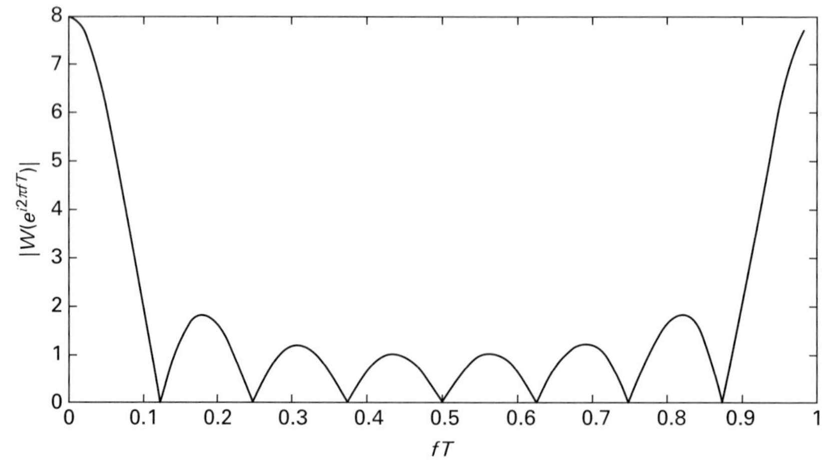

Figure 3.6 Spectrum magnitude of an $N = 8$-point rectangular window versus normalized frequency

Chebyshev Window

The Chebyshev window is specified in the frequency domain by

$$W(e^{i2\pi fT}) = \frac{e^{-i\pi(N-1)fT}\, T_{N-1}(\beta\cos(\pi fT))}{T_{N-1}(\beta)}, \quad 0 \le fT < 1, \tag{3.39a}$$

where the Chebyshev polynomial is defined by

$$T_{N-1}(x) = \begin{cases} (-1)^{N-1}\cosh((N-1)\cosh^{-1}(|x|)), & -\infty < x \le -1, & (3.39\text{b}) \\ \cos((N-1)\cos^{-1}(x)), & -1 \le x \le 1, & (3.39\text{c}) \\ \cosh((N-1)\cosh^{-1}(x)), & 1 \le x < \infty, & (3.39\text{d}) \end{cases}$$

and

$$1 < \beta = \cosh\left(\frac{\cosh^{-1}(\alpha)}{(N-1)}\right), \quad \alpha > 1. \tag{3.40}$$

The quantity α is the ratio of the mainlobe peak $|W[0]| = 1$ to the sidelobe peak,

$$\alpha = \frac{|W[0]|}{\text{SL peak}} = \frac{1}{|W(f_0T)|} > 1, \tag{3.41}$$

where f_0 is the smallest positive frequency such that $|W(f_0T)|$ equals the sidelobe peaks. Indeed, by using (3.40) in (3.39d),

$$\begin{aligned} T_{N-1}(\beta) &= \cosh((N-1)\cosh^{-1}(\beta)) \\ &= \cosh((N-1)\cosh^{-1}(\cosh(\tfrac{1}{N-1})\cosh^{-1}(\alpha))) = \alpha = \tfrac{1}{|W(f_0T)|}. \end{aligned} \tag{3.42}$$

But from (3.39a) with $f = f_0$ and (3.42), we have

$$|W(f_0T)| = \frac{T_{N-1}(\beta\cos(\pi f_0T))}{T_{N-1}(\beta)} = |W(f_0T)|T_{N-1}(\beta\cos(\pi f_0T)),$$

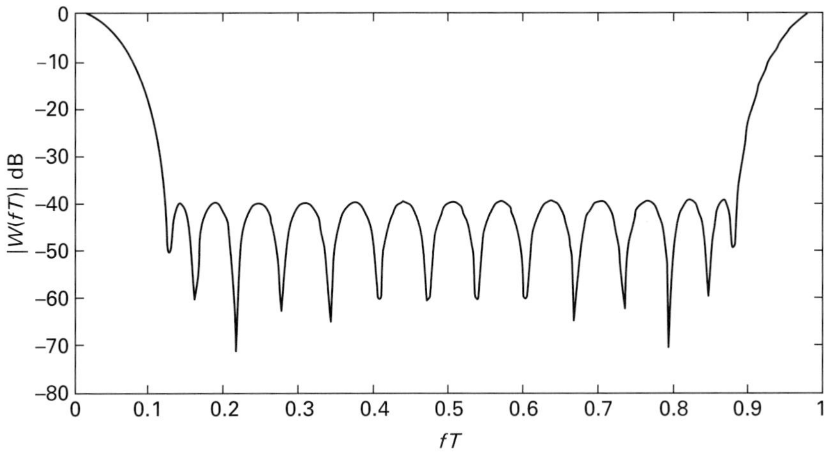

Figure 3.7 Normalized frequency response of a Chebyshev window

which shows $T_{N-1}(\beta\cos(\pi f_0 T)) = 1$. From (3.39c), this implies $\beta\cos(\pi f_0 T) = 1$. Thus, $f_0 T$ is the smallest positive number given by

$$f_0 T = \frac{1}{\pi}\cos^{-1}\left(\frac{1}{\beta}\right). \tag{3.43}$$

Furthermore, from (3.39a) and (3.39c), all the sidelobe peaks are equal to $|W(f_0 T)|$. The Chebyshev window has the optimum property that among all normalized window transform functions (i.e., $|W[0]| = 1$) with sidelobe peaks less than or equal to $|W(f_0 T)|$, all mainlobe widths are larger than or equal to $2f_0 T$.

Example 3.11 Consider a 40dB Chebyshev window with $N = 15$. From $40(\mathrm{dB}) = 20\log_{10}\alpha(\mathrm{dB})$ this implies $\alpha = 100 = 1/0.01$ or an SL peak $= |W(f_0 T)| = 0.01$. Then $\beta = \cosh((1/14)\cosh^{-1}(100)) = 1.0725$. Then $f_0 T = (1/\pi)\cos^{-1}(1/\beta) = 0.1177$. A plot of the normalized frequency response of this Chebyshev window is shown in Fig. 3.7.

3.2.3 Digital Prolate Window

The digital prolate window is a real-valued window sequence that has maximum normalized energy in the region of $[-f_0 T, f_0 T]$. Let $W(e^{i2\pi fT})$ given by (3.23) denote the window transform function and β denote the ratio of energy of $W(e^{i2\pi fT})$ in $[-f_0 T, f_0 T]$ to the total energy in $[-0.5, 0.5]$. Then

$$\begin{aligned}\beta &= \frac{\int_{-f_0 T}^{f_0 T} |W(e^{i2\pi fT})|^2\,df}{\int_{-0.5}^{0.5} |W(e^{i2\pi fT})|^2\,df}\\ &= \frac{\sum_{n=0}^{N-1}\sum_{m=0}^{N-1} w[n]w[m]a_{m,n}(f_0 T)}{\sum_{n=1}^{N}\sum_{m=1}^{N} w[n]w[m]a_{m,n}(0.5)},\end{aligned} \tag{3.44}$$

where

$$a_{m,n}(fT) = \int_{-fT}^{fT} e^{i2\pi(n-m)x}\,dx = \frac{\sin[2\pi(n-m)fT]}{\pi(n-m)}, \qquad m,n = 0,\ldots,N-1. \quad (3.45)$$

Let

$$A = \left[a_{m,n}(f_0T)\right], \qquad m,n = 0,\ldots,N-1, \quad (3.46)$$

then A is a real-valued symmetric positive-definite Toeplitz matrix. Furthermore, denote the window weights by the $N \times 1$ vector,

$$w = [w[0],\ldots,w[N-1]]^T.$$

Then β in (3.44) can be expressed in the form of the Rayleigh ratio

$$\beta = \frac{w^T A w}{w^T w}. \quad (3.47)$$

Thus, the digital prolate window weight vector, $\hat{w}$, attains the maximum of β among all real-valued weight vectors of length N in

$$\hat{\beta} = \max_{w \in \mathcal{R}^N} \frac{w^T A w}{w^T w} = \frac{\hat{w}^T A \hat{w}}{\hat{w}^T \hat{w}}. \quad (3.48)$$

Furthermore, this maximum fraction of energy $\hat{\beta}$ in $[-f_0T, f_0T]$, is given by the largest eigenvalue λ_1 and $\hat{w}$ is given by the eigenvector w_1 corresponding to λ_1 in the eigenvalue problem of

$$Aw_i = \lambda_i w_i, \qquad \lambda_1 \geq \lambda_2 \geq \cdots \geq \lambda_N > 0. \quad (3.49)$$

A simple computationally efficient method for the evaluation of λ_1 and w_1 is based on the power method algorithm.

Power Method Algorithm

Step 1 Consider any N-dimensional vector u in $\mathcal{R}^N$ with the first component normalized to unity.

Step 2 Evaluate $v = [v[1],\ldots,v(N)]^T = Au$.

Step 3 At the end of one iteration, $v[1]$ approximates the largest eigenvalue of (3.49) and v approximates the associated non-normalized eigenvector.

Step 4 Stop the algorithm when all the components of v in two successive iterations are within some acceptable tolerance. Otherwise, go to Step 5.

Step 5 Normalize v by taking $z = v/v[1]$ and set $u = z$. Then go to Step 2.

At the end of the algorithm, the resulting $v = w_1 = \hat{w}$ yields the digital prolate window weight vector and $v[1] = \lambda_1$.

Example 3.12 Consider an 8-point digital prolate window with $f_0T = 0.05$. Then $\beta = \lambda_1 = 0.681$ and the normalized $w_1 = \hat{w} = [1, 1.10, 1.17, 1.21, 1.21, 1.17, 1.10, 1]^T$ is obtained in four iterations using the power method algorithm.

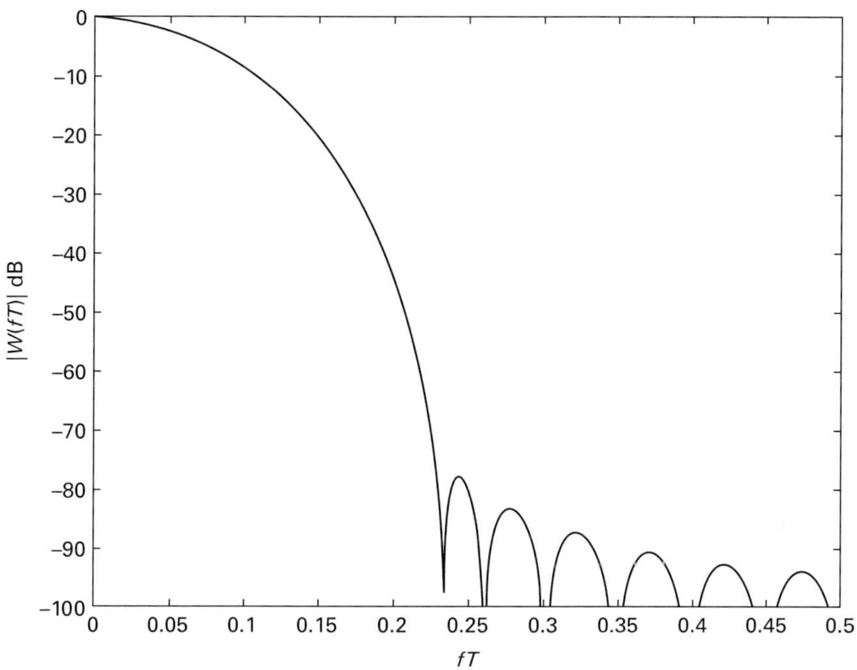

Figure 3.8 Normalized spectrum magnitude versus normalized frequency of a 16-point Kaiser window with $\alpha = 3.35$

3.2.4 Kaiser Window

A simple approximation to the digital prolate window is given by the Kaiser window of

$$w[n] = \frac{I_0\left(\pi\alpha\sqrt{1 - [\frac{2}{N}(n - \frac{N}{2})]^2}\right)}{I_0(\pi\alpha)}, \qquad n = 0, \ldots, N-1, \tag{3.50}$$

where α is related to the f_0 of the digital prolate window, and $I_0(\cdot)$ is the zeroth order modified Bessel function of the first kind.

Example 3.13 In Fig. 3.8, the magnitude of the normalized frequency response of a 16-point Kaiser window with $\alpha = 3.35$ is shown. The first null appears at 0.22. The first sidelobe peak has a value of -79dB.

3.2.5 Maximum Energy Window with Constrained Spectral Values

As discussed above, the digital prolate window is a real-valued window sequence that has maximum normalized energy in the region of $[-f_0T, f_0T]$. However, in various applications we may not only be interested in packing maximum energy in some desired frequency region, but also be interested in imposing precise spectral window values or spectral nulls over some specified frequencies in order to reduce unwanted large

Table 3.1 Parameters of non-constrained and constrained spectral value windows

Case	R	θ_0	Constrained θ_m	Constrained $W(\theta_m)/W(\theta)$	Constrained $\lvert W(\theta_m)/W(\theta)\rvert$dB	max β	Spectral nulls	Iter.	Characterization
1.1	7	0.1π	None	None	None	0.618	$(0.297,0.567,0.859)\pi$	4	Digital Prolate
1.2	7	0.1π	0	0.35π	$-\infty$	0.597	$(0.350,0.599,0.855)\pi$	3	Opt. const.
1.3	7	0.1π	0	$(0.35, 0.60)\pi$	$-\infty$	0.594983	$(0.350,0.600,0.851)\pi$	2	Opt. const.
1.4	7	0.1π	0	$(0.35, 0.60, 0.85)\pi$	$-\infty$	0.594981	$(0.350,0.600,0.850)\pi$	2	Opt. const.
2.1	8	0.1π	None	None	None	0.681	$(0.263,0.505,0.752,1)\pi$	4	Digital Prolate
2.2	8	0.1π	$(0.50, 0.70)\pi$	$(-0.01778, 0.01)\pi$	-35dB, -40dB	0.674	$(0.268,0.511,0.707,1)\pi$	4	Opt. const.
2.3	8	0.1π	–	–	-40 dB	0.558	$(0.460,0.589,0.783,1)\pi$	–	-40 dB Chebyshev
2.4	8	0.1π	–	–	0,-35dB, -40dB	0.565	$(0.396,0.695,1)\pi$	–	Finite F.S. app.

valued interferers at these frequencies. We can show that the unconstrained optimization problem of (3.48) can be modified to a constrained optimization problem imposing the specified spectral window values or nulls as shown in Table 3.1 from [8]. Specifically, in Case 1.1, a length 7 digital prolate window with $\theta_0 = 0.1\pi$ yielded the maximum $\beta = 0.618$ with no spectral value constraint but having spectral nulls at $(0.297, 0.567, 0.859)\pi$. In Case 1.2, suppose we impose a desired spectral null at 0.35π, then the maximum $\beta = 0.597$ only decreased slightly, but now yielded the desired spectral nulls of $(0.350, 0.599, 0.855)\pi$. Additional spectral value constraints were considered in Cases 1.3 and 1.4.

3.3 Processing Gain and Equivalent Noise Bandwidth

Consider an N-point window with time-domain weights given by $w[n], n = 0, 1, \ldots, N-1$, and the frequency-domain values given by $W[k]$ of (3.27). Let the input discrete-time data be

$$x[n] = s[n] + v[n], \qquad n = 0, 1, \ldots, N-1, \tag{3.51}$$

where $s[n]$ is a sequence taken from a complex sinusoidal waveform of normalized frequency k_1/N, $s[n] = Ae^{i2\pi k_1 n/N}, n = 0, 1, \ldots, N-1$, and $v[n]$ is a complex-valued zero-mean white Gaussian random sequence of variance σ^2. Let the output sequence of the window be denoted by

$$y[n] = x[n]w[n], \qquad n = 0, 1, \ldots, N-1. \tag{3.52}$$

The frequency-domain characterization of $y[n], n = 0, 1, \ldots, N-1$, is given by

$$Y[k] = \sum_{n=0}^{N-1} y[n]e^{-i2\pi nk/N} = Y_s[k] + Y_v[k], \qquad k = 0, 1, \ldots, N-1, \tag{3.53}$$

where the output signal component is

$$Y_s[k] = A\sum_{n=0}^{N-1} w[n]e^{i2\pi k_1 n/N}e^{-i2\pi nk/N}, \qquad k = 0, 1, \ldots, N-1, \tag{3.54}$$

and the output noise component is

$$Y_v[k] = \sum_{n=0}^{N-1} w[n]v[n]e^{-i2\pi nk/N}, \qquad k = 0, 1, \ldots, N-1. \tag{3.55}$$

The input signal-to-noise ratio (SNR) is given by

$$SNR_{in} = \frac{A^2}{\sigma^2}. \tag{3.56}$$

The SNR at the k_1th filter output is given by

$$SNR_{out} = \frac{|Y_s(k_1)|^2}{E\{|Y_v(k_1)|^2\}}, \tag{3.57}$$

$$= \frac{A^2}{\sigma^2}\frac{\left|\sum_{n=0}^{N-1} w[n]\right|^2}{\sum_{n=0}^{N-1} |w[n]|^2}. \tag{3.58}$$

The processing gain (PG) of a window is defined as the ratio of the output SNR to the input SNR. Therefore

$$PG = \frac{\left|\sum_{n=0}^{N-1} w[n]\right|^2}{\sum_{n=0}^{N-1} |w[n]|^2}. \tag{3.59}$$

For the problem of detecting a single complex sinusoid in white Gaussian noise, we clearly want to maximize the SNR_{out}. From Schwarz's inequality, we have

$$\frac{\left|\sum_{n=0}^{N-1} a[n]w[n]\right|^2}{\sum_{n=0}^{N-1} |w[n]|^2} \leq \sum_{n=0}^{N-1} |a[n]|^2. \tag{3.60}$$

In particular, if $a[n] = 1$ for $n = 0, 1, \ldots, N-1$, the upper bound is attainable when $w[n] = 1$ for $n = 0, 1, \ldots, N-1$. This shows that the maximum PG is given by N. Thus, for the optimum detection of a single complex sinusoid in white Gaussian noise, the optimum window is an uniform window. This conclusion is reached when nearby interferences in the frequency-domain are ignored.

In general, we use the term of equivalent noise bandwidth BW_{eq} to denote the effectiveness of the window in a noisy environment. An equivalent noise bandwidth BW_{eq} of a window is defined as the bandwidth of an equivalent flat-top filter in the frequency-domain such that the total output noise power of this equivalent filter equals the total noise power of the original filter. Thus

$$\sigma^2|W[0]|^2(BW_{eq}) = \sigma^2 \sum_{n=0}^{N-1} |w[n]|^2.$$

Hence BW_{eq} is given by

$$BW_{eq} = \frac{\sum_{n=0}^{N-1} |w[n]|^2}{\left|\sum_{n=0}^{N-1} w[n]\right|^2} = \frac{1}{PG}, \tag{3.61}$$

which is the reciprocal of the processing gain obtained before.

3.4 Conclusion

Section 3.1 considered some frequency resolution properties for simple continuous- and discrete-time signals. Section 3.2 introduced the influence of windowing on the resolution of continuous- and discrete-signals. Some details are given for the Chebyshev, the digital prolate, the constrained digital prolate, and the Kaiser windows. Section 3.3 treats the processing gain of the FFT viewed as a filter and the equivalent noise bandwidth issues.

3.5 References

Frequency resolutions of Fourier transform were of interest to physicists, astronomers, statisticians working in time-series [1] and to engineers [2]. Extensive spectral windows were tabulated in [3] and treated in [4]. More recent scholarly treatment of these issues appeared in [5]. The maximum energy window based on the prolate spheroidal wave function was introduced in [6] and the Kaiser window [7] can provide a good approximation. Maximum energy window with and without constrained spectral values was introduced in [8]. Processing gain and equivalent noise bandwidth concepts were well-known to sonar and radar engineers [9] and [10].

[1] M.S. Bartlett, "Smoothing Periodograms for Time-Series with Continuous Spectra," *Nature*, 1948, pp. 686–687.

[2] R.B. Blackman and J.W. Tukey, *The Measurement of Power Spectra*, Dover, 1958, pp. 95–100.

[3] F.J. Harris, "On the Use of Windows for Harmonic Analysis with the Discrete Fourier Transform," *Proceedings of the IEEE*, 1978, pp. 51–83.

[4] L. Marple, *Digital Spectral Analysis with Applications*, Prentice-Hall, 1987.

[5] P. Stoica and R. Moses, *Introduction to Spectral Analysis*, Prentice-Hall, 1997.

[6] H. Landau and H. Pollak, "Prolate-Spheroidal Wave Function, Fourier Analysis and Uncertainty-II," *Bell Tel. Syst. J.*, 1961, pp. 65–84.

[7] J.F. Kaiser, "Nonrecursive Digital Filter Design Using the I0-sinh Window Function," *Proc. 1974 IEEE Symp. Circuits and Syst.* 1974, pp. 20–23.

[8] K. Yao, "Maximum Energy Window with Constrained Spectral Values," *Signal Processing*, 1986, pp. 157–168.

[9] J.V. DiFranco and W.L. Rubin, *Radar Detection*, Artech House, 1980.

[10] G.W. Stimson, *Introduction to Airborne Radar*, Hughes Aircraft Company, 1983.

3.6 Exercises

1. Plot $|X(f)|$ with $\phi = 0, \pi/4, \pi/2, 3\pi/4$, and π using the parameters in Equation (3.11). What general conclusion can one draw about the resolution capability of Fourier transformation for these two equal amplitude complex sinusoids as a function of the relative phase angle ϕ?

2. For the following $N = 16$ windows, find $w[n]$, $n = 0,\ldots,N-1$, and $W[k]$, $k = 0,\ldots,N-1$. Plot the frequency response of the windows with few nulls, sidelobe peaks, etc. Find the processing gain of the window.

   a. Uniform window.

   b. Chebyshev window with -45 dB sidelobe peaks.

      i. Use $W(f)$ as defined in Section 3.2.2.

      ii. Use time-domain weights given by $w[0] = 1, w[(N/2)-1-i] = w[(N/2)+i], i = 0,\ldots,(N/2)-1$, for even valued N, where $w[n] = \frac{(N-1)}{(N-1-n)}\Sigma_{s=1}^{n} C_{s-1}^{n-1} C_{s}^{N-1-n} a^s$, $n = 1,\ldots,(N/2)-1$, $C_B^A = \frac{A}{(A-B)B}$, $a = \left(\frac{(e^{2x}-1)}{(e^{2x}+1)}\right)^2$, $x = \frac{\ln(\alpha+(\alpha^2-1)^{0.5})}{(N-1)}$.

      iii. Show (i) and (ii) yield equivalent numerical results.

3. Let

$$x[n] = e^{i(2\pi n f_1+\theta_1)} + e^{i(2\pi n f_2+\theta_2)} + u[n] + iv[n],\ n = 0,\ldots,31,$$

where $u[n]$ and $v[n]$ are uncorrelated real-valued WGN of zero-mean and equal variance σ_0^2. Plot the $M = 2$ AR spectral estimator (with $\sigma^2 = 1$) using first the theoretical ensemble-averaged autocorrelation values of $R[n]$ and then the time-averaged autocorrelation values of $\tilde{R}[n]$ for

   a.

$$f_1 = 0.3,\ f_2 = 0.35,\ \sigma_0 = 0.01;$$

   b.

$$f_1 = 0.3,\ f_2 = 0.35,\ \sigma_0 = 0.1;$$

   c.

$$f_1 = 0.3,\ f_2 = 0.4,\ \sigma_0 = 0.01;$$

   d.

$$f_1 = 0.3,\ f_2 = 0.4,\ \sigma_0 = 0.1.$$

In the above model, σ_0^2 is the variance of both the real part of the additive noise $u(n)$ as well as the imaginary part of the additive noise $v(n)$ in the observation equation. Are there two distinct peaks for these four cases? Plot the magnitudes of the FFT of these data (with sufficient zero paddings) and compare with your AR spectral estimators.

4. Solve the above Problem 3 based on the Pisarenko algorithm. Use both the theoretical ensemble-averaged and the time-averaged autocorrelations to solve for the four cases.

5. Solve the above Problem 3 based on the MUSIC algorithm. Use both the theoretical ensemble-averaged and the time-averaged autocorrelations to solve for the four cases.

6. Consider a noise-free model consisting of two sinusoids defined by $x[n] = s[n] = \cos(2\pi k_1 n/N) + \cos(2\pi(k_1+1)n/N)$, for some integer $0 \le k_1 \le N-2$, $n = 0, \ldots, N-1$, where N is an even integer.

   a. Consider a cosine window defined by $w[n] = \cos(\frac{2\pi}{N}(n - \frac{N}{2})), n = 0, \ldots, N-1$. Find this window spectral function $W[k], k = 0, \ldots, N-1$. Sketch $W[k]$.
   b. Consider the windowed sequence $y[n] = w[n]x[n], n = 0, \ldots, N-1$. Sketch $|Y[k]|, k = 0, \ldots, N-1$.
   c. From $|Y[k]|, k = 0, \ldots, N-1$, do we have spectral leakages? That is, if we only observe $|Y[k]|$ over the integers on $k = 0, \ldots, N-1$, do we observe any spectral values other than those in the original data of $s[n]$.

7. Consider a discrete-time sequence $x[n] = s[n] + q[n], 1 \le n \le N$, where $s[n] = \sin(2\pi f_1 n + \theta_1) + \sin(2\pi f_2 n + \theta_2)$, where f_1 and f_2 are in $[-0.5, 0.5)$, θ_1 and θ_2 are two uniformly distributed random variables on $[0, 2\pi]$, uncorrelated to each other and to the zero-mean WGN $q(t)$ with a spectral density $S_q(f) = 0.001, -0.5 \le f < 0.5$.

   a. Evaluate the ensemble-averaged autocorrelation values $R_x[n], n = 0, 1, 2, 3, 4,$ of $x[n]$.
   b. Let $f_1 = 1/4$ and $f_2 = -1/4$. Evaluate explicitly the values of $R_x[n]$, $n = 0, 1, 2, 3, 4$.
   c. Using the autoregressive spectral estimation method, find the optimal estimation coefficients $\{\hat{a}_1, \hat{a}_2, \hat{a}_3, \text{and } \hat{a}_4\}$. Hint: The four normal equations for this set of f_1 and f_2 parameter values can be reduced to a simpler form and the explicit evaluation (using no computer) of the approximate estimation coefficients can be performed.
   d. Show the AR spectral estimator using the values of $\{\hat{a}_1, \hat{a}_2, \hat{a}_3, \text{and } \hat{a}_4\}$ indeed yield large spectral values at $\hat{f} \approx \frac{1}{4}, \frac{-1}{4}$.

4 Parametric Spectral Analysis

In Chapter 3, the spectral analysis techniques based on DFT/FFT and windowing methods are considered to be non-parametric since no assumptions were made on the spectral sources under consideration. In this chapter, we consider various parametric spectral analysis techniques for spectral data generated from specific models characterized by sets of parameters. The spectral estimation problems reduce to the determination of these sets of parameters.

4.1 Maximum Entropy Spectral Analysis

4.1.1 *M*th-Order Autoregressive Model

Consider the use of an Mth-order autoregressive system to model the behavior and the spectral estimation of a complex-valued random sequence obtained by sampling $x(t)$ at time nT. Specifically, let $\{x[n], -\infty < n < \infty\}$ be a zero-mean complex-valued random sequence modeled as an Mth-order autoregressive (AR) sequence

$$x[n] - \sum_{m=1}^{M} a_m x[n-m] = u[n], \qquad -\infty < n < \infty, \tag{4.1}$$

where $\{a_1, a_2, \ldots, a_M\}$ is the set of complex-valued AR coefficients and $\{u[n], -\infty < n < \infty\}$ is a zero-mean complex-valued uncorrelated (i.e., white) sequence of variance σ^2.

Let the Z-transform of $\{x[n]\}$ and $\{u[n]\}$ be denoted by

$$\mathcal{Z}\{x[n]\} = X(z), \qquad \mathcal{Z}\{u[n]\} = U(z). \tag{4.2}$$

The Z-transform of (4.1) yields

$$X(z)\left[1 - \sum_{m=1}^{M} a_m z^{-m}\right] = U(z). \tag{4.3}$$

Let the transfer function $H(z)$ be defined as the ratio of the output $X(z)$ to the input $U(z)$. Then

$$H(z) = \frac{X(z)}{U(z)} = \frac{1}{1 - \sum_{m=1}^{M} a_m z^{-m}}. \tag{4.4}$$

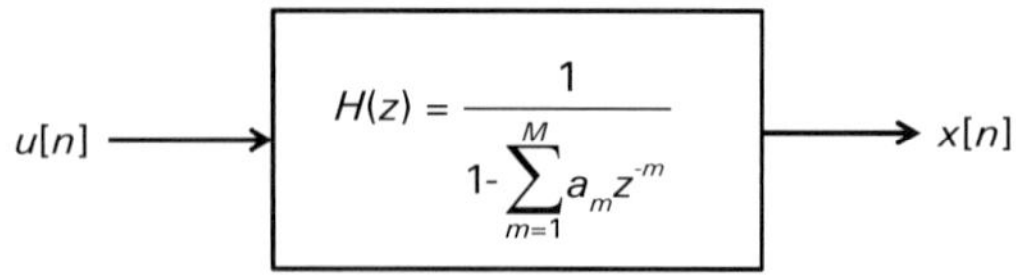

Figure 4.1 Generation of an Mth-order AR sequence

By setting $z = e^{i2\pi fT}$ in (4.4) and denoting $H(e^{i2\pi fT})$ by $H(f)$. Then

$$H(f) = \frac{1}{1 - \sum_{m=1}^{M} a_m e^{-im2\pi fT}}, \qquad 0 \le fT < 1. \tag{4.5}$$

While (4.1) is a time-domain characterization of the AR sequence, (4.4) and (4.5) show that if $\{x[n]\}$ is a wide-sense stationary random sequence with a power spectral density $S_x(f)$, then

$$S_x(f) = S_u(f)|H(f)|^2 = \frac{\sigma^2}{|1 - \sum_{m=1}^{M} a_m e^{-im2\pi fT}|^2}, \qquad 0 \le fT < 1, \tag{4.6}$$

where the input white spectral density $S_u(f) = \sigma^2,\ 0 \le fT < 1$.

An AR sequence $\{x[n]\}$ is characterized by its coefficients $\{a_1, a_2, \ldots, a_M\}$. Given the dependency of $x[n]$ on $\{\Sigma_{m=1}^{M} a_m x[n-m]\}$ in (4.1), we use the minimum mean-square error criterion to estimate these coefficients based on the observed $x[n]$. Let $\tilde{x}[n]$ be an approximation of $x[n]$ by using a linear combinations of its past M values as given by

$$\tilde{x}_n = \sum_{m=1}^{M} a_m x[n-m]. \tag{4.7}$$

The estimation error is given by

$$e[n] = x[n] - \tilde{x}[n] = x[n] - \sum_{m=1}^{M} a_m x[n-m]. \tag{4.8}$$

The mean-square error is given by

$$\begin{aligned} \epsilon[n] &= E\{|e[n]|^2\} = E\{e[n]e[n]^*\} \\ &= E\left\{(x[n] - \sum_{m=1}^{M} a_m x[n-m])(x^*[n] - \sum_{\ell=1}^{M} a_\ell^* x[n-\ell]^*)\right\}. \end{aligned} \tag{4.9}$$

Let us use the orthogonal principle method in minimizing the mean-square estimation.

$$\min E\{|\varepsilon[n]|^2\} = \min_{\{a_1,\ldots,a_M\}} E\left\{\left|x[n] - \sum_{m=1}^{M} a_m x[n-m]\right|^2\right\}.$$

Then we obtain

$$E\{|(x[n]-\sum_{m=1}^{M}a_m x[n-m])x_{n-j}^*|\}=0,\ j=1,\dots,M.$$

Denote $E\{x[i]x[j]H^* = R[i-j]\}$, then we obtain the normal equation of (4.11).

$$\mathbf{R}=\begin{bmatrix} R[0] & R[1] & \cdots & R[M-1] \\ R[1] & & & R[M-2] \\ \vdots & & & \vdots \\ R[M-1] & \dots & R[1] & R[0] \end{bmatrix}\begin{bmatrix} a_1 \\ a_2 \\ \vdots \\ a_M \end{bmatrix}=\begin{bmatrix} R[1] \\ R[2] \\ \vdots \\ R[M] \end{bmatrix}.$$

Then the optimum $\hat{a}_\ell$ are given by

$$\sum_{\ell=1}^{M}\hat{a}_\ell R[i-\ell]=R[i],\ i=1,\dots,M. \tag{4.10}$$

Equation (4.10) is called the normal equation (also called the Yule–Walker equation) solution of the AR coefficients.

Example 4.1 Let $M=1$. Then

$$\hat{a}_1=\frac{R[1]}{R[0]}. \tag{4.11}$$

Example 4.2 Let $M=2$. Then $\hat{a}_\ell$ are solved from

$$\begin{aligned}\hat{a}_1 R[0]+\hat{a}_2 R[-1]&=R[1],\\ \hat{a}_1 R[1]+\hat{a}_2 R[0]&=R[2].\end{aligned} \tag{4.12}$$

4.1.2 An Heuristic Interpretation of the AR Spectral Estimator for Pure Complex Exponentials

Consider

$$x(t)=s(t)+v(t), \qquad -\infty<t<\infty, \tag{4.13}$$

where

$$s(t)=Ae^{i(2\pi f_1 t+\theta)}, \tag{4.14}$$

is a pure complex exponential of frequency f_1, real-valued amplitude of A, θ is a uniformly distributed random variable (r.v.) and uncorrelated to the $v(t)$ zero-mean complex-valued white process. Upon sampling at time instants nT, (4.13) and (4.14) become

$$s[n] = Ae^{i(2\pi f_1 nT+\theta)},$$
$$x[n] = s[n] + v[n], \tag{4.15}$$

where $v[n]$ is a zero-mean complex-valued white random sequence of variance σ_0^2. Initially, let the noise $v(t)$ be negligible, then (4.15) becomes

$$\begin{aligned} x[n] &= A\cos(2\pi f_1 nT + \theta) + iA\sin(2\pi f_1 nT + \theta) \\ &= Ae^{i(2\pi f_1 nT+\theta)} = Ae^{i\phi[n]}. \end{aligned} \tag{4.16}$$

Then the phase of the sampled data $x[n]$ is given by

$$\phi[n] = 2\pi f_1 nT + \theta. \tag{4.17}$$

Clearly, one sample of $x[n]$ cannot determine the unknown frequency f_1 and the unknown initial phase θ in (4.17). Next, consider two adjacent sampled data. From (4.17),

$$\phi[n+1] = 2\pi f_1(n+1)T + \theta. \tag{4.18}$$

The phase difference is given by

$$\Delta\phi = \phi[n+1] - \phi[n] = 2\pi f_1 T, \tag{4.19}$$

and the frequency is given by

$$f_1 = \frac{\Delta\phi}{2\pi T}. \tag{4.20}$$

Thus, two adjacent samples can determine the frequency f_1 of a single complex exponential in the absence of noise.

Now, consider $\{x[n] = s[n] + v[n], n = 1, \ldots, N\}$ of (4.15) in the presence of noise. The autocorrelation sequence yields

$$R[0] = E\{|x[n]|^2\} = |A|^2 + \sigma_0^2, \tag{4.21}$$

$$R[1] = E\{x[n+1]x[n]^*\} = |A|^2 e^{i2\pi f_1 T} = |A|^2 e^{i\phi[1]}. \tag{4.22}$$

The phase of $R[1]$ is given by

$$\phi[1] = 2\pi f_1 T, \tag{4.23}$$

and the frequency f_1 is obtained from

$$f_1 = \frac{\phi[1]}{2\pi T}. \tag{4.24}$$

In practice, we do not measure the statistical (i.e., ensemble) averaged $R[0]$ and $R[1]$, but need to use the time sample averaged $\tilde{R}[0]$ and $\tilde{R}[1]$ to approximate $R[0]$ and $R[1]$, where

$$\tilde{R}[0] = \frac{1}{N}\sum_{n=1}^{N}|x[n]|^2, \tag{4.25}$$

$$\tilde{R}[1] = \frac{1}{N-1}\sum_{n=1}^{N-1}x[n+1]x[n]^*. \tag{4.26}$$

Consider a one-pole AR model of (4.5). Then

$$H(f) = \frac{1}{1 - a_1 e^{-i2\pi fT}}, \qquad 0 \le fT < 1. \tag{4.27}$$

From (4.6), the spectral density $S_x(f)$ becomes

$$S_x(f) = \frac{\sigma^2}{|1 - a_1 e^{-i2\pi fT}|^2}, \qquad 0 \le fT < 1. \tag{4.28}$$

We want to find an a_1 so that by using it in (4.28), $S_x(f)$ attains a large value at $f = f_1$. Thus, the crossing over the threshold of $S_x(f)$ at $f = f_1$ yields the estimation of the complex exponential at frequency f_1. Indeed, if

$$a_1 = e^{i2\pi f_1 T}, \tag{4.29}$$

by using (4.29) in (4.28), $S_x(f_1) \to \infty$. Thus, the AR spectral estimator has the form of

$$\tilde{S}_x(f) = \frac{\sigma^2}{|1 - \hat{a}_1 e^{-i2\pi fT}|^2} \approx \frac{\sigma^2}{|1 - e^{-i2\pi T(f-f_1)}|^2}, \qquad 0 \le fT < 1. \tag{4.30}$$

Fig. 4.2 shows a sketch of $\tilde{S}_x(f)$ with an unbounded value at $f = f_1$ if $\tilde{a}_1$ equals a_1 of (4.29).

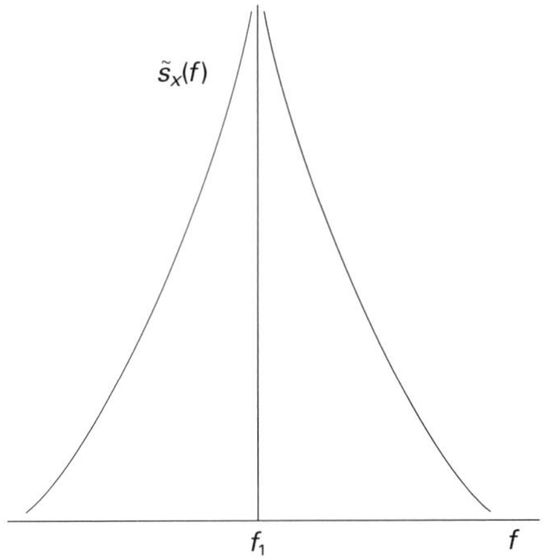

Figure 4.2 Estimated spectral density of a first-order AR estimator

If we consider the estimated $\hat{a}_1$ of (4.11) using $R[0]$ and $R[1]$ given by (4.21) and (4.22) respectively, then we have

$$\begin{aligned}\hat{a}_1 &= \frac{R[1]}{R[0]} = \left(\frac{|A|^2}{|A|^2+\sigma_0^2}\right) e^{i2\pi f_1 T} \\ &\approx e^{i2\pi f_1 T} = a_1,\end{aligned} \tag{4.31}$$

under the high SNR condition (i.e., $|A|^2 \gg \sigma_0^2$). Thus, the $\hat{a}_1$ of (4.31) provide the data dependent approximation of a_1 of (4.29) needed in the ideal first-order AR spectral estimator $\tilde{S}_x(f)$ of (4.30). In practice, we do not have $R[0]$ and $R[1]$, but need to use the time sample averaged $\tilde{R}[0]$ and $\tilde{R}[1]$ of (4.25) and (4.26). Thus, by setting

$$\hat{a}_1 = \frac{\tilde{R}[1]}{\tilde{R}[0]} = |\hat{a}_1| e^{i2\pi \hat{f}_1 T}, \tag{4.32}$$

we have

$$\text{phase}\left\{\frac{\tilde{R}[1]}{\tilde{R}[0]}\right\} = \text{phase}\left\{\tilde{R}[1]\right\} = 2\pi \hat{f}_1 T$$

and

$$\hat{f}_1 = \frac{\text{phase}\left\{\tilde{R}[1]\right\}}{2\pi T}. \tag{4.33}$$

Thus, $\hat{f}_1$ can be considered as a practical first-order AR estimate of f_1 for a single complex exponential in (4.15). By using $\tilde{a}_1$ of (4.32), $\tilde{S}_x(f)$ has a large but finite value at $f = f_1$.

Consider two complex exponentials in the presence of a complex-valued white sequence. Let

$$x[n] = A_1 e^{i(2\pi n f_1 T + \theta_1)} + A_2 e^{i(2\pi n f_2 T + \theta_2)} + v[n], \qquad n = 1, \ldots, N, \tag{4.34}$$

where θ_1 and θ_2 are two uncorrelated uniformly distributed r.v. also uncorrelated to the $v[n]$ zero-mean uncorrelated random sequence of variance σ_0^2. Its autocorrelation sequence values are given by

$$R[0] = |A_1|^2 + |A_2|^2 + \sigma_0^2, \tag{4.35}$$

$$R[1] = |A_1|^2 e^{i2\pi f_1 T} + |A_2|^2 e^{i2\pi f_2 T}, \tag{4.36}$$

$$R[-1] = |A_1|^2 e^{-i2\pi f_1 T} + |A_2|^2 e^{-i2\pi f_2 T}, \tag{4.37}$$

$$R[2] = |A_1|^2 e^{i4\pi f_1 T} + |A_2|^2 e^{i4\pi f_2 T}. \tag{4.38}$$

The desired optimum coefficients $\hat{a}_1$ and $\hat{a}_2$ of the second-order AR spectral estimator

$$\tilde{S}_x(f) = \frac{\sigma^2}{|1 - \hat{a}_1 e^{-i2\pi f T} - \hat{a}_2 e^{-i4\pi f T}|^2}, \qquad 0 \le fT < 1, \tag{4.39}$$

are given by the solution of (4.12) using $R[0]$, $R[1]$, $R[-1]$, and $R[2]$ given by (4.35) to (4.38). In practice, we need to use time-averaged $\tilde{R}[0]$ and $\tilde{R}[1]$ given by (4.25), (4.26),

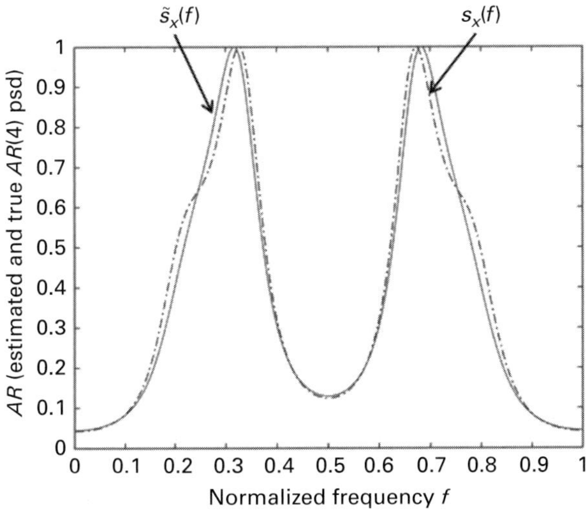

Figure 4.3 Fourth-order AR and estimated AR spectral densities

and

$$\tilde{R}[-1] = \frac{1}{N-1}\sum_{n=2}^{N} x[n-1]x[n]^*, \tag{4.40}$$

$$\tilde{R}[2] = \frac{1}{N-2}\sum_{n=1}^{N-2} x[n+2]x[n]^*. \tag{4.41}$$

The value of $\tilde{S}_x(f)$ at $f = f_1$ and $f = f_2$ using $\hat{a}_1$ and $\hat{a}_2$ based on $\tilde{R}[0]$, $\tilde{R}[1]$, $\tilde{R}[-1]$, and $\tilde{R}[2]$ has large but finite values at $f = f_1$ and $f = f_2$.

Example 4.3 Comparison of AR and FFT spectral estimations for complex exponential AR modeled data. Consider a 4th-order AR data sequence $\{x[n], n = 1, \ldots, 512\}$, generated from (4.1), where the AR model coefficients are given by

$$\mathbf{a} = [a_1, a_2, a_3, a_4]^T = [-0.3760, -0.6487, -0.0966, -0.1764]^T,$$

and the input sequence $\{u[n]\}$ is a zero-mean white sequence of unit variance. These AR model coefficients are obtained by setting the denominator term of (4.4) to

$$\begin{aligned} 1 - \sum_{m=1}^{4} a_m z^{-m} &= z^{-4}z^4(1 - a_1 z^{-1} - a_2 z^{-2} - a_3 z^{-3} - a_4 z^{-4}) \\ &= z^{-4}(-a_4 - a_3 z^1 - a_2 z^2 - a_1 z^3 + z^4) = 0, \end{aligned}$$

where the complex exponential frequencies and amplitudes are taken as $f_1 = 0.33 = -f_3$, $f_2 = 0.21 = -f_4$, $A_1 = A_3 = 0.7$, and $A_2 = A_4 = 0.5$ respectively.

Now, suppose we model the observed sequence $\{x[n]\}$ of length 512 as a 4th-order AR sequence. The time-averaged 4×4 correlation matrix $\tilde{\mathbf{R}}$ is given by

$$\tilde{\mathbf{R}} = \begin{pmatrix} 1.5960 & -0.3649 & -0.7606 & 0.3773 \\ -0.3649 & 1.5960 & -0.3649 & -0.7606 \\ -0.7606 & -0.3649 & 1.5960 & -0.3649 \\ 0.3773 & -0.7606 & -0.3649 & 1.5960 \end{pmatrix}$$

and the time-averaged correlation vector $\tilde{\mathbf{R}}$ is given by

$$\tilde{\mathbf{R}} = [\tilde{r}(1), \tilde{r}(2), \tilde{r}(3), \tilde{r}(4)]^T = [-0.3649, -0.7606, 0.3773, 0.1457]^T.$$

The normal equation solution of (4.10) yields the estimated AR coefficients

$$\hat{\mathbf{a}} = [\hat{a}_1, \hat{a}_2, \hat{a}_3, \hat{a}_4]^T = \tilde{\mathbf{R}}^{-1}\tilde{\mathbf{R}} = [-0.4250, -0.6966, -0.1662, -0.1782]^T.$$

As expected, the roots of the denominator of the AR transfer function with the true coefficients $\{a_1, a_2, a_3, a_4\}$ yield the frequencies of $f_1 = 0.33 = -f_3$ and $f_2 = 0.21 = -f_4$, while the roots of the estimated coefficients $\{\tilde{a}_1, \tilde{a}_2, \tilde{a}_3, \tilde{a}_4\}$ yield the estimated frequencies of $\tilde{f}_1 = 0.32 = -\tilde{f}_3$ and $\tilde{f}_2 = 0.2215 = -\tilde{f}_4$. We note, the estimated frequencies are quite close to the true frequencies. In Fig. 4.3, we plot the true AR spectral density estimator

$$\mathbf{S}_x(f) = \frac{1}{|1 - a_1 e^{-i2\pi f} - a_2 e^{-i4\pi f} - a_3 e^{-i6\pi f} - a_4 e^{-i8\pi f}|^2}, \qquad 0 \le f < 1,$$

as well as the estimated AR spectral density

$$\tilde{\mathbf{S}}_x(f) = \frac{1}{|1 - \hat{a}_1 e^{-i2\pi f} - \hat{a}_2 e^{-i4\pi f} - \hat{a}_3 e^{-i6\pi f} - \hat{a}_4 e^{-i8\pi f}|^2}, \qquad 0 \le f < 1.$$

We observe the estimated AR spectral density approximates closely the true AR spectral density. In Fig. 4.4, we plot the true AR spectral density $\mathbf{S}_x(f)$ as well as the absolute value of the FFT of the zero padded observed data, given by

$$\bar{\mathbf{S}}_x(k/2048) = \text{abs}(\text{fft}(\mathbf{x}, \text{zeros}(1536, 1))), \qquad k = 0, \ldots, 2047,$$

where $\mathbf{x} = [x_1, \ldots, x_{512}]^T$ is the 512 point observed data vector. The padding of 1,536 number of zeros in the FFT does not improve its resolution, but only yields additional observation values in the frequency domain. As can be seen in Fig. 4.4, the FFT estimated spectral density has great variations and appears to be less desirable than the estimated AR spectral density as compared to the true AR spectral density. Furthermore, while the AR model can yield quite good estimates of the frequencies of the complex exponentials as shown above, the FFT approach yields no information in this direction. For AR model generated data, the parametric AR spectral density approach, in particular when the order of the AR model is known or can be estimated correctly, generally yields a better approximation than the non-parametric FFT spectral density approach.

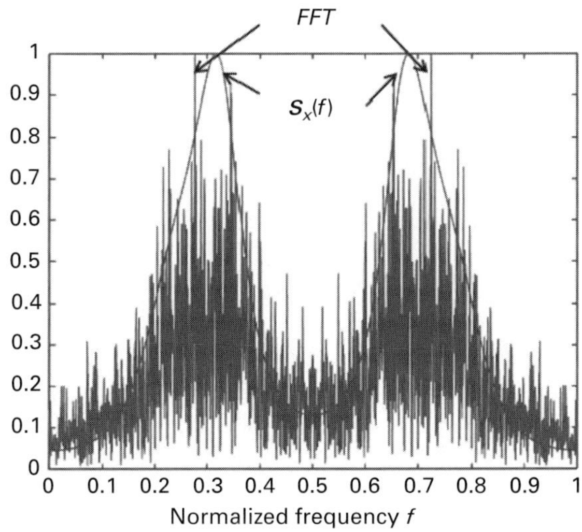

Figure 4.4 Fourth-order AR spectral density compared to FFT magnitude spectrum

4.2 Maximization of Entropy for an AR Spectral Estimator

Now we show that an AR spectral estimator actually maximizes the entropy of the estimated random sequence under the assumption that the sequence is Gaussian and wide-sense stationary. For ease of presentation, we restrict the sequence to be real-valued. Thus, the zero-mean Gaussian real-valued random sequence $\{x[n], -\infty < n < \infty\}$ has an autocorrelation sequence value of $R[m] = E\{x[n]x[n-m]\} = R[-m]$. Consider an $(M+2)$ dimensional vector $\mathbf{x}_{M+2} = [x_1, x_2, \ldots, x_{M+2}]^T$. Its autocorrelation matrix is given by

$$\mathbf{R}_{M+2} = \begin{pmatrix} R[0] & R[1] & \ldots & R[M] & R[M+1] \\ R[1] & R[0] & \ldots & R[M-1] & R[M] \\ \vdots & & \ddots & & \vdots \\ R[M+1] & R[M] & \ldots & R[1] & R[0] \end{pmatrix}. \tag{4.42}$$

The entropy of this Gaussian random vector is

$$H = \log\left[(\sqrt{2\pi e})^{M+2}|\mathbf{R}_{M+2}|^{\frac{1}{2}}\right], \tag{4.43}$$

where $|\mathbf{R}_{M+2}|$ is the determinant of $\mathbf{R}_{M+2}$.

The maximum entropy spectral estimation states that for fixed values of $R[m]$, $m = 0, 1, \ldots, M$, we want to vary $R[M+1]$ in order to maximize the entropy H of (4.43). Direct evaluation of

$$\frac{\partial|\mathbf{R}_{M+2}|}{\partial R[m+1]} = 0 \tag{4.44}$$

yields

$$\begin{vmatrix} R[1] & R[0] & \dots & R[M-2] & R[M-1] \\ R[2] & R[1] & \dots & R[M-3] & R[M-2] \\ \vdots & & \ddots & & \vdots \\ R[M+1] & R[M] & \dots & R[2] & R[1] \end{vmatrix} = 0\,. \tag{4.45}$$

The matrix in (4.45) is the lower left $(M+1)\times(M+1)$ submatrix of $\mathbf{R}_{M+2}$. Furthermore, direct evaluation of (4.44) also shows

$$\frac{\partial^2|\mathbf{R}_{M+2}|}{\partial^2 R[M+1]} = -2|\mathbf{R}_M| \leq 0. \tag{4.46}$$

The last inequality of (4.46) follows from the non-negative definiteness of the autocorrelation matrix $\mathbf{R}_M$. Thus, (4.45) can be used for verifying the maximization of the entropy of the Gaussian vector.

Example 4.4 Consider the case $M = 1$. Then

$$\mathbf{R}_3 = \begin{pmatrix} R[0] & R[1] & R[2] \\ R[1] & R[0] & R[1] \\ R[2] & R[1] & R[0] \end{pmatrix}, \tag{4.47}$$

$$\begin{aligned} |\mathbf{R}_3| &= R[0][R^2[0] - R^2[1]] - R[1][R[1]R[0] - R[1]R[2]] \\ &\quad + R[2][R^2[1] - R[0]R[2]]\,, \end{aligned} \tag{4.48}$$

$$\begin{aligned} \frac{\partial|\mathbf{R}_3|}{\partial R[2]} &= R^2(1) - R[2]R[0] + R^2(1) - R[0]R[2] \\ &= 2[R^2(1) - R[0]R[2]] = 0. \end{aligned} \tag{4.49}$$

Thus, (4.49) is equivalent to

$$\begin{vmatrix} R[1] & R[0] \\ R[2] & R[1] \end{vmatrix} = 0. \tag{4.50}$$

Furthermore,

$$\frac{\partial^2|\mathbf{R}_3|}{\partial^2 R[2]} = -2R[0] \leq 0. \tag{4.51}$$

Indeed (4.50) has the form of (4.45) and (4.51) has the form of (4.46).

Example 4.5 Consider the case $M = 2$. Then

$$\mathbf{R}_4 = \begin{pmatrix} R[0] & R[1] & R[2] & R[3] \\ R[1] & R[0] & R[1] & R[2] \\ R[2] & R[1] & R[0] & R[1] \\ R[3] & R[2] & R[1] & R[0] \end{pmatrix} \tag{4.52}$$

$$\begin{aligned}
\frac{\partial|\mathbf{R}_4|}{\partial R[3]} &= -R[1]\frac{\partial}{\partial R[3]}\begin{vmatrix} R[1] & R[1] & R[2] \\ R[2] & R[0] & R[1] \\ R[3] & R[1] & R[0] \end{vmatrix} + R[2]\frac{\partial}{\partial R[3]}\begin{vmatrix} R[1] & R[0] & R[2] \\ R[2] & R[1] & R[1] \\ R[3] & R[2] & R[0] \end{vmatrix} \\
&\quad - R[3]\frac{\partial}{\partial R[3]}\begin{vmatrix} R[1] & R[0] & R[1] \\ R[2] & R[1] & R[0] \\ R[3] & R[2] & R[1] \end{vmatrix} - \begin{vmatrix} R[1] & R[0] & R[1] \\ R[2] & R[1] & R[0] \\ R[3] & R[2] & R[1] \end{vmatrix} \\
&\quad - -R[1]\begin{vmatrix} R[1] & R[2] \\ R[0] & R[1] \end{vmatrix} + R[2]\begin{vmatrix} R[0] & R[2] \\ R[1] & R[1] \end{vmatrix} - R[3]\begin{vmatrix} R[0] & R[1] \\ R[1] & R[0] \end{vmatrix} \\
&\quad - \begin{vmatrix} R[1] & R[0] & R[1] \\ R[2] & R[1] & R[0] \\ R[3] & R[2] & R[1] \end{vmatrix}
\end{aligned}$$

$$= -2\begin{vmatrix} R[1] & R[0] & R[1] \\ R[2] & R[1] & R[0] \\ R[3] & R[2] & R[1] \end{vmatrix} = 0\,, \tag{4.53}$$

$$\begin{aligned}
\frac{\partial^2|\mathbf{R}_4|}{\partial^2 R[3]} &= -2\begin{vmatrix} R[0] & R[1] \\ R[1] & R[0] \end{vmatrix} \\
&= -2\mathbf{R}_2\,.
\end{aligned} \tag{4.54}$$

Thus, (4.53) has the form of (4.45) and (4.54) has the form of (4.46).

Consider the Mth-order AR sequence of (4.1). Let $b_m = -a_m, m = 1, \ldots, M$. Then

$$x[n] + b_1 x[n-1] + \cdots + b_M x[n-M] = u[n], \qquad -\infty < n < \infty. \tag{4.55}$$

Since $\{u[n], -\infty < n < \infty\}$ is an uncorrelated sequence, $E\{u[n]u[n-k]\} = 0, k \geq 1$, and

$$E\{u[n]x[n-k]\} = 0, \qquad k \geq 1. \tag{4.56}$$

By multiplying (4.55) by x_{n-k} and taking expectation, we obtain

$$R[k] + b_1 R[k-1] + \cdots + b_M R[k-M] = 0, \qquad k = 1, \ldots, M+1. \tag{4.57}$$

The right-hand side of (4.57) follows from (4.56). Then (4.57) can be expressed as

$$\begin{pmatrix} R[1] & R[0] & \dots & R[M-2] & R[M-1] \\ R[2] & R[1] & \dots & R[M-3] & R[M-2] \\ \vdots & & \ddots & & \vdots \\ R[M+1] & R[M] & \dots & R[2] & R[1] \end{pmatrix} \begin{pmatrix} 1 \\ b_1 \\ b_2 \\ \vdots \\ b_M \end{pmatrix} = \begin{pmatrix} 0 \\ 0 \\ 0 \\ \vdots \\ 0 \end{pmatrix}. \tag{4.58}$$

For the homogeneous set of equations in (4.58) to hold, we must have

$$\begin{vmatrix} R[1] & R[0] & \dots & R[M-2] & R[M-1] \\ R[2] & R[1] & \dots & R[M-3] & R[M-2] \\ \vdots & & \ddots & & \vdots \\ R[M+1] & R[M] & \dots & R[2] & R[1] \end{vmatrix} = 0. \tag{4.59}$$

However, (4.59) is equivalent to the expression of (4.45). This shows a Gaussian Mth-order AR spectral estimator maximizes the entropy of the sequence.

4.3 Pisarenko Spectral Estimation Method

The Pisarenko method is one of the earliest eigendecomposition methods used for time-frequency-domain as well as spatial/transform domain spectral analysis of pure complex exponentials. In this section, we consider spectral estimation on the time-frequency model. Consider

$$x[n] = s[n] + v[n], \qquad n = 1, 1, \dots, N, \tag{4.60}$$

where $s[n]$ is given by

$$s[n] = \sum_{j=0}^{M-1} A_j e^{i(2\pi f_j n + \theta_j)}, \tag{4.61}$$

where A_j are unknown real-valued amplitudes, f_j are unknown distinct frequencies in $[0, 1)$, θ_j are uncorrelated and uniformly spaced random variables in $[0, 2\pi)$, and $q[n]$ are complex-valued, zero-mean, white Gaussian random variables of variance σ_0^2 and also uncorrelated to θ_j. Denote its autocorrelation sequence values by

$$R[n-m] = E\{x[n]x^H[m]\} .$$

Then

$$R[n-m] = \begin{cases} \sigma_0^2 + \sum_{j=0}^{M-1} A_j^2, & n = m , \\ \sum_{j=0}^{M-1} A_j^2 e^{i2\pi f_j (n-m)}, & n \neq m . \end{cases} \tag{4.62}$$

The $N \times N$ autocorrelation matrix of $\mathbf{x} = [x(0), x(1), \dots, x(N-1)]^T$ is given by

$$\mathbf{R} = E\{\mathbf{x}\mathbf{x}^H\} = \mathbf{P} + \sigma_0^2 \mathbf{I}_N , \tag{4.63}$$

where

$$\mathbf{P} = \sum_{j=0}^{M-1} \mathbf{P}_j, \tag{4.64}$$

$$\mathbf{P}_j = A_j^2 \mathbf{e}_j(N)\mathbf{e}_j^H(N), \tag{4.65}$$

$$\mathbf{e}_j(\ell) = [1, e^{i2\pi f_j}, e^{i4\pi f_j}, \ldots, e^{i2\pi(\ell-1)f_j}]^T,$$

$$j = 0, 1, \ldots, M-1, \qquad M \le \ell \le N. \tag{4.66}$$

From the ensemble-averaged **R** matrix, we want to find the frequencies $\{f_0, f_1, \ldots, f_{M-1}\}$, the amplitudes $\{A_0, A_1, \ldots, A_{M-1}\}$, and the noise variance σ_0^2.

Lemma 4.1 *Let* $\{f_0, f_1, \ldots, f_{M-1}\}$ *be distinct real numbers on* $[0, 1)$. *Then* $\{\mathbf{e}_0(N), \mathbf{e}_1(N), \ldots, \mathbf{e}_{M-1}(N)\}$ *are M linearly independent vectors in* $\mathcal{C}^N$ *for* $M \le N$.

Proof: Let $a_0, a_1, \ldots, a_{M-1}$ be M arbitrary but distinct complex coefficients. Then

$$\mathbf{A} = \begin{pmatrix} a_0^0 & a_1^0 & \ldots & a_{M-1}^0 \\ a_0^1 & a_1^1 & & a_{M-1}^1 \\ a_0^2 & a_1^2 & & a_{M-1}^2 \\ \vdots & \vdots & & \vdots \\ a_0^{M-1} & a_1^{M-1} & & a_{M-1}^{M-1} \end{pmatrix} \tag{4.67}$$

is an $M \times M$ Vandermonde matrix and thus is non-singular. Denote the P columns of the $M \times 1$ vector $e_j(M), j = 0, \ldots, P-1$, by

$$V(M, P) = [\mathbf{e}_0(M), \mathbf{e}_1(M), \ldots, \mathbf{e}_{P-1}(M)]. \tag{4.68}$$

Thus, $\mathbf{V}(M, M)$ is an $M \times M$ Vandermonde matrix of the form of **A** in (4.67) with

$$a_j = e^{i2\pi f_j}, \qquad j = 0, \ldots, M-1. \tag{4.69}$$

Then the non-singularity of $\mathbf{V}(M, M)$ implies

$$\mathbf{V}(M, M) \begin{pmatrix} c_0 \\ c_1 \\ \vdots \\ c_{M-1} \end{pmatrix} = 0, \tag{4.70}$$

or

$$\sum_{j=0}^{M-1} c_j \mathbf{e}_j(M) = 0,$$

which shows that $c_j, j = 0, \ldots, M-1$, must be zero. Thus, $\{\mathbf{e}_0(M), \mathbf{e}_1(M), \ldots, \mathbf{e}_{M-1}(M)\}$ are linearly independent in $\mathcal{C}^M$. But if (4.70) implies $[c_0, c_1, \ldots, c_{M-1}]^T = 0$, then for $N \ge M$,

$$\mathbf{V}(N,M)\begin{pmatrix} c_0 \\ c_1 \\ \vdots \\ c_{M-1} \end{pmatrix} = \begin{pmatrix} 1 & 1 & 1 & 1 \\ e^{i2\pi f_0} & e^{i2\pi f_1} & \ldots & e^{i2\pi f_{M-1}} \\ e^{i4\pi f_0} & e^{i4\pi f_1} & \ldots & e^{i4\pi f_{M-1}} \\ \vdots & & \ddots & \vdots \\ e^{i2(N-1)\pi f_0} & e^{i2(N-1)\pi f_1} & \ldots & e^{i2(N-1)\pi f_{M-1}} \end{pmatrix} \begin{pmatrix} c_0 \\ c_1 \\ \vdots \\ c_{M-1} \end{pmatrix} \tag{4.71}$$

$$= 0$$

still implies $[c_0, c_1, \ldots, c_{M-1}]^T = 0$. This means $\{\mathbf{e}_0(N), \mathbf{e}_1(N), \ldots, \mathbf{e}_{M-1}(N)\}$ are linearly independent in $\mathcal{C}^N$.

From (4.65), each $\mathbf{P}_j, j = 0, 1, \ldots, M-1$, is a dyadic matrix of $\mathbf{e}_j(N)$ and thus the range of $\mathbf{P}_j$ is a one-dimensional space spanned by $\mathbf{e}_j(N)$. Furthermore, for any $x \in \mathcal{C}^N$,

$$\mathbf{x}^H\mathbf{P}_j\mathbf{x} = A_j^2\mathbf{x}^H\mathbf{e}_j(N)\mathbf{e}_j^H(N)\mathbf{x} \geq 0, \;\; j = 0, 1, \ldots, M-1.$$

This shows $\mathbf{P}_j$ is a non-negative matrix. Then $\mathbf{P}$ of (4.64) is a rank M non-negative matrix. We summarize these results in the next lemma.

Lemma 4.2 *The matrix* $\mathbf{P}$ *defined in (4.64) is an* $N \times N$ *non-negative matrix of rank* M*. The null space of* $\mathbf{P}$ *is of dimension* $(N-M)$*. The determinant of* $\mathbf{P}$ *is zero.*

Let $\lambda_j, j = 0, 1, \ldots, N-1$, be the eigenvalues of the autocorrelation matrix $\mathbf{R}$, and $\{\phi_j, j = 0, \ldots, N-1\}$ be the associated eigenvectors. The λ_j are the solutions of

$$|\mathbf{R} - \lambda\mathbf{I}_N| = |\mathbf{P} - (\lambda - \sigma_0^2)\mathbf{I}_N| = 0\,, \tag{4.72}$$

or equivalently

$$\mathbf{R}\phi_j = \lambda_j\phi_j, \qquad j = 0, 1, \ldots, N-1. \tag{4.73}$$

Equation (4.72) shows σ_0^2 is an eigenvalue of $\mathbf{R}$. Let $\mathbf{y}$ be the associated eigenvector. Then

$$\mathbf{R}\mathbf{y} = \mathbf{P}\mathbf{y} + \sigma_0^2\mathbf{y} = \sigma_0^2\mathbf{y}. \tag{4.74}$$

Therefore, $\mathbf{y}$ belongs to the null space of $\mathbf{P}$. From Lemma 4.2, since the dimension of the null space of $\mathbf{P}$ is $(N-M)$, the eigenvalue σ_0^2 has a multiplicity of $\nu = (N-M)$. Furthermore, from Lemma 4.2, since $\mathbf{P}$ is non-negative definitive of rank M,

$$\mathbf{P}\mathbf{w}_j = \mu_j\mathbf{w}_j, \qquad j = 0, 1, \ldots, M-1, \tag{4.75}$$

$$\mu_j \geq 0, \qquad j = 0, 1, \ldots, M-1, \tag{4.76}$$

where the μ_j are eignevalues of $\mathbf{P}$ and the $\mathbf{w}_j$ are the associated eigenvectors. From (4.72), (4.75), and (4.76), the eigenvalues of $\mathbf{R}$ are given by

$$\lambda_j - \sigma_0^2 = \begin{cases} \mu_j \geq 0, & j = 0, 1, \ldots, M-1, \\ 0, & j = M, \ldots, N-1, \end{cases} \tag{4.77}$$

and the eigenvectors of **R** are given by

$$\phi_j = \begin{cases} \mathbf{w}_j, & j = 0, 1, \ldots, M-1, \\ \mathbf{y}_j, & j = M, \ldots, N-1. \end{cases} \tag{4.78}$$

In (4.78), $\mathbf{w}_j$ satisfies (4.75) and $\mathbf{y}_j$ satisfies (4.74). This shows that the minimum eigenvalue of **R** is σ_0^2. From (4.73), all the eigenvectors $\{\phi_j\}$ are mutually orthogonal. Since the linear span of $\{\mathbf{w}_0, \mathbf{w}_1, \ldots, \mathbf{w}_{M-1}\}$ equals the linear span of

$$\{\mathbf{e}_0(N), \mathbf{e}_1(N), \ldots, \mathbf{e}_{M-1}(N)\},$$

then each $\mathbf{y}_j$ is orthogonal to the linear span of

$$\{\mathbf{e}_0(N), \mathbf{e}_1(N), \ldots, \mathbf{e}_{M-1}(N)\}.$$

Lemma 4.3 *The $N \times N$ autocorrelation matrix* **R** *of (4.63) with $\sigma_0^2 > 0$ is a positive definitive matrix with the minimum eigenvalue σ_0^2 of multiplicity $\nu = N - M$. The eigenspace corresponding to σ_0^2 is orthogonal to the subspace spanned by $\{\mathbf{e}_0(N), \mathbf{e}_1(N), \ldots, \mathbf{e}_{M-1}(N)\}$.*

Consider an eigenvector $\mathbf{y} = [y_0, y_1, \ldots, y_{N-1}]^T$ in the eigenspace corresponding to σ_0^2. Then from Lemma 4.3,

$$0 = \mathbf{e}_j(N)\mathbf{y} = \sum_{k=0}^{N-1} y_k e^{-i2\pi f_j k}, \qquad j = 0, 1, \ldots, M-1. \tag{4.79}$$

Now, consider the $(N-1)$st degree polynomial of z defined by

$$S(z) = \sum_{k=0}^{N-1} y_k z^{N-1-k} = z^{(N-1)} \sum_{k=0}^{N-1} y_k z^{-k} = 0\,. \tag{4.80}$$

By setting $z = e^{i2\pi f}$, we see all M frequencies of $\{f_0, f_1, \ldots, f_{M-1}\}$ of (4.79) are roots of $S(z)$ of (4.80) on the unit circle. The other $(N-M-1)$ roots of $S(z)$ are not located on the unit circle.

Example 4.6 Consider the case of $M = 1$ and $N = 3$. Then $s[n] = A_0 e^{i(2\pi f_0 n + \theta_0)}$, $n = 0, 1, 2$. From (4.62), the autocorrelation matrix is given by

$$\mathbf{R} = \begin{pmatrix} A_0^2 + \sigma_0^2 & A_0^2 e^{-i2\pi f_0} & A_0^2 e^{-i4\pi f_0} \\ A_0^2 e^{i2\pi f_0} & A_0^2 + \sigma_0^2 & A_0^2 e^{-i2\pi f_0} \\ A_0^2 e^{i4\pi f_0} & A_0^2 e^{i2\pi f_0} & A_0^2 + \sigma_0^2 \end{pmatrix}. \tag{4.81}$$

Direct evaluation shows that the three eigenvalues of **R** are σ_0^2, σ_0^2, and $\sigma_0^2 + 3A_0^2$. The eigenvectors associated with σ_0^2 are

$$\mathbf{y}_1 = \begin{pmatrix} 0 \\ 1 \\ -e^{i2\pi f_0} \end{pmatrix} \quad \text{and} \quad \mathbf{y}_2 = \begin{pmatrix} 1 \\ 0 \\ -e^{i4\pi f_0} \end{pmatrix}.$$

By using $\mathbf{y} = \mathbf{y}_1$ in (4.80), we obtain

$$0 \cdot z^2 + z - e^{i2\pi f_0} = 0. \tag{4.82}$$

This shows $z = e^{i2\pi \hat{f}} = e^{i2\pi f_0}$. Thus, $\hat{f} = f_0$ yields the correct frequency estimate. Similarly, if we use $\mathbf{y} = \mathbf{y}_2$ in (4.80), we obtain

$$z^2 + 0 \cdot z - e^{i4\pi f_0} = 0. \tag{4.83}$$

This shows $z^2 = e^{i4\pi \hat{f}} = e^{i4\pi f_0}$. Thus, $\hat{f} = f_0$ also yields the correct frequency estimate.

Once the frequencies $\{f_0, f_1, \ldots, f_{M-1}\}$ of (4.79) are determined, the unknown amplitudes $\{A_0, A_1, \ldots, A_{M-1}\}$ are available from the linear system of equations

$$\mathbf{V}(M, M)\mathbf{A} = \mathbf{r}, \tag{4.84}$$

where $\mathbf{V}(M, M)$ is the non-singular Vandermonde matrix given by (4.68),

$$\mathbf{A} = \left[A_0^2, A_1^2, \ldots, A_{M-1}^2\right]^T, \tag{4.85}$$

$$\mathbf{r} = \left[R[0] - \sigma_0^2, R[1], \ldots, R[M-1)\right]^T. \tag{4.86}$$

If $R[m], m = 0, 1, \ldots, M-1$, as well as the noise variance σ_0^2 can be estimated from the observed data $x[n]$, then the unknown amplitudes in $\mathbf{A}$ can also be obtained from

$$\mathbf{A} = \mathbf{V}^{-1}(M, M)\mathbf{r}. \tag{4.87}$$

In practice, we may need to use time-averaged autocorrelation values

$$\tilde{R}[m] = \frac{1}{N-m} \sum_{n=1}^{N-m} x[n+m]x[n]^* \tag{4.88}$$

to approximate the true $R[m]$. An estimate $\hat{\sigma}_0^2$ of σ_0^2 can be obtained by using

$$\tilde{\sigma}_0^2 = \frac{1}{N} \sum_{n=1}^{N} |x[n]|^2, \tag{4.89}$$

when $x[n]$ in (4.60) is known to consist only of the noise $v[n]$ with no signal $s[n]$.

4.4 MUSIC for Direction-of-Arrival Estimation

4.4.1 Source and Array Model

Consider the determination of the direction-of-arrival (DOA) of M point sources located in spatial directions $\{\theta_0, \theta_1, \ldots, \theta_{M-1}\}$ from $[-\pi, \pi)$ using K sensors arranged in a linear array manner by the use of the multiple signal classification method. We assume $K \geq M$ and the sensors are all in the far fields of the sources. Then the mth source plane wave impinges relative to the normal of the array at angle θ_m as shown in Fig. 4.5. Let the spacing of all the sensors be uniform with an inter-sensor spacing value of d. Then the propagation path distance of the wavefront between each sensor is given by $d \sin \theta_m$. Assume the source frequencies are located in a narrowband and modeled by a single frequency of f_o and wavelength $\lambda = c/f_o$, where c is the speed of propagation.

The output of the kth sensor due to all M sources at time t is given by

$$x_k(t) = \sum_{m=1}^{M-1} a_k(\theta_m) S_m(t) e^{-i(2\pi f_o t + k 2\pi \frac{d}{\lambda} \sin \theta_m + \psi_m)} + N_k(t), \qquad -\infty < t < \infty, \quad (4.90)$$

where $a_k(\theta_m)$ is the kth sensor response at direction θ_m, $S_m(t)$ is the mth source amplitude, ψ_m is a uniformly distributed random variable on $[0, 2\pi)$, and $N_k(t)$ is a WGN of zero-mean and power spectral density σ^2 also assumed to be uncorrelated to ψ_m. We also assume that the sensor spacing d is taken to be the half wavelength, resulting in $d = \lambda/2$. We further assume that each sensor is sampled every T seconds. Then at time nT, (4.90) becomes

$$x_k[n] = \sum_{m=1}^{M-1} a_k(\theta_m) S_m[n] e^{-i(2\pi f_o nT + k\pi \sin \theta_m + \psi_m)} + N_k[n], \quad (4.91)$$

where $x_k[n] = x_k(nT)$, $S_m[n] = S_m(nT)$, and $N_k[n] = N_k(nT)$.

The "time snap-shot" of all the sensor outputs due to all the sources at time index n is denoted as a K-dimensional column vector of

$$\begin{aligned} \mathbf{x}_n &= [x_0[n], x_1[n], \ldots, x_{K-1}[n]]^T \\ &= \tilde{\mathbf{A}} \mathbf{s}_n e^{-i2\pi f_o nT} + \mathbf{N}_n \,, \end{aligned} \quad (4.92)$$

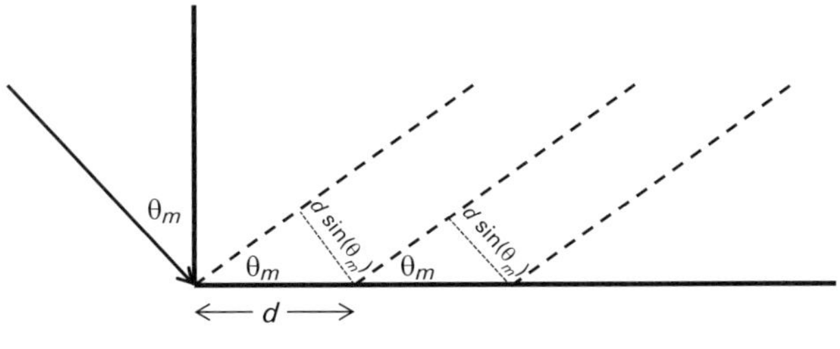

Figure 4.5 Propagation path delay

where $\tilde{\mathbf{A}}$ and $\mathbf{A}$ are both $K \times M$ matrices given by

$$\tilde{\mathbf{A}} = [\tilde{\mathbf{a}}(\theta_0), \ldots, \tilde{\mathbf{a}}(\theta_{M-1})] , \qquad \mathbf{A} = [\mathbf{a}(\theta_0), \ldots, \mathbf{a}(\theta_{M-1})] , \tag{4.93}$$

and

$$\begin{aligned} \tilde{\mathbf{a}}(\theta_m) &= e^{-i\psi_m} \cdot [a_0(\theta_m), a_1(\theta_m)e^{-i\pi \sin\theta_m}, \ldots, a_{K-1}(\theta_m)e^{-i(K-1)\pi \sin\theta_m}]^T \\ &= e^{-i\psi_m}\mathbf{a}(\theta_m) , \end{aligned} \tag{4.94}$$

$$\mathbf{S}_n = [S_0[n], S_1[n], \ldots, S_{M-1}[n]]^T , \tag{4.95}$$

and

$$\mathbf{N}_n = [N_0[n], N_1[n], \ldots, N_{K-1}[n]]^T . \tag{4.96}$$

Each $\mathbf{a}(\theta_m)$ is called a "steering vector" in the direction θ_m. In particular, if each sensor response is uniform spatially, then all $a_k(\theta_m)$ are identical and can be taken to have unity value. The steering vector $\mathbf{a}(\theta_m)$ in (4.94) then reduces to

$$\mathbf{a}(\theta_m) = [1, e^{-i\pi \sin\theta_m}, e^{-i2\pi \sin\theta_m}, \ldots, e^{-i(K-1)\pi \sin\theta_m}]^T . \tag{4.97}$$

Let the $M \times M$ autocorrelation matrix of $\mathbf{S}_n$ be denoted by

$$\mathbf{R}_S = E\{\mathbf{S}_n\mathbf{S}_n^H\}, \tag{4.98}$$

then the $K \times K$ autocorrelation matrix of $\mathbf{x}_n$ becomes

$$\mathbf{R} = E\{\mathbf{x}_n\mathbf{x}_n^H\} = \mathbf{A}\mathbf{R}_S\mathbf{A}^H + \sigma^2\mathbf{I}_K. \tag{4.99}$$

Example 4.7 Let the amplitudes $\{\mathbf{S}_m[n], m = 0, \ldots, M-1\}$ be modeled by uncorrelated Gaussian random variables with $\mathcal{N}(\mathbf{A}_m, \sigma_m^2)$, where $\mathbf{A}_m$ and σ_m^2 are real and positive. Then

$$E\{S_m[n]S_{m'}^*[n]\} = \begin{cases} E\{|S_m[n]|^2\} = \mathbf{A}_m^2 + \sigma_m^2, & m = m' \\ E\{S_m[n]\}E\{S_{m'}^*[n]\} = \mathbf{A}_m\mathbf{A}_{m'}, & m \neq m' \end{cases} . \tag{4.100}$$

Thus, the autocorrelation matrix $\mathbf{R}_S$ is given by

$$\mathbf{R}_S = \alpha\alpha^H + \mathrm{diag}(\sigma_0^2, \sigma_1^2, \ldots, \sigma_{M-1}^2), \tag{4.101}$$

where

$$\alpha = [\mathbf{A}_0, \mathbf{A}_1, \ldots, \mathbf{A}_{M-1}]^T , \tag{4.102}$$

is the vector of the mean amplitude values. We note, while $\alpha\alpha^H$ is of rank one, the $M \times M$ autocorrelation matrix $\mathbf{R}_S$ is a full rank M matrix.

Example 4.8 Let the amplitudes $\{\mathbf{S}_m[n], m = 0, \ldots, M-1\}$ be modeled by a set of deterministic positive numbers. Then the autocorrelation matrix $\mathbf{R}_S$ is given by

$$\mathbf{R}_S = \alpha\alpha^H , \tag{4.103}$$

where α is defined by (4.102). Now, $\mathbf{R}_S$ of (4.103) is of rank one. Indeed, this deterministic model in Example 4.8 is included as a special case of Example 4.7 if all the variances σ_m^2 becomes zero.

4.4.2 Signal and Noise Subspaces

Consider the eigendecomposition of the $K \times K$ autocorrelation matrix $\mathbf{R}$ of (4.99) as given by

$$\mathbf{R} = \mathbf{Q}\Lambda\mathbf{Q}^H = \sum_{k=0}^{K-1} \lambda_k \mathbf{V}_k \mathbf{V}_k^H, \tag{4.104}$$

where $\mathbf{Q} = [\mathbf{V}_0, \mathbf{V}_1, \ldots, \mathbf{V}_{K-1}]$ is a $K \times K$ unitary matrix with columns given by the orthonormal eigenvectors $\mathbf{V}_k$ corresponding to eigenvalues

$$\lambda_0 \geq \lambda_1 \geq \ldots \geq \lambda_{K-1} \geq 0$$

of the matrix $\mathbf{R}$. With M distinct spatial locations, the rank of the $K \times K$ positive definite matrix $\mathbf{A}\mathbf{R}_S\mathbf{A}^H$ is M assuming $K \geq M$. Since the $K \times K$ matrices $\sigma^2\mathbf{I}_k$ and $\mathbf{R} = \mathbf{A}\mathbf{R}_S\mathbf{A}^H + \sigma^2\mathbf{I}_k$ are of rank K, then just as in the Pisarenko method problem, in order for $\mathbf{A}\mathbf{R}_S\mathbf{A}^H = \mathbf{R} - \sigma^2\mathbf{I}_k$ to be positive-definitive, the minimum eigenvalues of $\mathbf{R}$ need to be σ^2 of multiplicity $(K - M)$. Thus, the eigenvalues of R satisfy

$$\lambda_0 \geq \lambda_1 \geq \ldots \geq \lambda_{M-1} > \lambda_M = \lambda_{M+1} = \ldots = \lambda_{K-1} = \sigma^2 > 0.$$

The space $\mathcal{L}(\mathbf{V}_0, \mathbf{V}_1, \ldots, \mathbf{V}_{M-1})$ spanned by the M eigenvectors corresponding to the M largest eigenvalues $\{\lambda_0, \lambda_1, \ldots, \lambda_{M-1}\}$ of $\mathbf{R}$ is called the signal subspace. Denote the $K \times M$ matrix

$$\mathbf{V}_\mathbf{S} = [\mathbf{V}_0, \mathbf{V}_1, \ldots, \mathbf{V}_{M-1}].$$

Furthermore, each steering vector can be expressed as

$$\mathbf{a}(\theta_m) = \sum_{n=0}^{M-1} b_{m,n}\mathbf{V}_n, \qquad m = 0, \ldots, M-1, \tag{4.105}$$

where the coefficients in $\{b_{m,0}, b_{m,1}, \ldots, b_{m,M-1}\}$ are not all zero for $m = 0, \ldots, M-1$. The space

$$\mathcal{L}(\mathbf{V}_M, \mathbf{V}_{M+1}, \ldots, \mathbf{V}_{K-1})$$

spanned by the $(K - M)$ eigenvectors corresponding to the minimum eigenvalues of $\mathbf{R}$ is called the noise subspace. Denote the $K \times (K - M)$ matrix

$$\mathbf{V}_\mathbf{N} = [\mathbf{V}_M, \mathbf{V}_{M+1}, \ldots, \mathbf{V}_{K-1}].$$

Then

$$\mathbf{Q} = [\mathbf{V}_\mathbf{S}, \mathbf{V}_\mathbf{N}] = [\mathbf{V}_0, \mathbf{V}_1, \ldots, \mathbf{V}_{M-1}, \mathbf{V}_M, \mathbf{V}_{M+1}, \ldots, \mathbf{V}_{K-1}].$$

The orthogonality of the eigenvectors can be expressed in the form of $\mathbf{Q}^H\mathbf{Q} = \mathbf{I}_K$, ${\mathbf{V_S}}^H\mathbf{V_S} = \mathbf{I}_M$, ${\mathbf{V_N}}^H\mathbf{V_N} = \mathbf{I}_{K-M}$, as well as the orthogonality of the subspace by ${\mathbf{V_N}}^H\mathbf{V_S} = \mathbf{O}_{(K-M),M}$.

Let the eigendecomposition of $\mathbf{A}\mathbf{R}_S\mathbf{A}^H$ be expressed as

$$\mathbf{A}\mathbf{R}_S\mathbf{A}^H = \sum_{m=0}^{M-1} \mu_m \mathbf{V}_m \mathbf{V}_m^H, \tag{4.106}$$

where

$$\mu_0 \geq \mu_1 \geq \ldots \geq \mu_{M-1} > 0$$

and $\{\mathbf{V}_0, \mathbf{V}_1, \ldots, \mathbf{V}_{M-1}\}$ are the eigenvalues and eigenvectors. Then (4.104) can be re-expressed as

$$\begin{aligned}\mathbf{R} &= \sum_{k=0}^{K-1} \lambda_k \mathbf{V}_k \mathbf{V}_k^H = \sum_{m=0}^{M-1} \mu_m \mathbf{V}_m \mathbf{V}_m^H + \sigma^2 \sum_{k=0}^{K-1} \mathbf{V}_k \mathbf{V}_k^H \\ &= \sum_{m=0}^{M-1} (\mu_m + \sigma^2)\mathbf{V}_m \mathbf{V}_m^H + \sigma^2 \sum_{m=M}^{K-1} \mathbf{V}_m \mathbf{V}_m^H.\end{aligned} \tag{4.107}$$

This shows

$$\lambda_k = \begin{cases} \mu_k + \sigma^2, & k = 0, 1, \ldots, M-1, \\ \sigma^2, & k = N, \ldots, K-1. \end{cases} \tag{4.108}$$

From (4.107) and (4.108), we see that the K leading eigenvalues of $\mathbf{R}$ depend on the powers of both signals and noise. Those eigenvalues in the noise subspace depend only on the noise.

4.4.3 MUSIC Spatial DOA Estimator

In the MUSIC approach, the orthogonality of the signal and noise subspaces are exploited explicitly to determine the DOA of the sources. From (4.105) to (4.108), we know

$$\mathcal{L}(\mathbf{a}(\theta_0), \mathbf{a}(\theta_1), \ldots, \mathbf{a}(\theta_{M-1})) = \mathcal{L}(\mathbf{V}_0, \mathbf{V}_1, \ldots, \mathbf{V}_{M-1}) \perp \mathcal{L}(\mathbf{V}_M, \mathbf{V}_{M+1}, \ldots, \mathbf{V}_{K-1}). \tag{4.109}$$

The orthogonality of (4.109) implies

$$\mathbf{a}^H(\theta_m)\mathbf{V}_k = 0, \quad 0 \leq m \leq M-1, \quad M \leq k \leq K-1, \tag{4.110}$$

or equivalently

$$\mathbf{a}^H(\theta_m)\mathbf{V}_k\mathbf{V}_k^H\mathbf{a}(\theta_m) = |\mathbf{a}^H(\theta_m)\mathbf{V}_k|^2 = 0, \quad 0 \leq m \leq M-1, \ M \leq k \leq K-1. \tag{4.111}$$

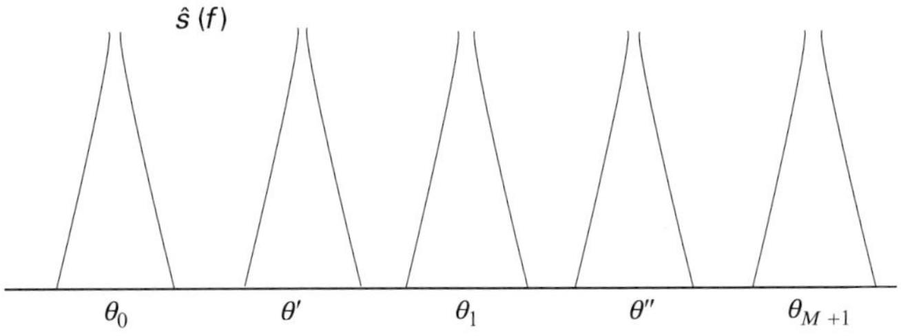

Figure 4.6 Spatial DOA estimator $\hat{\mathbf{S}}(\theta)$

Then in the spirit of the estimated AR spectral estimator, a MUSIC spatial DOA estimator $\hat{\mathbf{S}}(\theta)$ can be taken to be

$$\hat{\mathbf{S}}(\theta) = \frac{1}{\mathbf{a}^H(\theta)\mathbf{V}_k\mathbf{V}_k^H\mathbf{a}(\theta)}, \quad -\pi \le \theta \le \pi. \tag{4.112}$$

From (4.111), $\hat{\mathbf{S}}(\theta)$ has large values at angles of $\theta = \theta_m, 0 \le m \le M-1$, as shown in Fig. 4.6. Thus, by observing the peaks of $\hat{\mathbf{S}}(\theta)$, we can determine the DOA at $\theta = \theta_m, 0 \le m \le M-1$.

Unfortunately, at $\theta = \theta'$, $\theta = \theta''$, etc., it is also possible for $\hat{\mathbf{S}}(\theta)$ to have high peaks as shown in Fig. 4.6. Indeed, for a given $\theta' \ne \theta_m, m = 0, 1, \ldots, M-1$, $\mathbf{a}(\theta') \in \mathcal{L}(\mathbf{V}_M, \mathbf{V}_{M+1}, \ldots, \mathbf{V}_{K-1})$. Thus, for some $k \in \{M, M+1, \ldots, K-1\}$, we may have

$$|\mathbf{a}^H(\theta')\mathbf{V}_k|^2 = 0, \qquad M \le k \le K-1, \tag{4.113}$$

but there must be some other $k' \in \{M, M+1, \ldots, K-1\}$ such that

$$|\mathbf{a}^H(\theta')\mathbf{V}_{k'}|^2 \ne 0, \qquad M \le k \le K-1. \tag{4.114}$$

This means if k in (4.112) is selected such that (4.113) is true, then indeed $\hat{\mathbf{S}}(\theta)$ has the undesired peak at θ' as shown in Fig. 4.6. In light of the observations of (4.113) and (4.114), we note for any given $\theta' \ne \theta_m, m = 0, 1, \ldots, M-1$,

$$|\mathbf{a}^H(\theta')\mathbf{V}_{k'}|^2 \ne 0, \qquad \text{for some } k' \in \{M, \ldots, K-1\}, \tag{4.115}$$

but from (4.110), we still have

$$|\mathbf{a}^H(\theta_m)\sum_{k=M}^{K-1}\mathbf{V}_k|^2 = 0, \quad 0 \le m \le M-1. \tag{4.116}$$

Finally, the MUSIC spatial DOA estimator is defined by

$$\hat{\mathbf{S}}_{MUSIC}(\theta) = \frac{1}{\mathbf{a}^H(\theta)\sum_{k=M}^{K-1}\mathbf{V}_k\mathbf{V}_k^H\mathbf{a}(\theta)}, \quad -\pi \le \theta \le \pi. \tag{4.117}$$

From (4.115) and (4.116), we note that $\hat{\mathbf{S}}_{MUSIC}(\theta)$ has the desired peak at $\theta = \theta_m$, $m = 0, 1, \ldots, M-1$, but no additional spurious peaks as compared to $\hat{\mathbf{S}}(\theta)$ of (4.112). In practice, the ensemble-averaged autocorrelation matrix $\mathbf{R}$ of the data $\mathbf{x}_n$ is not available but only some time-averaged version of $\mathbf{R}$ as given by

$$\tilde{\mathbf{R}}_n = \frac{1}{N} \sum_{i=n-N+1}^{n} \mathbf{x}_i \mathbf{x}_i^H, \tag{4.118}$$

where (4.118) uses N "time snap-shots" of the sensor outputs of the form in (4.92). Of course, the eigendecomposition of $\tilde{\mathbf{R}}_n$ in (4.118), in contrast to the time correlation $\mathbf{R}$ of (4.104), degrades the performance of the $\hat{\mathbf{S}}_{MUSIC}(\theta)$ estimator, and infinite peaks at $\theta = \theta_m$ do not occur.

4.5 Conclusion

In this chapter, we considered spectral analysis for sources defined by models characterized by some parameters. In Section 4.1, we considered the maximum entropy spectral analysis for sources characterized by Mth order autoregressive time-series. Section 4.2 provided some background information on the entropy of a Gaussian wide-sense stationary random AR source. Section 4.3 treated the Pisarenko spectral estimation problem. The MUSIC DOA estimation algorithm was considered in Section 4.4.

4.6 References

Early works on parametric spectral estimation for sources with continuous spectra were performed by Yule–Walker [1], [2] with detailed discussion in Chapter 3 of [3]. The concept of ME in statistical mechanics and information theory was introduced by Jaynes in [4]. ME spectral analysis was studied by Burg in [5] and used for speech applications by Flanagan [6] and his colleagues at Bell Lab. Pisarenko [7] pioneered the use of the eigendecomposition method for spectral estimation of pure complex exponentials. Schmidt [8] introduced the concept of MUSIC method for DOA estimation.

[1] G.U. Yule, "On a Method of Investigating Periodicities in Disturbed Series," *Philos. Trans. Royal Society of London*, 1927, pp. 267–298.

[2] G. Walker, "On Periodicities in Series of Related Terms," *Proceedings of the Royal Society of London*, 1931, pp. 518–532.

[3] P. Stoica and R. Moses, *Introduction to Spectral Analysis*, Prentice-Hall, 1997.

[4] E.T. Jaynes, "Information Theory and Statistical Mechanics," *Physical Review, Ser. II,* 1957, pp. 171–190.

[5] J.P. Burg, "Maximum Entropy Spectral Analysis," Ph.D. thesis, Stanford University, 1975.

[6] J. Flanagan, *Speech Analysis Synthesis and Perception*, Springer Verlag, 1983.

[7] V.F. Pisarenko, "The Retrieval of Harmonics from a Covariance Function," *Geophys. J. Roy. Astron. Soc.*, 1973, pp. 347–366.

[8] R.O. Schmidt, "Multiple Emitter Location and Signal Parameters Estimation," *Proc. RADC Spectral Estimation Workshop*, 1979, pp. 243–258.

4.7 Exercises

1. Consider a noise-free model consisting of two sinusoids defined by $x[n] = s[n] = \cos(2\pi k_1 n/N) + \cos(2\pi(k_1 + 1)n/N)$, for some integer $0 \leq k_1 \leq N - 2$, $n = 0, \ldots, N - 1$, where N is an even integer.

   a. Consider a cosine window defined by $w[n] = \cos(\frac{2\pi}{N}(n - \frac{N}{2})), n = 0, \ldots, N - 1$. Find this window spectral function $W[k], k = 0, \ldots, N - 1$. Sketch $W[k]$.
   b. Consider the windowed sequence $y[n] = w[n]x[n], n = 0, \ldots, N - 1$. Sketch $|Y[k]|, k = 0, \ldots, N - 1$.
   c. From $|Y[k]|, k = 0, \ldots, N - 1$, do we have spectral leakages? That is, if we only observe $|Y[k]|$ over the integers on $k = 0, \ldots, N - 1$, do we observe any spectral values other than those in the original data of $s[n]$.

2. Consider a discrete-time sequence $x[n] = s[n] + q[n], 1 \leq n \leq N$, where $s[n] = \sin(2\pi f_1 n + \theta_1) + \sin(2\pi f_2 n + \theta_2)$, where f_1 and f_2 are in $[-0.5, 0.5)$, θ_1 and θ_2 are two uniformly distributed r.v. on $[0, 2\pi]$, uncorrelated to each other and to the zero-mean WGN $q(t)$ with a spectral density $S_q(f) = 0.001, \; -0.5 \leq f < 0.5$.

   a. Evaluate the ensemble-averaged autocorrelation values $R_x[n], n = 0, 1, 2, 3, 4$, of $x[n]$.
   b. Let $f_1 = 1/4$ and $f_2 = -1/4$. Evaluate explicitly the values of $R_x[n], n = 0, 1, 2, 3, 4$.
   c. Using the autoregressive spectral estimation method, find the optimal estimation coefficients $\{\hat{a}_1, \hat{a}_2, \hat{a}_3, \text{and}\, \hat{a}_4\}$. Hint: The four normal equations for this set of f_1 and f_2 parameter values can be reduced to a simpler form and the explicit evaluation (using no computer) of the approximate estimation coefficients can be performed.
   d. Show the AR spectral estimator using the values of $\{\hat{a}_1, \hat{a}_2, \hat{a}_3, \text{and}\, \hat{a}_4\}$ indeed yield large spectral values at $\hat{f} \approx \frac{1}{4}, \frac{-1}{4}$.

3. Solve the above Exercise 1 based on the Pisarenko algorithm. Use both the theoretical ensemble-averaged and the time-averaged autocorrelations to solve for the four cases.
4. Solve the above Exercise 1 based on the MUSIC algorithm. Use both the theoretical ensemble-averaged and the time-averaged autocorrelations to solve for the four cases.
5. Consider N r.v. $\{X_1, \ldots, X_n\}$ taken from a zero-mean real-valued wide-sense stationary random sequence. Denote the sample spectral density function by

$$P(\omega) = \frac{1}{N} \left| \sum_{n=-(N-1)}^{N-1} x[n] e^{-j\omega n} \right|^2 ,$$

and the time-averaged autocorrelation function by

$$\hat{r}[k] = \hat{r}[-k] = \frac{1}{N} \sum_{n=-(N-1)}^{N-1} x[k+n]x[n] , \; k = 0, \; \ldots, \; N-1 .$$

Prove

$$P(\omega) = \frac{1}{N} \sum_{n=-(N-1)}^{N-1} \tilde{r}[n] e^{-j\omega n} , \; 0 \leq \omega \leq 2\pi .$$

5 Time-Frequency Spectral Analysis

5.1 Time-Frequency Signal Representation

Many applications ranging from radar/sonar to speech processing, from seismic surveying, bio-acoustical analysis, and condition-based maintenance, all deal with signals with time-varying spectral characteristics. Such signals are usually referred to as *non-stationary*. The Fourier transform $X(f)$ of a time signal $x(t)$ provides information about its spectral contents, but any information about the *time-localization* of the frequency components of $x(t)$ are obscurely hidden in the phase spectrum $\arg\{X(f)\}$. No information contained in the signal $x(t)$ is lost when its Fourier tranform $X(f)$ is computed, but it is an arduous task to extract from $X(f)$ any information regarding the temporal behavior of the different frequency components. The following two examples consider the limitation of the Fourier analysis for time-varying signals.

Example 5.1 Consider a real-valued sequence of unit magnitude sinusoid of frequency 20 Hz over time interval [280, 450] followed by another real-valued sequence of unit magnitude sinusoid of frequency 100 Hz over time interval [451, 780] as shown in Fig. 5.1(a). The magnitudes of their Fourier transform are shown as four spikes at ± 20 Hz and ± 100 Hz in Fig. 5.1(b). The spectrogram of these two waveforms is displayed in Fig. 5.1(c). Clearly, Fig. 5.1(a) does not provide any frequency-domain information, while Fig. 5.1(b) does not provide any time-domain information about these two waveforms. However, the spectrogram in Fig. 5.1(c) shows some frequency information around 20 Hz and 100 Hz and some time duration information.

Example 5.2 Consider the following two signals defined by

$$x_1(t) = u_T(t)\cos\left(2\pi(f_0 t + \alpha t^2/2)\right), \quad -\infty < t < \infty, \tag{5.1}$$

$$x_2(t) = K\, u_T(t)\,\frac{\sin(\pi B t)}{\pi t}\cos(2\pi f_0 t), \quad -\infty < t < \infty, \tag{5.2}$$

where $u_T(t)$ has unit amplitude over the observation interval $[-T/2, T/2]$ and zero elsewhere as shown in Figs. 5.2(a) and (b) respectively. The signal $x_1(t)$ is a "chirp" signal (i.e., its frequency changes linearly in time), whereas $x_2(t)$ is a modulated sinc signal. Although $x_1(t)$ and $x_2(t)$ are very different, for essentially all values of the

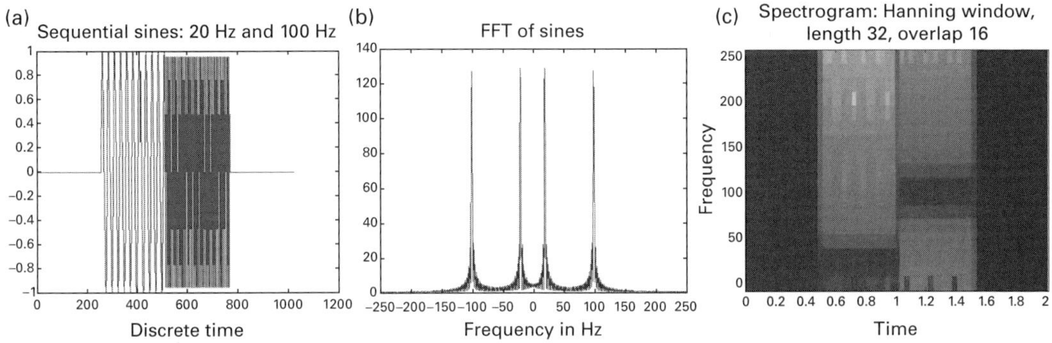

Figure 5.1 (a) Two sinusoids of frequencies 20 Hz and 100 Hz; (b) magnitudes of the Fourier transform of these two sinusoids; (c) spectrogram of these two waveforms

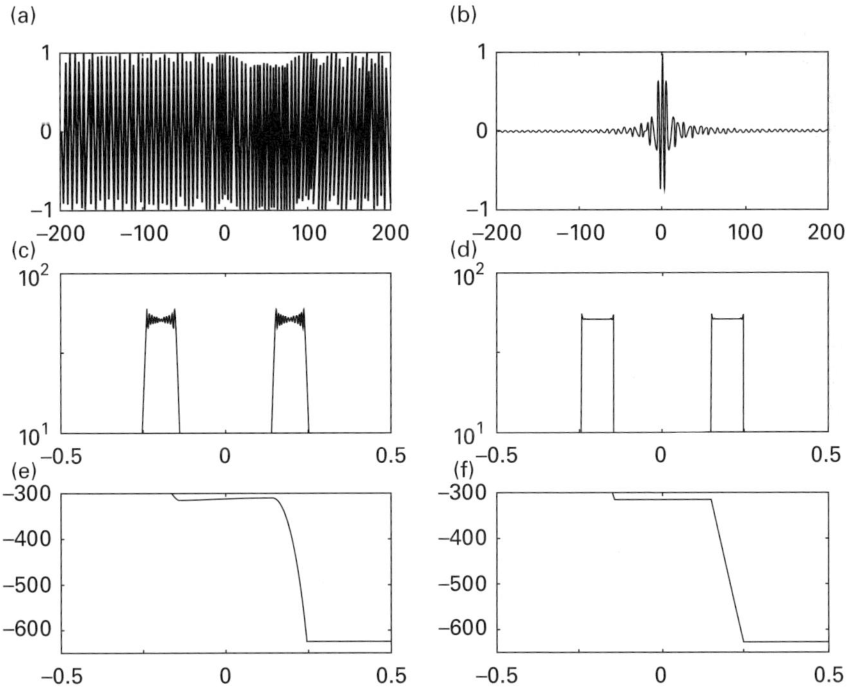

Figure 5.2 (a) Chirp signal of (5.1); (b) modulated sinc signal of (5.2); (c) magnitude spectrum of the chirp signal; (d) magnitude spectrum of the modulated sinc signal; (e) phase spectrum of the chirp signal; and (f) phase spectrum of the modulated sinc signal

parameters f_0, α, B, K, and T, they have somewhat similar magnitude spectra, as can be seen in Figs. 5.2(c) and (d) and also somewhat similar phase spectra in Figs. 5.2(e) and (f), for $f_s = 1024$, $T = 1/f_s$, $f_0 = 200/f_s$, $\alpha = 100/(f_s)^2$, $B = 100/f_s$, $K = 10$. Thus, properties in the time-domain of a signal manifest in the frequency-domain in complex manners.

Figure 5.3 J.S. Bach Partita No.1 in B Flat Major

Combined time- and frequency-domain representations of signals can yield valuable information in analysis and synthesis of time-varying signals and non-stationary systems, as well as time-varying filter design, noise suppression, etc. Time-frequency representation (TFR) has been devised to deal with non-stationary signals and may be considered as an extension of the classical Fourier analysis. A TFR $F_x(t, f)$ is a two-dimensional function of time and frequency which shows what spectral components are present at any given time. The oldest form of time-frequency representation is a musical score, see Fig. 5.3, where (discrete) time runs horizontally across the page and (discrete) frequency (pitch) is shown by vertical superposition of notes.

For a given real-valued signal $x(t)$, it is possible to define a corresponding analytic signal $z(t)$ and then characterize its *instantaneous frequency*, $f_z(t)$ and its *group delay*, $\tau_z(f)$. Both quantities characterize some aspects of the time and frequency attributes of $x(t)$. In particular, the instantaneous frequency measures the localization in time of the frequency components of $x(t)$, while the group delay describes the localization in frequency of its time components.

Define the Hilbert transform of $x(t), \quad -\infty < t < \infty$, as

$$\begin{aligned}\hat{x}(t) = \mathcal{H}\{x(t)\} &= \frac{1}{\pi}\,\text{p.v.}\left\{\int_{-\infty}^{\infty}\frac{x(t-\tau)}{\tau}\,d\tau\right\} = \frac{1}{\pi}\,\text{p.v.}\left\{\int_{-\infty}^{\infty}\frac{x(\tau)}{t-\tau}\,d\tau\right\}\\ &= [x(t)] * \left[\frac{1}{\pi t}\right], \quad -\infty < t < \infty, \end{aligned} \tag{5.3}$$

where "p.v." denotes the Cauchy principal value. In (5.3), since $\hat{x}(t)$ is the convolution of $x(t)$ with $h(t) = 1/(\pi t)$, then $\hat{X}(f) = \mathcal{F}\{\hat{x}(t)\} = \mathcal{F}\{x(t)\} \times \mathcal{F}\{h(t)\}$. Denote $X(f) = \mathcal{F}\{x(t)\}$ and $H(f) = \mathcal{F}\{h(t)\}$. By computation,

$$\begin{aligned}H(f) = \mathcal{F}\{h(t)\} &= \mathcal{F}\left\{\frac{1}{\pi t}\right\} = \int_{-\infty}^{\infty}\frac{1}{\pi t}\exp(-i2\pi ft)dt\\ &= \frac{1}{\pi}\int_{-\infty}^{\infty}\frac{\cos(2\pi ft) - i\sin(2\pi ft)}{t}dt = \frac{-i}{\pi}\int_{-\infty}^{\infty}\frac{\sin(2\pi ft)}{t}dt.\end{aligned}$$

From direct integration, we know

$$\int_{-\infty}^{\infty}\frac{\sin(at)}{t}dt = \begin{cases}\pi, & a > 0,\\ -\pi, & a < 0.\end{cases}$$

Thus,

$$H(f) = \begin{cases} -i, & f > 0, \\ i, & f < 0. \end{cases}$$

Example 5.3 Consider $x(t) = \cos(2\pi f_0 t), \ -\infty < t < \infty$. Then $\mathcal{F}\{x(t)\} = X(f) = (\delta(f - f_0) + \delta(f + f_0))/2$ and $\hat{X}(f) = X(f) \times H(f) = (-i\delta(f - f_0) + \delta(f + f_0))/2 = (\delta(f - f_0) - \delta(f + f_0))/(2i)$. Thus, $\hat{x}(t) = \mathcal{F}^{-1}\{\hat{X}(f)\} = (\exp(i2\pi f_0 t) - \exp(-i2\pi f_0 t))/(2i) = \sin(2\pi f_0 t)$.

The analytic signal $z(t)$ of the real-valued $x(t)$ is defined by

$$z(t) = x(t) + i\hat{x}(t) = a(t)e^{i\phi(t)}, \quad -\infty < t < \infty. \tag{5.4}$$

Equation (5.4) shows the analytic signal is complex-valued. In particular, the spectrum is given by

$$\mathcal{F}\{z(t)\} = Z(f) = X(f) + i\hat{X}(f) = X(f) + iX(f) \times H(f) \tag{5.5}$$

$$= \begin{cases} X(f) + (-i) \times i \times X(f) = 2X(f), & f > 0, \\ X(f) + i \times i \times X(f) = 0, & f < 0. \end{cases} \tag{5.6}$$

Equation (5.6) shows that the spectrum of an analytic signal vanishes for negative-valued frequency.

Example 5.4 Consider $x(t) = \cos(2\pi f_0 t), \ -\infty < t < \infty$. Since $X(f) = (\delta(f - f_0) + \delta(f + f_0))/2$, $X(f)$ is non-zero for both positive and negative frequencies. However, its analytic signal $z(t) = \exp(i2\pi f_0 t)$ has a spectrum $Z(f) = \delta(f - f_0), \ -\infty < f < \infty$, which has zero content for negative frequencies.

Denote the Fourier transform of $z(t)$ by $Z(f) = A(f)e^{i\theta(f)}, \ -\infty < f < \infty$. Then the instantaneous frequency and the group delay are defined by

$$f_z(t) = \frac{1}{2\pi}\frac{d\phi(t)}{dt}, \tag{5.7}$$

$$\tau_z(f) = -\frac{1}{2\pi}\frac{d\theta(f)}{df}. \tag{5.8}$$

These quantities implicitly assume that at each time instant, t, there exists only one single frequency component, or that only one frequency is concentrated around a single time instant. These assumptions are sensible if $x(t)$ is *monocomponent*, and its *time-bandwidth product* BT is large enough. Fig. 5.4 shows examples of mono and multi-component signals and gives an intuitive definition of T, the duration, and B, the bandwidth. Note that the requirement that $BT \gg 1$ is equivalent to saying that the signal is "long enough" for its bandwidth, that is, $x(t)$ bears enough information to define an instantaneous frequency: no instantaneous frequency can be reasonably defined for a very short (less than one period) chunk of a sinusoidal signal. Signals with large BT

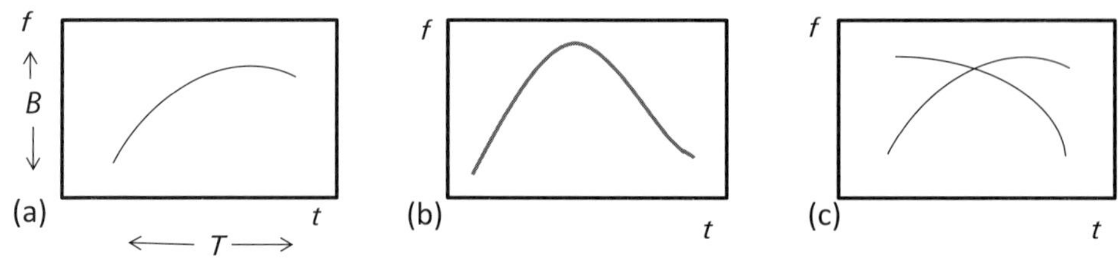

Figure 5.4 (a) Monocomponent time-frequency signal; (b) Monocomponent time-frequency signal; (c) Multi-component time-frequency signal

product are sometimes referred to as *asymptotic*. It can be shown that for asymptotic monocomponent signals

$$f_z(t) \cong \frac{1}{\tau_z(f)}. \tag{5.9}$$

For multi-component signals no intuitive physical meaning can be associated with instantaneous frequency and group delay. A general time-frequency tool is then needed which can separate the components of a signal and determine for each component its instantaneous frequency and the energy spread around it. In the following sections, several different time-frequency representations with their desired properties are presented and compared.

5.1.1 The Spectrogram

The most intuitive approach to time-frequency representation is to apply a *moving window* to the signal and then compute the Fourier transform. If the window $w(t)$ is a signal significantly different from zero only in a symmetric interval $[-\Delta t/2, \Delta t/2]$ about the origin, the short-time Fourier transform (STFT) of the signal $z(t)$ is defined as follows

$$S_z^{(w)}(t, f) = \int_{-\infty}^{\infty} z(\tau)\, w^*(t-\tau)\, e^{-j2\pi f\tau}\, d\tau. \tag{5.10}$$

The *spectrogram* is defined as the magnitude squared of the STFT of the signal. In order to obtain "short-time" characteristics, the aperture Δt of the window should be made considerably smaller than the total duration T of the signal under analysis. The width of the window cannot be made too small, though, because of the "uncertainty principle," since this would produce an "excessive smearing" of the spectral peaks. Denote the Fourier transforms of $z(t)$ and $w(t)$ in (5.10) as $Z(f)$ and $W(f)$ respectively, an equivalent formulation of the STFT can be given by

$$S_z^{(w)}(t, f) = e^{-j2\pi ft} \int_{-\infty}^{\infty} Z(\nu)\, W^*(\nu - f)\, e^{j2\pi\nu t}\, d\nu. \tag{5.11}$$

This formulation suggests a practical way to compute the STFT. The signal can be passed through a bank of bandpass filters with frequency responses $W^*(\nu_i - f)$, centered around consecutive discretized frequencies ν_i. This approach, which incidentally is used

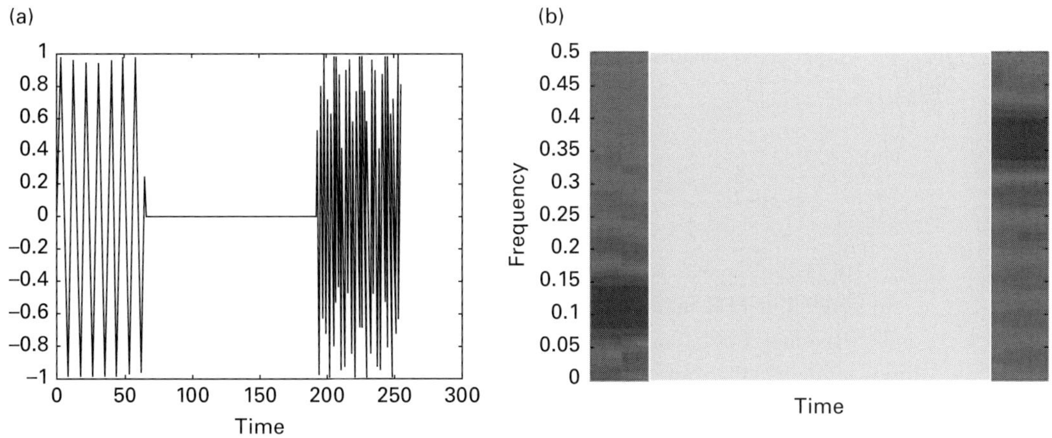

Figure 5.5 (a) Two sinusoidal signals with components well separated in time and frequency; (b) Spectrogram of the signal in (a)

to model the auditory response of the cochlea in human inner ears, is referred to as *sonogram*. Note that from the two equivalent definitions of the STFT in time and in frequency, it is clear that the time and frequency resolutions, defined by the duration of the window $w(t)$ and its bandwidth, respectively, are always fixed quantities.

Example 5.5 Fig. 5.5 shows the spectrogram of a signal composed of two portions of unit-amplitude sinusoidal signals, of normalized frequency 0.11 for samples 0 through 64, followed by no signal for samples 65 through 192, and a sinusoid of frequency 0.37 for samples 193 through 255, using a window aperture of 32 samples.

5.1.2 Quadratic Time-Frequency Representations

The spectrogram belongs to a broad class of TFRs called quadratic TFRs. These TFRs can be loosely interpreted as time-frequency energy distributions. The following properties are required from a "good" quadratic TFR:

1. Since the TFR should represent an energy distribution, it shoud be *real* and *positive-valued*.
2. The *marginal properties* should be satisfied, namely

$$\int_{-\infty}^{\infty} F_x(t,f)\,df = |x(t)|^2 \text{(instantaneous power)}, \tag{5.12}$$

$$\int_{-\infty}^{\infty} F_x(t,f)\,dt = |X(f)|^2 \text{(energy spectral density)}. \tag{5.13}$$

3. The integral of $F_x(t,f)$ over time and frequency should equal the energy of the signal E_x as given by

$$\int_{-\infty}^{\infty}\int_{-\infty}^{\infty} F_x(t,f)\,df\,dt = E_x. \tag{5.14}$$

4. The first moments of $F_x(t, f)$ over f and t should yield the instantaneous frequency and the group delay of $x(t)$ as given by

$$\frac{\int_{-\infty}^{\infty} f\, F_x(t, f)\, df}{\int_{-\infty}^{\infty} F_x(t, f)\, df} = f_x(t), \tag{5.15}$$

$$\frac{\int_{-\infty}^{\infty} t\, F_x(t, f)\, dt}{\int_{-\infty}^{\infty} F_x(t, f)\, dt} = \tau_x(f). \tag{5.16}$$

In general, a TFR may be negative-valued and the first moment properties in (5.15) and (5.16) may not be valid. Thus, a TFR cannot always be considered as a pointwise time-frequency distribution function of energy. It is possible though to define the energy of a localized time-frequency interval $(\Delta t, \Delta f)$ centered around time t and frequency f as

$$E_x(\Delta t, \Delta f) = \int_{t-\Delta t/2}^{t+\Delta t/2} \int_{f-\Delta f/2}^{f+\Delta f/2} F_x(t, f)\, dt\, df, \tag{5.17}$$

satisfying the uncertainty relation $\Delta t \Delta f \geq (4\pi)^{-1}$. With this restriction, $E_x(\Delta t, \Delta f)$ is positive for the following large class of signals under consideration.

Due to the nature of the quadratic TFRs, the superposition principle is invalid in the sense that the TFR of a weighted sum of signals does not translate to the same weighted combination of the TFRs of the individual signals. For two signals $x_1(t)$ and $x_2(t)$,

$$F_{c_1 x_1 + c_2 x_2}(t, f) = |c_1|^2 F_{x_1}(t, f) + |c_2|^2 F_{x_2}(t, f) + I_{x_1, x_2}(t, f), \tag{5.18}$$

where the $I_{x_1, x_2}(t, f)$ term represents the cross-term (also called the *interference term*, or the *artifact*). This interference term is usually characterized by a highly oscillatory structure. It is essentially zero for the spectrogram, whenever the two signals are well separated in time and frequency. For other TFRs, this is not necessarily true and smoothing techniques are used to eliminate such undesirable components.

5.1.3 Shift-Invariant Time-Frequency Distribution

Among the quadratic TFRs, a special class of distributions satisfies the basic property of *time-frequency shift invariance*. If the signal is shifted in time and/or frequency, the TFRs of this class will be shifted by the same amount, both in time and in frequency

$$x'(t) = x(t - t_0)e^{j2\pi f_0 t} \quad \Rightarrow \quad F_{x'}(t, f) = F_x(t - t_0, f - f_0). \tag{5.19}$$

This class of TFRs is also referred to as the *Cohen class*. The spectrogram belongs to this class of TFRs. The general definition of a TFR in the Cohen class for an analytic signal $z(t)$ is given by

$$F_z^{(g)}(t, f) = \int_{-\infty}^{\infty} \int_{-\infty}^{\infty} \int_{-\infty}^{\infty} g(\nu, \tau) z\left(u + \frac{\tau}{2}\right) z^*\left(u - \frac{\tau}{2}\right) e^{j2\pi \nu (u-t)}\, e^{-j2\pi f \tau}\, d\nu\, du\, d\tau, \tag{5.20}$$

where $g(\nu, \tau)$ is the weighting function which defines the particular TFR of interest. An alternative definition is given by

$$F_z^{(g)}(t,f) = \mathcal{F}_{\tau \to f}\left\{G(t,\tau) \overset{t}{*} K_z(t,\tau)\right\}, \tag{5.21}$$

where

$$G(t,\tau) = \mathcal{F}^{-1}_{\nu \to t}\{g(\nu,\tau)\}, \tag{5.22}$$

$$K_z(t,\tau) = z\left(t+\frac{\tau}{2}\right) z^*\left(t-\frac{\tau}{2}\right). \tag{5.23}$$

The convolution integral in t is denoted by $\overset{t}{*}$, $\mathcal{F}_{\tau\to f}$ denotes the Fourier transformation with time variable τ and frequency variable f, and $\mathcal{F}^{-1}_{\nu\to t}$ denotes the inverse Fourier transformation with the frequency variable ν and time variable t.

We note that the choice of the weighting function $g(\nu,\tau)$ cannot be totally arbitrary. In particular, each of the properties 1–4 of Section 5.1.2 can be translated into specific constraints on the weighting function.

- The integrals (5.20) converge if $|g(\nu,\tau)| \le 1$.
- $F_z^{(g)}(t,f)$ is real if and only if $g(\nu,\tau) = g^*(-\nu,-\tau)$.
- The integral of $F_z^{(g)}(t,f)$ over t and f equals the energy of $z(t)$ if and only if $g(0,0) = 1$.
- The requirements on the first moments of $F_z^{(g)}(t,f)$ (5.15) and (5.16) are satisfied if

$$\left.\frac{\partial g(\nu,\tau)}{\partial \tau}\right|_{(0,0)} = \left.\frac{\partial g(\nu,\tau)}{\partial \nu}\right|_{(0,0)} = 0, \tag{5.24}$$

$$g(\nu,0) = \text{constant } \forall \nu, \tag{5.25}$$

$$g(0,\tau) = \text{constant } \forall \tau. \tag{5.26}$$

A TFR is a function of two variables, generally denoting *time* and *frequency*. No information is lost if the Fourier transform of this two-dimensional function is taken on either one or both of its variables. Various alternative representations related by Fourier transformations are possible, but will be omitted here.

5.1.4 The Wigner–Ville Distribution

The most well-known member of the Cohen class of TFRs is the Wigner–Ville distribution (WVD) obtained by setting $g(\nu,\tau) = 1$ in (5.20) and is defined by

$$W_z(t,f) = \int_{-\infty}^{\infty} z\left(t+\frac{\tau}{2}\right) z^*\left(t-\frac{\tau}{2}\right) e^{-j2\pi f\tau}\, d\tau, \tag{5.27}$$

where $z(t)$ is an analytic signal. Every quadratic time-frequency shift-invariant TFR can be considered as a smoothed or filtered version of the WVD. The WVD is always real (but may assume negative values), satisfies the marginal properties, has the highest energy concentration (i.e., in the second moment sense) around the instantaneous frequency of the signal, and its first moments yield the instantaneous frequency and the group delay of the signal. Moreover, the time-frequency support of the WVD is exactly equal to the duration and the bandwidth of the signal. The signal can be recovered from its WVD, to within a complex scaling c, as follows

$$z(t) = c \int_{-\infty}^{\infty} W_z(t, f)\, e^{j2\pi ft}\, df. \tag{5.28}$$

Note that since $z(0) = c|z(0)|^2$, because of (5.12), it follows that $c = 1/z^*(0)$. The WVD is the ideal tool for monocomponent signals, since it displays a high concentration of energy around the instantaneous frequency of the signal. In the presence of multi-component signals, the artifacts of the WVD are present even if the components are well separated in time and frequency. Such interference terms are highly oscillatory and concentrate in regions of the time-frequency plane that lie between every two components. Smoothing can be applied in order to attenuate these terms at the expense of resolution. Time smoothing and frequency smoothing can be decoupled, by carefully choosing the weighting function, yielding the so-called smoothed pseudo-WVD (the term "pseudo-WVD" is reserved for a short-time WVD, a WVD smoothed in time only). In the following four figures in Fig. 5.6, various TFRs of two unit-magnitude sinusoids

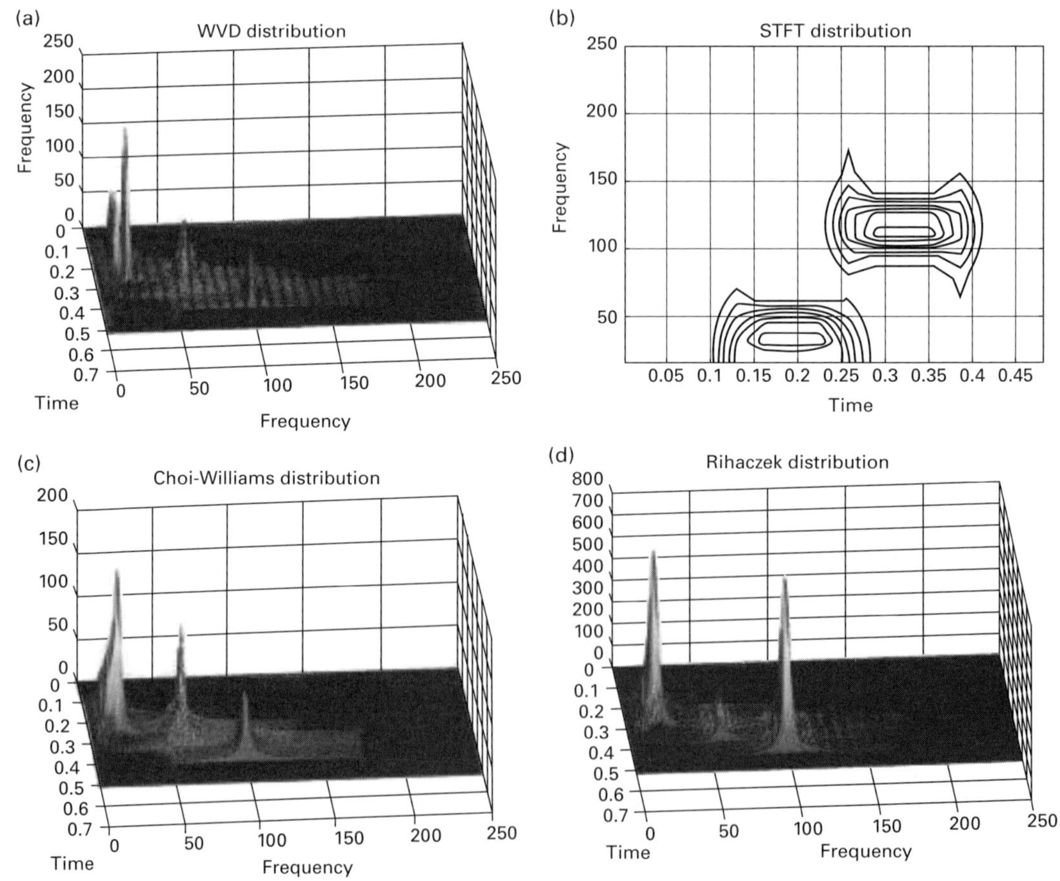

Figure 5.6 (a) WVD of the signal; (b) STFT of the signal; (c) Choi–Williams distribution with $\sigma = 5$; (d) Rihaczek distribution.

of frequency $f_1 = 20$ Hz and $f_2 = 100$ Hz each of length 256 after sampling with a $f_s = 500$ Hz are shown. Note that the WVD is defined on the analytic signal. This is due to the fact that when the definition is applied to a real signal, interference between negative and positive frequency components results in disturbing oscillatory artifacts.

5.2 Conclusion

Section 5.1 introduced various basic aspects of the time-frequency properties of a time-varying signal. Section 5.1.1 considered the short-time Fourier transform and spectrogram by using various possible windowings in conjunction with a Fourier transform to provide the time-localization property of the signal. Section 5.1.2 treated the quadratic TFR distributions. Section 5.1.3 considered various properties of the shift-invariant TFD. Finally, Section 5.1.4 treated the WVD, considered to be a commonly encountered and important class of the TFD.

5.3 References

In Section 5.1.1, more details on the two waveforms considered in Example 5.1 can be found in [1]. The musical score of J.S. Bach is shown in Fig. 5.3. Some time-frequency analysis books include [2], [3], [4], [5], [6]. The WVD was originated by E. Wigner in quantum mechanics [7] and in signal analysis by J. Ville [8].

[1] B. Boashash and V. Sucic, "High Performance Time-Frequency Distributions for Practical Applications," L. Debnath, ed., *Wavelets and Signal Processing*, Birkhauser, 2003, Chapter 6.

[2] B. Boashash, *Time-Frequency Signal Analysis*, Wiley, 1992.

[3] S. Qian and D. Chen, *Joint Time-Frequency Analysis*, Prentice-Hall, 1996.

[4] S.L. Marple, *Digital Spectral Analysis*, Prentice-Hall, 1987.

[5] L. Cohen, *Time-Frequency Analysis*, Prentice-Hall, 1995.

[6] B. Boashash, *Time Frequency Signal Analysis and Processing: A Comprehensive Reference*, Elsevier, 2016.

[7] E. Wigner, "On the Quantum Correction for Thermodynamic Equilibrium," *Physics Review*, 1932, p. 749.

[8] J. Ville, "Theorie et Application de la Notion de Signal Analytique," *Cables et Transmission*, 1948, pp. 61–74.

5.4 Exercises

1. Consider an exponentially decaying oscillating function

$$f_1(t) = \begin{cases} e^{-\gamma t} e^{iw_0 t}, & t \in [0, \infty), \\ 0, & t \in (-\infty, 0]. \end{cases}$$

Denote the Fourier transform of $f_1(t)$ by $F_1(\omega)$. Show

$$S_1(\omega) = |F_1(\omega)|^2 = \frac{1}{4\pi^2[\gamma^2 + (\omega - \omega_0)^2]}, \quad \omega \in (-\infty, \infty).$$

Consider a Gaussian decaying oscillatory function

$$f_2(t) = e^{-t^2/(2\sigma^2)} e^{i\omega_0 t}, \ t \in (-\infty, \infty).$$

Denote the Fourier transform of $f_2(t)$ by $F_2(\omega)$. Show

$$S_2(\omega) = |F_2(\omega)|^2 = e^{-(\omega-\omega_0)^2\sigma^2}, \quad \omega \in (-\infty, \infty).$$

Show that $S_1(\omega)$ is a unimodal peaked function about ω_0 with a one-sided width of γ and $S_2(\omega)$ is a unimodal peaked function about ω_0 with a one-sided width of $1/\sigma$.

2. There are several definitions and notations for the Wigner–Ville distribution (WVD) (also called the cross-WVD). One definition is given by

$$\mathrm{WVD}_{s,g}^{(1)}(t,\omega) = \int_{-\infty}^{\infty} s\left(t + \frac{\tau}{2}\right) g^*\left(t - \frac{\tau}{2}\right) \exp(-j\omega\tau) d\tau \,,$$

then the auto-WVD can be defined

$$\mathrm{WVD}_{s}^{(1)}(t,\omega) = \int_{-\infty}^{\infty} s\left(t + \frac{\tau}{2}\right) s^*\left(t - \frac{\tau}{2}\right) \exp(-j\omega\tau) d\tau \,.$$

Show

$$\mathrm{WVD}_{s,g}^{(1)}(t,\omega) = \mathrm{WVD}_{g,s}^{(1)*}(t,\omega),$$

which implies $\mathrm{WVD}_{s}^{(1)}(t,\omega)$ is real-valued.

3. Consider the WVD of a complex-valued square-integrable function $z(t)$ on $(-\infty, \infty)$ defined by

$$W_z(t,f) = \int_{-\infty}^{\infty} z\left(t + \frac{\tau}{2}\right) z^*\left(t - \frac{\tau}{2}\right) \exp(-j2\pi f\tau) d\tau.$$

a. Find the WVD of $az(t)$;
$z_1(t) + z_2(t)$;
$z(t - a)$;
$z(t)\exp(j2\pi f_0 t)$;
$z(t)\exp(j2\pi f_0 t^2)$;
$z(t/a), a > 0$.

b. Express the WVD of $z(t)$ in terms of the $Z(f) = FT\{z(t)\}$.

c. What is the Fourier transform of $W_z(t,f)$?

4. Consider the analysis of the chirp signal $x[n] = e^{i2\pi 6(n/N)^2}$, $n = 1, \ldots, N$, by FFT, STFT, and Wigner–Ville TFD. Pick your own parameters to illustrate the features of this signal using these three spectral estimation methods.

6 Wavelets and Subband Decomposition

6.1 Introduction

In this chapter, we consider various aspects of the wavelet transform (WT). WT is a relatively new analytical method of performing signal decomposition. Just as in the Fourier series, sets of complete orthonormal functions are used in performing signal decomposition. However, WT techniques have found applications for theoretical and practical data compressions. We will approach WT first from the affine and linear time-frequency representations point of view, then we will consider it as a form of multi-resolution subband decomposition.

6.2 Affine Time-Frequency Representations

In contrast to the previously considered quadratic TFRs, now we consider the affine class of TFRs, which preserve time scaling and time shifts of a real-valued signal $x(t)$, $-\infty < t < \infty$, in the sense that

$$x'(t) = \sqrt{|a|}\, x\,(a(t - t_0)) \quad \Rightarrow \quad F_{x'}(t, f) = F_x\left(a(t - t_0), \frac{f}{a}\right). \tag{6.1}$$

Any member of the affine class of TFRs can be written in terms of the WVD of the signal, as given by

$$F_z(t, f) = \int_{-\infty}^{\infty} \int_{-\infty}^{\infty} \chi\left(f(t - u), \frac{v}{f}\right) W_z(u, v)\, du\, dv, \tag{6.2}$$

where $\chi(\alpha, \beta)$ is a two-dimensional kernel function, independent of the analytic signal $z(t)$ of $x(t)$. We note that the Cohen shift-invariant and the affine classes of TFRs are not disjoint. The most commonly used affine TFRs, which are *not* members of the Cohen shift-invariant class, possess the *constant-Q* property. From basic circuit and system analysis, recall that the parameter Q of an oscillatory system is defined as the ratio of the center frequency to the 3dB bandwidth. For a constant-Q property TFR, as seen in Fig. 6.1 (Right), the -3dB to +3dB frequency band about f_2 is much larger than the -3dB to +3dB frequency band about f_1. In other words, for a constant-Q TRF, if $f_2 = 2 \times f_1$, then the 3dB bandwidth at f_2 must be twice as much as the 3dB bandwidth at f_1. Thus, the aperture of the frequency window that defines the frequency smoothing

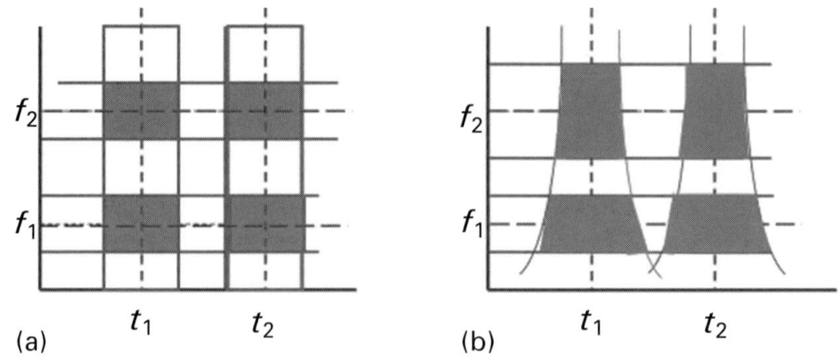

Figure 6.1 Comparison of resolutions of STFT (a) and WT (b)

increases as the frequency increases. The amount of time smoothing is in turn inversely proportional to the frequency of interest. An example of a constant-Q smoothing TRF is represented by the *scalogram*, defined by

$$C_z^{(g)}(t,f) = \int_{-\infty}^{\infty}\int_{-\infty}^{\infty} W_g\left(\frac{f}{f_0}(t-u), \frac{v}{f} f_0\right) W_z(u,v)\, du\, dv, \tag{6.3}$$

where the kernel function denoted now by $W_g(-\alpha/f_0, \beta f_0)$ is itself the WVD of a finite support bandpass signal $g(t)$, centered around its bandpass frequency f_0.

As mentioned earlier, there are TFRs which share the properties of shift invariance of the Cohen class and the affine property of the affine class. These TFRs are referred to as *shift-scale-invariants*. Because of the shift-invariant property, these TFRs do not allow constant-Q analysis. A characteristic of this subclass of TFRs is that their smoothing window $g(v,\tau)$ in (5.20) takes the form of a *product kernel*,

$$g(v,\tau) = s(v\tau), \tag{6.4}$$

where $s(\cdot)$ denotes the one-dimensional function of the form of $g(\cdot)$ in (6.3). The characteristics required from such a kernel are (i) $s(0) = 1$ in order to satisfy the marginal properties of (5.12) and (5.13), and (ii) $s(\cdot)$ be concentrated around the origin and tend to zero for the absolute value of the argument tending to infinity in order to achieve sufficient smoothing of the interference terms.

The *Choi–Williams distribution* belongs to the shift-scale-invariant class of TFRs and its kernel has the form of a Gaussian pdf as given by

$$s(\alpha) = e^{-(2\pi\alpha)^2/\sigma}, \tag{6.5}$$

where σ is a parameter that trades greater energy concentration with a smaller σ versus greater smoothing with a larger σ. The Choi–Williams distribution is often considered as the best compromise between resolution and artifact reduction in the presence of multi-component signals. The product form of the kernel introduces some restrictions on interference attenuation, since the support of the function $s(v\tau)$ on the (v,τ) plane is a region bounded by hyperboles ($v = \text{constant}/\tau$), and thus the $s(v\tau)$ is constant along the axes. This implies that whenever two or more signals occur at the same time or at the same frequency, substantial artifacts may show up.

6.3 Linear Time-Frequency Representations

Whenever the superposition principle holds for a given TFR, $F_x(t,f)$ is said to be *linear* so that

$$x(t) = a_1 x_1(t) + a_2 x_2(t) \quad \Rightarrow \quad F_x(t,f) = a_1 F_{x_1}(t,f) + a_2 F_{x_2}(t,f). \tag{6.6}$$

Linearity is desirable in multi-component signal analysis. Two well-known and important linear TFRs are the short-time Fourier transform (STFT) and the wavelet transform (WT).

6.3.1 The Short-Time Fourier Transform

The short-time Fourier transform, already defined in (5.10), is the most classical example of a linear TFR. The use of the window function $g(t)$ (previously denoted by $w(t)$ in (5.10)) has the result of suppressing all the signal lying outside of its support, thus retaining the spectral features of the portion of signal included in the aperture of the window. The STFT has the property of time and frequency shift invariance in the sense that

$$x'(t) = x(t - t_0)e^{i2\pi f_0 t} \quad \Rightarrow \quad S_{x'}^{(g)}(t,f) = S_x^{(g)}(t - t_0, f - f_0). \tag{6.7}$$

It is well known that the Fourier analysis imposes a dependence between time and frequency resolutions. The two extreme choices for $g(t)$, $g(t) = \delta(t)$ and $g(t) = 1$, $\infty < t < \infty$ illustrate such a trade-off. If $g(t) = \delta(t)$, then $S_x^{(\delta)}(t,f) = x(t)e^{-i2\pi f t}$, which means that the STFT preserves all the time fluctuations of the signal, at the cost of no frequency resolution. If on the other hand, $g(t) = 1$, then $S_x^{(1)}(t,f) = X(f)$, where all time resolution is completely lost.

The STFT keeps all the information of the signal $x(t)$, in the sense that the signal can be completely restored from its STFT,

$$x(t) = \int_{-\infty}^{\infty}\int_{-\infty}^{\infty} S_x^{(g)}(u,v)\gamma(t-u)\, e^{i2\pi v t}\, du\, dv, \tag{6.8}$$

where $\gamma(t)$, the *synthesis window function*, is any function such that

$$\int_{-\infty}^{\infty} g(t)\, \gamma^*(t)\, dt = 1. \tag{6.9}$$

We note that with adequate normalizations, $\gamma(t) = g(t)$, $\gamma(t) = \delta(t)$, and $\gamma(t) = 1$ are all good choices for the synthesis window.

6.3.2 The Wavelet Transform

Another prominent example of a linear TFR is the time-frequency version of the WT defined by

$$V_x^{(g)}(t,f) = \int_{-\infty}^{\infty} x(u) \sqrt{\left|\frac{f}{f_0}\right|}\, g^*\left(\frac{f}{f_0}(t-u)\right) du, \tag{6.10}$$

where $g(t)$ is a function with local support (is significantly non-zero in a limited interval around the origin) and is a bandpass signal centered on $f = f_0$. The function $g(t)$ is commonly referred to as the *mother wavelet*. Observe that the WT preserves time shifts and time scalings, but does not preserve frequency shifts as given by

$$x'(t) = x(t-t_0) \quad \Rightarrow \quad V_{x'}^{(g)}(t,f) = V_x^{(g)}(t-t_0,f), \tag{6.11}$$

$$x''(t) = \sqrt{|a|}x(at) \quad \Rightarrow \quad V_{x''}^{(g)}(t,f) = V_x^{(g)}(at,f/a). \tag{6.12}$$

We also note that the scaling in (6.12) preserves the energy of the signal.

The WT can be seen as the expansion of the signal $x(t)$ on the set of functions $\mathcal{G} = \{g_{(a,b)}(t)\}$, where $g_{(a,b)}(t) = |a|^{-1/2}g((t-b)/a)$. In our case, the scaling factor is $a = f_0/f$ and the translation parameter is $b = t_0$. The set $\mathcal{G}$ need not be orthonormal and may indeed be a redundant set (i.e., collection of functions from the set may be linearly dependent). Note that all the functions of $\mathcal{G}$ are obtained from the mother wavelet using simple translations and scalings. Technically, in the case of wavelets with discretized time-scale parameters, $\mathcal{G}$ constitutes a *frame*, that is, a set of functional elements such that any non-zero function has non-zero projection on at least one of these elements and any non-infinite function has finite projection. For a given analysis frequency, the signal $x(t)$ is convolved with a shifted and scaled version of the mother wavelet. As the analysis frequency increases, the time resolution increases (the window shrinks), and the bandwidth of the window broadens proportionally to the frequency. The WT can, for these reasons, be interpreted as a *constant-Q* analysis of the signal. The WT thus analyzes higher frequencies with better time resolutions but poorer frequency resolution than lower frequencies. Note, by contrast, that the time-frequency resolution of the STFT is constant for all times and frequencies. See Fig. 6.1 for a comparison of the resolution characteristics of the STFT and the WT.

The signal can be reconstructed from its WT according to the following equation

$$x(t) = \frac{1}{c_g} \int_{-\infty}^{\infty} \int_{-\infty}^{\infty} V_x^{(g)}(u,v) \sqrt{\left|\frac{v}{f_0}\right|}\, g\left(\frac{v}{f_0}(t-u)\right) \frac{du\, dv}{f_0}, \tag{6.13}$$

where

$$c_g \triangleq \int_{-\infty}^{\infty} \frac{|G(f)|^2}{|f|}\, df < \infty \tag{6.14}$$

is the *admissibility constant* of the mother wavelet, and $G(f) = \mathcal{F}\{g(t)\}$. Note that the WT of a signal is uniquely defined, but the inverse WT is *non-unique*. Different WT representations can be inverted with respect to the same mother wavelet and generate the *same* time function. Consider, for instance, the time-frequency function

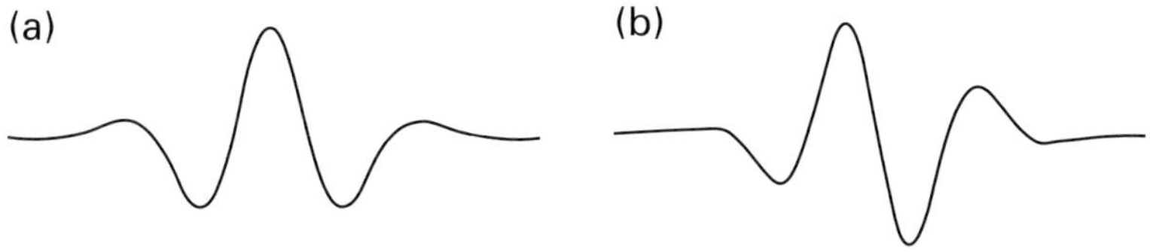

Figure 6.2 (a) The real-part of the Morlet mother wavelet. (b) The imaginary-part of the Morlet mother wavelet

$$V'(t, f) = c_g f_0 \, \delta(t, f - f_0). \tag{6.15}$$

If $V_x^{(g)}(t, f)$ in (6.13) is replaced with $V'(t, f)$, then the integral equals $g(t)$. This implies that both time-frequency representations $V_x^{(g)}(t, f)$, the WT of the mother wavelet with respect to itself, and $V'(t, f) = c_g f_0 \, \delta(t, f - f_0)$ correspond to the same inverse WT.

Usually, complex mother wavelets are used. If a real mother wavelet were employed, then the WT evaluated at the points which are nulls of $g(t)$ would be very sensitive to noise and prone to uncertainties. For this reason, it is customary to choose complex-valued oscillating mother wavelets without nulls. The properties of the WT are dictated by the mother wavelet of choice. Statements valid for certain mother wavelets can be totally incorrect for other choices of $g(t)$. The criterion of selection of a specific wavelet is intimately connected to the application. A classic example of a wavelet is the so-called *Morlet wavelet*, see Fig. 6.2, which is a Gaussian modulated tone,

$$g(t) = e^{-i2\pi f_0 t} h(t), \tag{6.16}$$

$$h(t) = C \, \sigma \, e^{-t^2/(2\sigma^2)}, \tag{6.17}$$

where C is a normalizing constant. The scalogram defined in (6.3) is in fact equal to the magnitude square of the WT whose mother wavelet is equal to the kernel $g(t)$. The scalogram therefore inherits all the properties of resolution and constant-Q analysis from the relative WT.

6.3.3 Relationship between STFT and WT

The weighting function $w(\cdot)$ in the STFT of (5.8) is commonly taken to be the unit value rectangular window defined on $[-\Delta t/2, \Delta t/2]$ and zero elsewhere to indicate an interest in this interval. We can also use a Gaussian-like weighting function $g(t) = C_o e^{-t^2/2\sigma^2}$ in (5.8) to indicate strong emphasis over a small symmetric interval about the origin and low emphasis for regions far from the origin. Then (5.8) becomes

$$S_z^{(g)}(t, f) = \int_{-\infty}^{\infty} z(\tau) \, C_o e^{-(t-\tau)^2/2\sigma^2} \, e^{-i2\pi f \tau} \, d\tau.$$

This STFT can then be interpreted as the projection of the waveform $z(\cdot)$ onto the Gaussian dampened complex exponential wavelet function, which is often called the Gabor wavelet function. A sketch of the real-part of this wavelet function for three

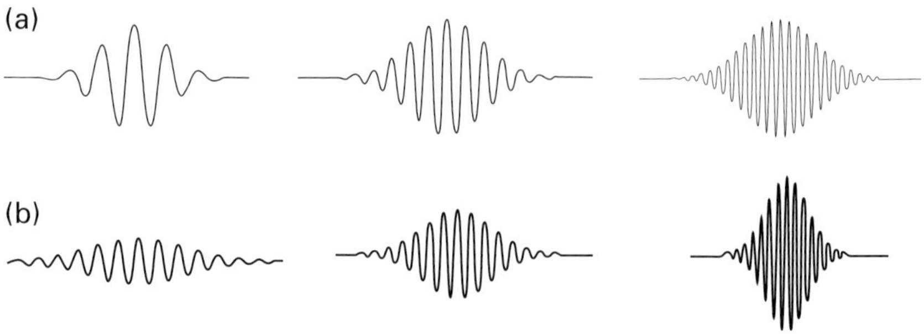

Figure 6.3 Transform wavelet functions for three increasing transformed frequency variables. (a) STFT; (b) Morlet WT

increasing values are given in Fig. 6.3(a). In Fig. 6.3(b), we sketch the corresponding real-part of the wavelet function for the Morlet WT given by (6.10), (6.16), and (6.17).

From direct inspection of the waveforms in Fig. 6.3(a) and Fig. 6.3(b), clearly there are similarities between these two wavelet functions. Both of these wavelet functions are Gaussianly tapered in their envelope values and the frequency of oscillation increases for higher transformed frequency variables. However, there are important differences. In the STFT cases of Fig. 6.3(a), the effective window length is invariant with the transformed frequency variable, while in the WT cases of Fig. 6.3(b), the effective window decreases with the transform frequency variable. These observations are consistent with the known constant time-frequency resolution of the STFT for all times and frequencies and the loss of frequency resolution for higher frequencies due to the constant-Q property of WT as seen in Fig. 6.1.

Example 6.1 We compare the time-frequency resolution of STFT and WT by considering two sinusoidal cases. In both cases, continuous-time waveforms of 256 ms are sampled at $f_s = 2{,}000$ Hz resulting in two sequences of length 512. Case 1 consists of the sum of two sinusoids of equal amplitude with $f_1 = 64$ Hz and $f_2 = 192$ Hz as shown in Fig. 6.4(a). Case 2 consists of a sinusoid of $f_1 = 128$ Hz with a gap of 64 samples in the middle of the sequence as show in Fig. 6.4(b). Fig. 6.5(a) shows the magnitude of the STFT for Case 1 with two sinusoids of Fig. 6.4(a) using a rectangular unit-valued window of length 128 samples (i.e., $\Delta t = 64$ ms), while Fig. 6.5(b) shows the magnitude using a window of length 32 samples (i.e., $\Delta t = 16$ ms). As can be seen, the larger windowed STFT can resolve the two sinusoidal frequencies, while the shorter windowed STFT can not. Similarly, Fig. 6.6(a) and (b) show the corresponding STFT magnitude results for Case 2 of a sinusoid with a gap in Fig. 6.4(b). We note, the gap in time is well detected by the 32 sample windowed STFT in Fig. 6.6(b) and is bearly detectable by the 128 sample windowed STFT in Fig. 6.6(a).

Fig. 6.7(a) shows the magnitude and phase plots of the Morlet WT for Case 1 of Fig. 6.4(a). Similar plots are shown in Fig. 6.7(b) for Case 2 of Fig. 6.4(b). We note, the same WT can resolve the two frequencies of Case 1 and the gap in the time-domain of Case 2. In contrast, the STFT needs to use two different appropriate window lengths

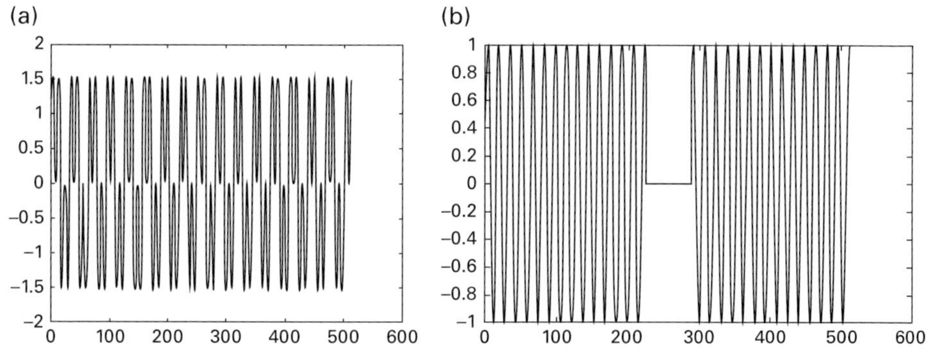

Figure 6.4 Time-domain sampled waveforms (a) Sum of two sinusoids at 64 Hz and 192 Hz (b) One sinusoid at 128 Hz with a center 64-sample gap

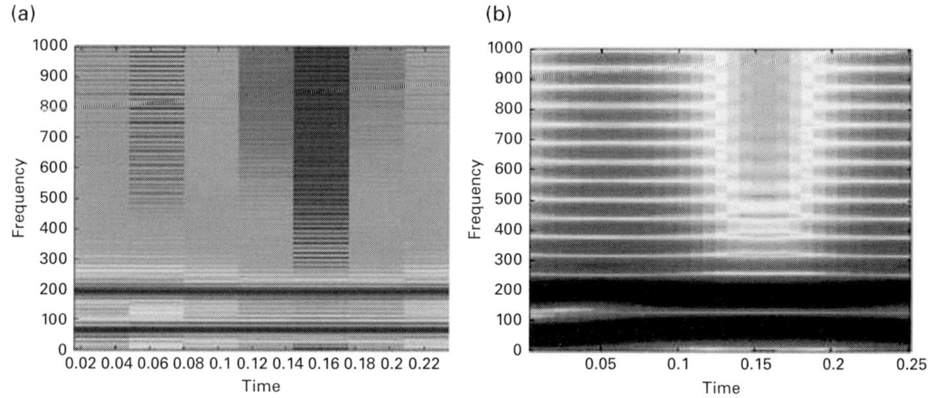

Figure 6.5 Magnitudes of STFT of Fig. 6.4(a) (a) 128-sample window (b) 32-sample window

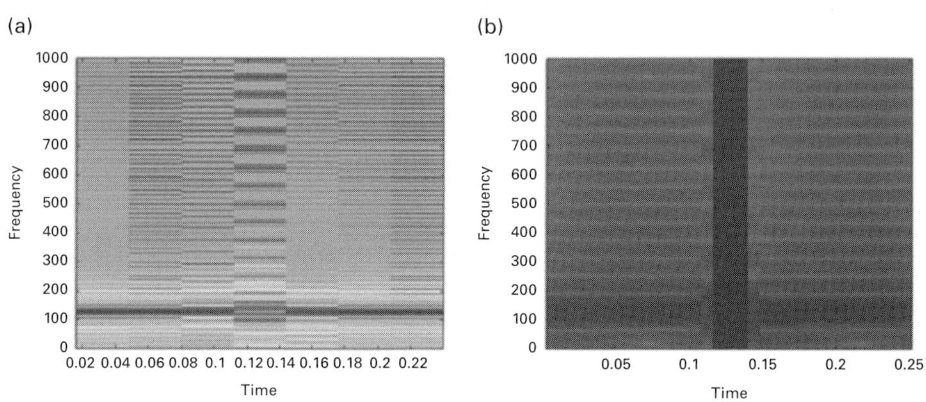

Figure 6.6 Magnitudes of STFT of Fig. 6.4(b) (a) 128-sample window (b) 32-sample window

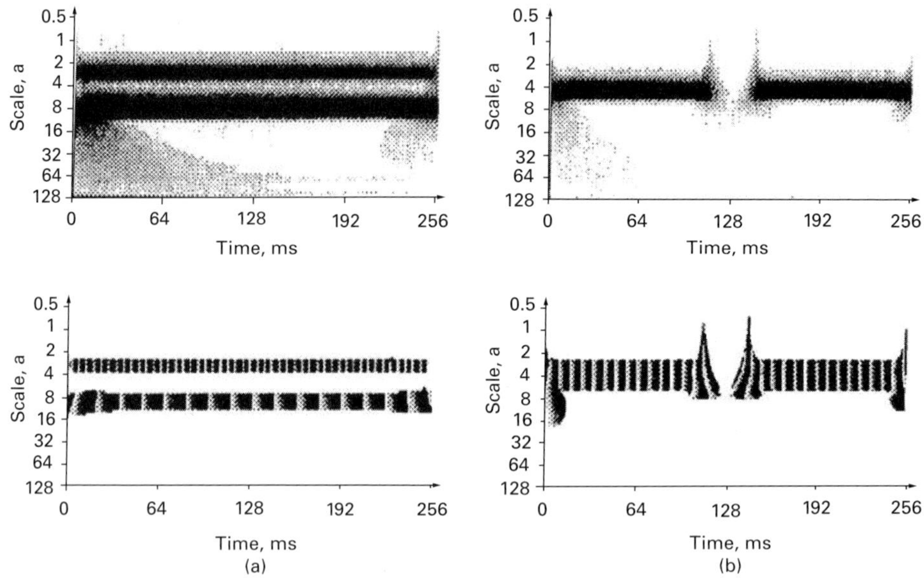

Figure 6.7 Magnitude and phase plots of the Morlet WT (a) Sampled waveform of Fig. 6.4(a) (b) Sampled waveform of Fig. 6.4(b). (IET Permission for using these figures from a 1994 Bentley-McDonnell paper.)

(i.e., 128 sample window for Case 1 and 32 sample window for Case 2) to perform the frequency-domain and time-domain resolutions. This example illustrates the robustness of the Morlet WT as compared to the STFT.

6.4 The Discrete Wavelet Transform

The continuous-time wavelet transform, $V_x^{(g)}(t, f)$, as described in (6.10), is defined in terms of dilations and translations of the mother wavelet

$$g_{(a,b)}(t) \equiv \frac{1}{\sqrt{a}} g\left(\frac{t-b}{a}\right), \tag{6.18}$$

for any pair of *scale* and *shift* parameters, a and b. The wavelet function in a continuous-time WT is highly redundant and its formulation in integral form is not practical for digital computation. In the discretized version of the continuous wavelet transform, called the *discrete-time wavelet transform* (DWT), the scale and shift parameters are sampled according to

$$a = a_0^m, \quad b = nb_0 a_0^m.$$

The *affine* wavelets, $g_{mn}(t)$, are defined by

$$g_{mn}(t) \equiv a_0^{-m/2} g(a_0^{-m} t - nb_0), \quad m, n \in \mathbf{Z}, \tag{6.19}$$

where the factor $a_0^{m/2}$ is a normalization factor.

In order to consider some basic properties of the DWT, we briefly review a few basic concepts on expansions in function space. We recall that $L^2(\mathbf{A})$ is the set of complex-valued functions $f(t)$ on the set $\mathbf{A}$ on the real-line, such that the energy $\int_{\mathbf{A}} |f(t)|^2\, dt < \infty$. The set $\mathbf{A}$ can be a finite interval, or a semi-infinite interval $[0,\infty)$, or the infinite interval $(-\infty,\infty)$ on the real-line. We also recall ℓ^2 is the set of complex-valued sequence $\{f_n, n \in \mathbf{Z}\}$, such that the energy $\Sigma_n |f_n|^2 < \infty$. A set $\{\phi_n(t)\}$ in $L^2(\mathbf{A})$ is said to be *complete* if every $f(t) \in L^2(\mathbf{A})$ has an expansion

$$f(t) = \Sigma_n f_n \phi_n(t),\ t \in \mathbf{A}. \tag{6.20}$$

The expansion in (6.20) need not be unique since the elements in a complete set may be redundant. A set $\{\phi_n(t)\}$ in $L^2(\mathbf{A})$ is a *basis* if all the elements of the set are linearly independent and the set is complete. The expansion using basis functions is unique. A set $\{\phi_n(t)\}$ in $L^2(\mathbf{A})$ is *orthonormal* if $(\phi_n(t), \phi_m(t)) = \delta_{m,n}$. Then for every complete orthonormal set $\{\phi_n(t)\}$ in $L^2(\mathbf{A})$, every $f(t) \in L^2(\mathbf{A})$ has the unique representation of (6.20) with the expansion coefficient given by

$$f_n = (f(t), \phi_n(t)) = \int_{\mathbf{A}} f(t)\phi_n^*(t)dt,\ n \in \mathbf{Z}, \tag{6.21}$$

satisfying

$$\Sigma_n |f_n|^2 = \|f\|^2 = \int_{\mathbf{A}} |f(t)|^2 dt. \tag{6.22}$$

Due to the complete orthonormality property of these basis functions, these expansion coefficients are unique as seen in (6.21). We note, the first expression of (6.22) gives the energy of the transform coefficients of the expansion, while the last expression of (6.22) gives the energy of the function in the time-domain. The Parseval theorem given by (6.22) states that the energy in the two domains are equal for every complete orthonormal set $\{\phi_n(t)\}$ in $L^2(\mathbf{A})$.

The concept of a "frame" is a generalization of the concept of a complete set. A set $\{\phi_n(t)\}$ in $L^2(\mathbf{A})$ is a *frame* if there exists $0 < A$ and $0 < B < \infty$ such that for every $f(t) \in L^2(\mathbf{A})$

$$A\|f\|^2 \leq \Sigma_n |f_n|^2 \leq B\|f\|^2. \tag{6.23}$$

Thus, for any $\{\phi_n(t)\}$ of a frame, every $f(t) \in L^2(\mathbf{A})$ has an expansion of the form of (6.20). In general, the coefficients of the expansion of a frame need not be unique. A frame is tight if $A = B$. If $\{\phi_n(t)\}$ in $L^2(\mathbf{A})$ is a tight frame with $A = B = 1$ and $\|\phi_n(t)\| = 1$, then $\{\phi_n(t)\}$ is an orthonormal basis.

The affine wavelet function $g_{mn}(t)$ of (6.19) is a frame if the mother wavelet function of (6.18) satisfies the admissibility constant condition of (6.14). Furthermore, an orthonormal set of wavelets satisfy the orthonormality condition in both shift and scale indices,

$$\int g_{mn}(t) g_{m'n'}(t)\, dt = \delta_{m,m'}\delta_{n,n'},$$

where the integral is zero everywhere, except for $m = m'$, $n = n'$ when it is equal to unity. Conditions on the mother wavelet can be imposed such that the wavelets satisfy the orthonormality condition. A common choice for the scale and shift parameters of DWT is to take $a_0 = 2$ and $b_0 = 1$, which define a *dyadic* decomposition.

6.5 Multi-resolution Decomposition

A practical implementation of the dyadic DWT is provided by the *multi-resolution analysis* or *pyramidal subband decomposition*. Both subband decomposition and DWT can be implemented by using a tree structure as in Fig. 6.8, where the input signal is passed through a low-pass filter, $\tilde{h}_0$ and a high-pass filter, $\tilde{h}_1$. The output of the low-pass filter, downsampled by two, is in turn the input of a second set of low- and high-pass filters, $\tilde{h}_0$ and $\tilde{h}_1$, and so on. The procedure is repeated for L stages. Note that the operation of decimation by two in this case does not reduce the amount of information, since the equivalent bandwidth of the filter output is half that of the original signal. The downsampled outputs of the low-pass filters at stage i, $1 \leq i \leq L$, are referred to as *approximation sequences at level* i, while the downsampled high-pass filter outputs are called *detail sequences at level* i. In short, if we define $D_i f_n$ as the detail sequence at level i and $A_i f_n$ the approximation sequence at level i, one has that

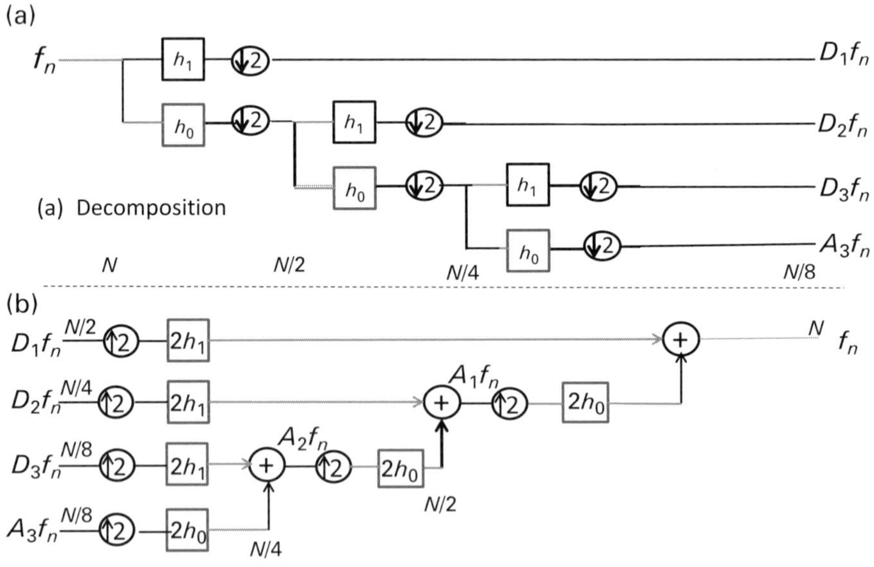

Figure 6.8 Block diagram of the DWT implemented as subband decomposition (a) and reconstruction using a filter bank tree (b)

$$D_i f_n = \sum_k A_{i-1} f_k \tilde{h}_{1,2n-k},$$

$$A_i f_n = \sum_k A_{i-1} f_k \tilde{h}_{0,2n-k}.$$

The complete decomposition includes the L detail sequences at levels 1 through L and the approximation sequence at level L. Note that at any time the number of samples remains constant. Suppose that the original signal has N samples, and for simplicity, assume that N is a power of 2. Both the detail and the approximation sequences at level 1 have $N/2$ samples. At level 2, the detail and approximation sequences have $N/4$ samples, which when added to the $N/2$ samples of the detail sequence at level 1 produces N samples, and so on.

During reconstruction, the detail sequences and the approximation sequences at level i are upsampled by two, by simply adding one zero between any two samples. The upsampled detail sequence is then passed through a reconstruction filter $h_{1,n} \equiv \tilde{h}_{1,-n}$, the upsampled approximation sequence is passed through a filter with coefficients $h_{0,n} \equiv \tilde{h}_{0,-n}$, and the two resulting sequences are summed to form the approximation sequence at level $(i-1)$. This can be represented as follows

$$A_{i-1} f_n = 2\left(\sum_k A_i f_n h_{0,n-2k} + \sum_k D_i f_n h_{1,n-2k}\right).$$

Now consider the selection of the filter coefficients and its relationship to the DWT. From the theory of *multi-resolution approximation* for $L^2(\mathbf{A})$, it is possible to show that the filter coefficients, $h_{0,n}$ and $h_{1,n}$, and their Fourier transforms, $H_0(f) \equiv \sum_n h_{0,n} e^{-jn2\pi f}$ and $H_1(f) \equiv \sum_n h_{1,n} e^{-jn2\pi f}$, satisfy the following conditions:

$$h_{0,n} = O(n^{-2}), \tag{6.24}$$

$$|H_0(0)| = 1, \tag{6.25}$$

$$|H_0(f)|^2 + |H_0(f+1/2)|^2 = 1, \tag{6.26}$$

$$h_{1,n} = (-1)^{1-n} h_{0,1-n}. \tag{6.27}$$

In general, filters satisfying (6.26) are named *conjugate*, while (6.27) defines *quadrature mirror filters*.

It can be proved that given a wavelet function, $g_{mn}(t)$, the filter coefficients are uniquely determined. This relationship can be conveniently described in terms of the *scaling function*, $\phi(t)$. The requirement on $\phi(t)$ is that the set of functions

$$\{\phi_{mn}(t) = 2^{-m/2}\phi(2^{-m}t - n)\}$$

be an orthonormal basis for a closed subspace of $L^2(\mathbf{A})$, that is,

$$\int \phi_{mn}\phi_{mn'}\, dt = \delta_{n,n'}.$$

It is customary to require that $\phi(t)$ be continuously differentiable, with asymptotic decay given by $|\phi(t)| = O(t^{-2})$, $d\phi/dt = O(t^{-2})$. Then we have that

$$\phi(t) = 2\sum_n h_{0,n}\phi(2t-n),$$

$$g(t) = 2\sum_n h_{1,n}\phi(2t-n),$$

and $\{g_{mn}(t) = 2^{-m/2}g(2^{-m}t-n)\}$ is a set of orthonormal wavelets. The coefficients of h_0 and h_1 can also be computed as

$$h_{0,n} = 2^{-1}\int \phi(2^{-1}u)\phi(u-n)\,du,$$

$$h_{1,n} = 2^{-1}\int g(2^{-1}u)\phi(u-n)\,du.$$

It can also be shown that the detail sequences actually coincide with the wavelet coefficients f_{mn}, that is,

$$D_i f_n \equiv f_{in}.$$

The reconstructed signal is then equal to

$$\begin{aligned} f(t) &= \sum_{n=-\infty}^{\infty} A_L f_n 2^{-L/2}\phi(2^{-L}t-n) + \sum_{m=1}^{L}\sum_{n=-\infty}^{\infty} D_m f_n 2^{-m/2} g(2^{-L}t-n) \\ &= \sum_{n=-\infty}^{\infty} A_L f_n 2^{-L/2}\phi(2^{-L}t-n) + \sum_{m=1}^{L}\sum_{n=-\infty}^{\infty} f_{mn} 2^{-m/2} g(2^{-L}t-n). \quad (6.28) \end{aligned}$$

The pure wavelet expansion, $f(t) = \sum_m \sum_n f_{mn}2^{-m/2}g(2^{-m}t-n)$ requires infinite resolution levels in order to fully reconstruct the signal. Clearly, the form of (6.28) is more practical for a realistic implementation.

6.5.1 The Haar and the Shannon Wavelets

The Haar transform provides the simplest possible multi-resolution analysis as well as DWT. The scaling function is defined by

$$\phi(t) = \begin{cases} 1, & 0 \le t \le 1, \\ 0, & \text{otherwise.} \end{cases}$$

The basic requirements are satisfied, although $\phi(t)$ is clearly discontinuous. The corresponding wavelet is given by

$$g(t) = \begin{cases} 1, & 0 \le t < 1/2 \\ -1, & 1/2 \le t < 1 \\ 0, & \text{otherwise} \end{cases}.$$

The following holds

$$\phi(t) = \phi(2t) + \phi(2t-1),$$
$$g(t) = \phi(2t) - \phi(2t-1).$$

Moreover, the orthogonality conditions hold

$$\int \phi(t-n)\phi(t-n')\,dt = \delta_{n,n'},$$
$$\int g_{mn}(t) g_{m'n'}(t)\,dt = \delta_{m,m'}\delta_{n,n'}.$$

The filter coefficients are then

$$h_{0,n} = \frac{1}{2}\left(\delta_{n,0} + \delta_{n,1}\right),\ h_{1,n} = \frac{1}{2}\left(\delta_{n,0} - \delta_{n,1}\right).$$

The Haar scaling functions constitute the simplest orthonormal family. The main drawback of these functions is that their discontinuity in time makes their frequency resolution very poor. The Shannon wavelets are at the other extreme, discontinuous in frequency and spread out in time. The scaling function is defined by

$$\phi(t-n) = \frac{\sin \pi(t-n)}{\pi(t-n)}, \qquad n \in \mathbf{Z},$$

and it is well known that the set of its shifted replicas constitutes an orthonormal basis. From the sampling theorem, it is known that

$$f(t) = \sum_k f_k \frac{\sin \pi(t-k)}{\pi(t-k)}.$$

The Fourier transform of the scaling function is a rectangular window in the frequency interval $(-1/2, 1/2)$. It turns out that the wavelet function is given by

$$g(t) = \frac{\sin \frac{\pi}{2} t}{\frac{\pi}{2} t} \cos \frac{3\pi}{2} t.$$

The filter coefficients are given by

$$h_{0,n} = \frac{1}{2} \frac{\sin(n\pi/2)}{n\pi/2},$$
$$h_{1,n} = (-1)^{n-1} h_{0,1-n}.$$

6.6 Wavelets and Compression

An important goal in certain communication systems is to reduce the amount of redundancy inherently present in a signal, in order to reduce the required transmission bandwidth. There are two kinds of compression, *lossless* and *lossy*. A compression scheme is lossless if the signal can be reconstructed with zero error. Higher compression ratios can usually be obtained if at the reconstruction level some distortion is allowed (lossy compression). The conventional strategy in lossy compression is to discard small data and all the data that have minimal contribution to the perception of the received signal. For instance, in image and sound signals, very high frequency components can be safely filtered out, since human eyes and ears are not very sensitive to highly oscillatory signals. In the method of transform coding, an invertible transform is usually applied to the signal in order to produce an alternative but equivalent representation. Then some or all of the transformed data are quantized or modified in order to minimize the number of transmitted parameters. At the receiving end, the signal is reconstructed by applying the inverse transform. A transform is suited for compressing a given class of signals if it produces close to zero data for most signals in the class. The wavelet transform has been extensively used for the purpose of compression. The critical step is the selection of the appropriate wavelet basis.

The Haar DWT is one of the oldest and simplest DWTs. The input signal, x_n, $n = 0, \ldots, N-1$, is passed through the two filters h_0 and h_1. At the output of the low-pass filter, one has the sequence given by $x_0 + x_1, x_2 + x_3, \ldots$, while the sequence output by the high-pass filter is given by $x_0 - x_1, x_2 - x_3, \ldots$ It is clear that if the original signal is almost constant or slowly varying, then one expects the detail sequence to be very small. The new representation, that is, the approximation sequence plus the detail sequence values, is suitable for compression. In fact, only a few of the detail sequence samples need to be retained and one can still have a very faithful signal reconstruction. Of course, the Haar transform can be applied in turn to the approximation sequence, iteratively, in a pyramidal scheme. The Haar transform is simple to understand and implement, but it lacks sophistication. In particular, non-constant signals may lead to a large number of non-zero coefficients, thereby reducing the compression capabilities.

Now, consider the Haar DWT in more detail. Denote the original data column vector $x = [x_0, x_1, x_2, x_3, \ldots, x_{N-1}]^T$. In matrix form, the Haar transform H can be described by the following matrix

$$H = \begin{pmatrix} 1 & 1 & & & & & \\ 1 & -1 & & & & & \\ & & 1 & 1 & & & \\ & & 1 & -1 & & & \\ & & & & \vdots & & \\ & & & & & 1 & 1 \\ & & & & & 1 & -1 \end{pmatrix}.$$

Then the transformed vector y is given by

$$y = Hx = \begin{pmatrix} 1 & 1 & & & & & \\ 1 & -1 & & & & & \\ & & 1 & 1 & & & \\ & & 1 & -1 & & & \\ & & & & \vdots & & \\ & & & & & 1 & 1 \\ & & & & & 1 & -1 \end{pmatrix} \begin{bmatrix} x_0 \\ x_1 \\ x_2 \\ x_3 \\ \vdots \\ x_{N-2} \\ x_{N-1} \end{bmatrix} = \begin{bmatrix} x_0 + x_1 \\ x_0 - x_1 \\ x_2 + x_3 \\ x_2 - x_3 \\ \vdots \\ x_{N-2} + x_{N-1} \\ x_{N-2} - x_{N-1} \end{bmatrix},$$

where the odd components correspond to the approximation sequence, and the even components correspond to the detail sequence. Next, note that the inverse Haar transform is given by $H^{-1} = \frac{1}{2}H$, since

$$H^{-1}H = \frac{1}{2}\begin{pmatrix} 1 & 1 & & & & & \\ 1 & -1 & & & & & \\ & & 1 & 1 & & & \\ & & 1 & -1 & & & \\ & & & & \vdots & & \\ & & & & & 1 & 1 \\ & & & & & 1 & -1 \end{pmatrix}\begin{pmatrix} 1 & 1 & & & & & \\ 1 & -1 & & & & & \\ & & 1 & 1 & & & \\ & & 1 & -1 & & & \\ & & & & \vdots & & \\ & & & & & 1 & 1 \\ & & & & & 1 & -1 \end{pmatrix}$$

$$= \frac{1}{2}\begin{pmatrix} 2 & 0 & 0 & & & & \\ 0 & 2 & 0 & & & & \\ & & 2 & 0 & 0 & & \\ & & & 2 & 0 & 0 & \\ & & & & \vdots & & \\ & & & & 0 & 2 & 0 \\ & & & & 0 & 0 & 2 \end{pmatrix} = I_N.$$

Then taking the inverse Haar DWT on y yields

$$x = H^{-1}y = \frac{1}{2}Hy = \frac{1}{2}\begin{pmatrix} 1 & 1 & & & & & \\ 1 & -1 & & & & & \\ & & 1 & 1 & & & \\ & & 1 & -1 & & & \\ & & & & \vdots & & \\ & & & & & 1 & 1 \\ & & & & & 1 & -1 \end{pmatrix}\begin{bmatrix} x_0 + x_1 \\ x_0 - x_1 \\ x_2 + x_3 \\ x_2 - x_3 \\ \vdots \\ x_{N-2} + x_{N-1} \\ x_{N-2} - x_{N-1} \end{bmatrix}$$

$$= \frac{1}{2}\begin{bmatrix} x_0 + x_1 + x_0 - x_1 \\ x_0 + x_1 - x_0 + x_1 \\ x_2 + x_3 + x_2 - x_3 \\ x_2 + x_3 - x_2 + x_3 \\ \vdots \\ x_{N-2} + x_{N-1} + x_{N-2} - x_{N-1} \\ x_{N-2} + x_{N-1} - x_{N-2} + x_{N-1} \end{bmatrix} = \begin{bmatrix} x_0 \\ x_1 \\ x_2 \\ x_3 \\ \vdots \\ x_{N-2} \\ x_{N-1} \end{bmatrix}.$$

We note the joint Haar DWT and inverse Haar DWT are lossless transformations.

Much research has been devoted to the development of smooth sets of wavelet bases for signal representation. If, with the other conditions, we impose our high-pass filter

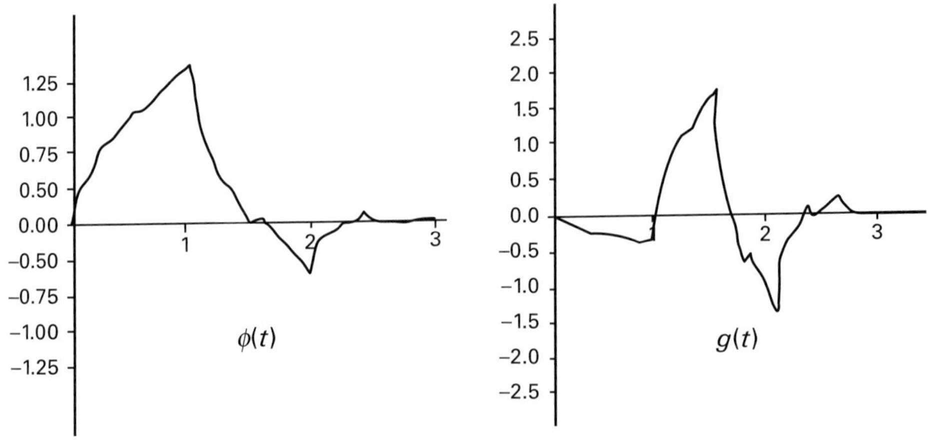

Figure 6.9 Daubechies scaling and wavelet functions of order 2

transforms any linear function to zero, then we obtain the following filter coefficients (for simplicity written as row vectors)

$$h_1 = \frac{1}{8}\left(-1+\sqrt{3} \quad -3+\sqrt{3} \quad 3+\sqrt{3} \quad -1-\sqrt{3}\right),$$
$$h_0 = \frac{1}{8}\left(1+\sqrt{3} \quad 3+\sqrt{3} \quad 3-\sqrt{3} \quad 1-\sqrt{3}\right).$$

As an example, verify that both sequences $1, 1, 1, 1, \ldots$ and $1, 2, 3, 4, \ldots$ produce the zero sequence at the output of the high-pass filter. This transform results in better compression ratios than the Haar transform, since it produces near zero detail sequences for a larger class of signals. The form of the scaling function can be obtained from the basic equation

$$\phi(t) = 2h_{0,0}\phi(2t) + 2h_{0,1}\phi(2t-1) + 2h_{0,2}\phi(2t-2) + 2h_{0,3}\phi(2t-3).$$

The values $\phi(1)$ and $\phi(2)$ can be obtained by setting $t = 1$ and $t = 2$. Iteratively, one can obtain all the remaining values. What one finally obtains is a continuous but "rough" looking function, as seen in Fig. 6.9, where the corresponding wavelet is also shown. Incidentally, this wavelet is referred to as Daubechies wavelet of order 2. The Daubechies wavelet of order 1 corresponds to the Haar wavelet.

6.7 Conclusion

Section 6.2 introduced the concept of affine time-frequency representations. Fig. 6.1 compared the resolutions of the STFT and that of a wavelet transform. Section 6.3 discussed the linear time-frequency representations, including the STFT and the wavelet transform and their relationship. Section 6.4 defined the discrete wavelet transform. Section 6.5 introduced the concept of multi-resolution decomposition, including the classical Haar and Shannon wavelets. Section 6.6 considered the use of wavelets in data compression and the advantage of the Daubechies wavelets.

6.8 References

An extensive number of publications on wavelet transforms include: tutorial papers [1], technical papers [2], [5], [6], and detailed books [3], [7], [8], [9].

[1] P.M. Bentley and J.T.E. McDonnel, "Wavelet Transforms: An Introduction," *Electronics & Communication Journal*, 1994, pp. 175–186.

[2] S.G. Mallat, "Wavelets and Signal Processing," *IEEE Trans. on Pattern Analysis and Machine Intelligence*, 1989.

[3] S. Mallat, *Wavelet Tour of Signal Processing*, Academic Press, 1998.

[4] C.S. Burrus, *et al., Computer-Based Exercises for Signal Processing*, Prentice-Hall, 1994.

[5] G. Strang, "Wavelet Transforms versus Fourier Transforms," *Bulletin American Mathematical Society*, 1993, pp. 288–305.

[6] I. Daubechies, *Ten Lectures on Wavelets*, SIAM, 1992.

[7] A. Aldroubi and M. Unser, *Wavelets in Medicine and Biology*, CRC, 1996.

[8] G. Strang and T. Nguyen, *Wavelets and Filter Banks*, Wellesley–Cambridge University Press, 1996.

[9] J.H. Davis, *Methods of Applied Mathematics with a Matlab Overview*, 2001, Chapters 8–9.

6.9 Exercises

1. When you have time, read the historical development of wavelets written by a non-mathematician in the book, *The World According to Wavelets*, second edition, by B.B. Hubbard, A.K. Peters, 1998.
2. Construct the multi-resolution using the classical Haar functions as outlined in *The World According to Wavelets*, second edition, by B.B. Hubbard, A.K. Peters, 1998, pp. 173–174.
3. Prove the Heisengberg's uncertainty principle in signal processing. Start with the definition of the variance as a measure of concentration of $f(t)$ in the time-domain. We assume the first moment is zero and $||f(t)||_2 = 1$. Then

$$\sigma^2 = \int_{-\infty}^{\infty} t^2 |f(t)|^2 dt\,.$$

Similarly, define the concentration of the signal in the frequency-domain by

$$\widehat{\sigma}^2 = \frac{1}{2\pi}\int_{-\infty}^{\infty} \omega^2 |F(\omega)|^2 d\omega\,.$$

Show

$$\sigma\hat{\sigma} \geq (1/2).$$

Hint: Apply the Schwarz inequality to

$$\left| \int_{-\infty}^{\infty} (t\, f(t)\, f'(t))dt \right|^2.$$

4. Consider the creation of a scaling function as outlined on pages 176–181 in *The World According to Wavelets,* second edition, by B.B. Hubbard, A.K. Peters, 1998.
5. Plot the frequency responses using FFT (with some zero paddings) to show that h_0 is a low-pass filter and h_1 is a high-pass filter for the following cases:

   a. The Haar wavelet.
   b. The Shannon wavelet.
   c. The Daubechies wavelet of order 2.

6. For the mathematically sophisticated students, relate the multi-resolution theory of the wavelet by S. Mallat to the operator-theoretic approach to the scattering theory work of P. Lax and R.S. Phillips. S. Mallat, "Multiresolution Approximation and Wavelet Orthonormal Bases of $L^2(R)$," *Trans. American Mathematical Society*, 1989, pp. 69–87. P. Lax and R.S. Phillips, *Scattering Theory,* Academic Press, 1967. More details on this issue were given in the doctoral thesis, "Operator Theoretic Approach to Wavelet Theory and Algorithm," by J.N.Q. Pham, Electrical Engineering Department, UCLA, 1998.

7 Beamforming and Array Processing

7.1 Introduction

Consider an RF transmitter sending a waveform

$$s(t) = A\cos(2\pi f_0 t), \quad -\infty < t < \infty,$$

over an ideal free-space propagation medium. Also consider a receiver in the far-field of the transmitter at a distance of r from the transmitter. Let the input of a receiver be denoted by

$$x(t) = (A/r)\cos(2\pi f_0 t + t_0) + n(t), \quad -\infty < t < \infty,$$

where t_0 denotes the transmission delay time from the transmitter to the receiver, A/r denotes the attenuated received amplitude due to $1/r$ propagation spatial loss, and $n(t)$ is modeled by a zero-mean white Gaussian noise (WGN) of variance σ_0^2. Then the signal-to-noise ratio (SNR) of the receiver is given by

$$SNR_1\,(\text{dB}) = 10\log_{10}(A/r)^2/\sigma_0^2\,.$$

Suppose we have N such receivers all at a distance r from the transmitter and each having a WGN $n_i(t)$ with its noise variance of $\sigma_i^2 = \sigma_0^2$, then denote the sum of all N receivers by $Y(t) = (NA/r)\cos(2\pi f_0 + t_0) + n_i(t), \quad -\infty < t < \infty,\ i = 1, \ldots, N$. Then the SNR of the combined system is given by

$$SNR_N\,(\text{dB}) = (NA/r)^2/(N\sigma_0^2) = N * SNR_1\,(\text{dB})\,.$$

Thus, under the above ideal conditions, this "beamformer" utilizing N such receivers has enhanced the SNR of a single receiver by a factor of N. In practice, due to all sorts of real-life constraints, the full SNR gain factor of N may not be achievable by such a simple beamformer, but nevertheless an effective SNR gain larger than a single receiver is quite possible.

Given the possible SNR gain of an array of receivers (sensors) over a single receiver (sensor), beamforming of array processing has motivated diverse applications in the past. Early works in hydrophone array processing appeared in [1] and in RF array processing in [2] and [3]. Since then, array processing has been utilized for acoustic/seismic processing in sensor networks [4] and [5], and more recently for cellular telephony MIMO systems.

Section 7.2 considers some early simple sum and delay linear array processing and minimum-variance distortionless array processing for narrowband waveforms. Section 7.3 treats three wideband beamformers. In Section 7.4, array system performance analysis by CRB, simulations, and field measurements are treated. Some recent array processing methodologies, including the robust beamforming method, random finite set method for array design, and wideband radar beamformers are considered in Section 7.5. Section 7.6 includes the conclusion.

7.2 Early Array Processing Systems

In this section, we will first deal with a uniform linear array and then with a minimum-variance distortionless response (MVDR) array. As before, first consider a narrowband source waveform, where the information signal is modulated on some RF frequency, f_0, for practical transmission purpose. The bandwidth of the signal over $[0, f_s]$ is generally much less than the RF frequency. Thus, the ratio of the highest to lowest transmitted frequency, $(f_0 + f_s)/(f_0 - f_s)$, is typically near unity. In the 802.11b ISM wireless LAN system, the ratio is 2.5 GHz/2.4 GHz = 1.04, which is denoted as narrowband. Narrowband waveforms have a nominal wavelength and time delays can be compensated by simple phase shifts. The classical narrowband beamformer operating on these waveforms is a spatial extension of the matched filter or coherent demodulator. In classical time-domain filtering, the time-domain signal is linearly combined with the filtering weight to achieve the desired coherent processing. This narrowband beamformer also combines the spatially distributed sensor collected array data linearly with the beamforming weight to achieve spatial filtering. Beamforming enhances the signal from the desired spatial direction and reduces the signal(s) from other direction(s) in addition to possible time-frequency filtering. On the other hand, most acoustical waveforms are not narrowband, and thus more intricate processing has utilized for these wideband waveforms.

Next, consider the RF transmission of a narrowband waveform over an ideal free-space propagation medium as before. Except now consider an N-sensor linear array beamformer as shown in Fig. 7.1. In this linear array model, the far-field wavefront is planar and is incident at an angle of θ relative to the normal of the array, and all the N receiving sensors are assumed to have omni-directional sensitivities, with an adjacent inter-sensor spacing of $\lambda/2$, where $\lambda = c/f_0$. Thus, the wavefront at sensor n has to travel an additional distance of $d \sin(\theta)$ relative to the wavefront at sensor $(n - 1)$. The relative time delay to travel this additional distance is then given by $d \sin(\theta)/c = d \sin(\theta)/(f_0\lambda)$, where c is the speed of propagation of the wavefront and λ is the wavelength corresponding to frequency f_0. Then the time delays for all the sensors are given by

$$\tau_n = \tau_1 + (n-1)d \sin(\theta)/(f_0\lambda), \; n = 1, \ldots, N, \tag{7.1}$$

and the expression for all the received waveforms is given by

$$x_n(t) = A \exp(i2\pi f_0 t) \exp(-i2\pi f_0 \tau_1) \exp(-i2\pi d(n-1) \sin(\theta)/\lambda) + n_n(t),$$
$$n = 1, \ldots, N. \tag{7.2}$$

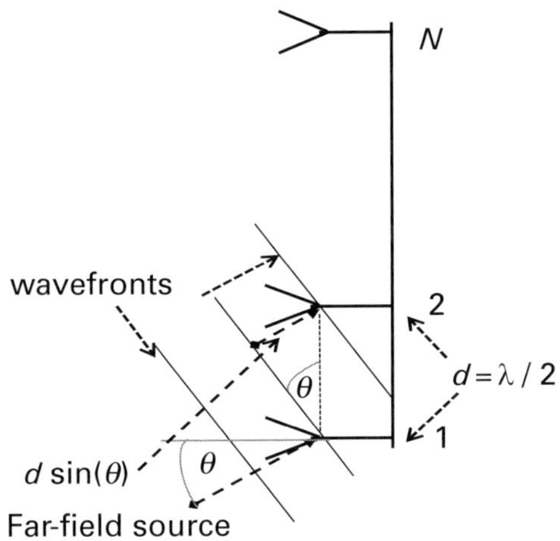

Figure 7.1 Uniform linear array of N sensors with inter-sensor spacing $d = \lambda/2$

If we use this uniform linear array, τ_1 is still known, but all the other $\tau_n, n = 2, \ldots, N$, are fully known relative to τ_1 as given by (7.1). Then the ideal beamformer output expression becomes

$$y(t) = \exp(i2\pi f_0 \tau_1) \left[N\, A \exp(i2\pi f_0 t) + \sum_{n=1}^{N} \{\exp(i2\pi d \ \sin(\theta)/\lambda)\, n_n(t)\} \right], \qquad (7.3)$$

which achieves the desired $SNR_N = N \times SNR_1$.

Now, suppose each sensor has an omni-directional response in all angular directions (i.e., isotropic over $[-\pi, \pi]$). The beamformer angular transfer function for a uniform linear array with inter-sensor spacing of $d = \lambda/2$ is given by

$$\begin{aligned} H(\theta) &= \sum_{n=1}^{N} \exp(-i\pi(n-1)\sin(\theta)) = \frac{1 - \exp(-i\pi N \sin(\theta))}{1 - \exp(-i\pi \sin(\theta))} \\ &= \frac{\exp(-i(\pi/2(N-1)\sin(\theta)) \ \sin((N\pi/2)\sin(\theta)))}{\sin((\pi/2)\sin(\theta))}, \quad -\pi \le \theta < \pi. \qquad (7.4) \end{aligned}$$

A polar plot of $|H(\theta)|$ displayed from 0° to 360° for $N = 4$ is shown in Fig. 7.2(a) and for $N = 8$ is shown in Fig. 7.2(b). In this figure, the linear array lies on the 90° to −90° line (in these two figures 270° = −90°), and the gain is symmetric about this line. Thus, there is a high gain at the 0° direction (the "forward broadside" of the array) as well as at the 180° direction (the "backward broadside") with various sidelobes in other directions. In some applications, we may assume all the desired and unwanted sources are known to be in the forward sector (i.e., in the −90° to 0° to +90° sector). If the desired source is at the high gain 0° direction, then other unwanted sources in the directions of the sidelobes form interferences to the reception of the desired source. Thus, an array with a mainlobe having high gain over a narrow angular sector plus sidelobes with small values is considered desirable. From (7.4) and Fig. 7.2, as the number of array elements

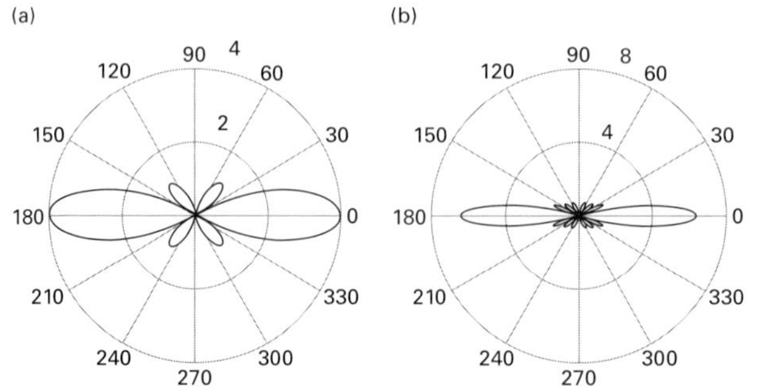

Figure 7.2 Polar plot of $|H(\theta)|$ vs θ for $d = \lambda/2$ and $N = 4$ (a) and $N = 8$ (b)

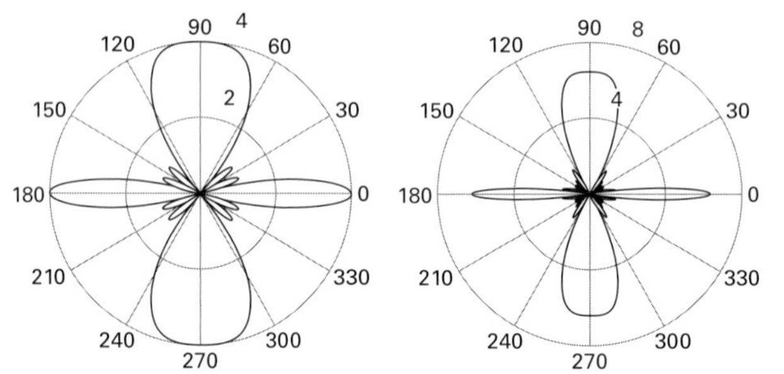

Figure 7.3 Polar plot of $|H(\theta)|$ vs θ for $d = \lambda$ and $N = 4$ (a) and $N = 8$ (b)

N increases, the mainlobe of the beamformer angular response becomes narrower and thus able to provide a better angular resolution, while the sidelobe peak values stay the same relative to the beam peak.

On the other hand, if we set $d = \lambda$, then (7.4) has the form of

$$H(\theta) = \frac{\exp(-i(\pi(N-1)\sin(\theta))\ \sin((N\pi)\sin(\theta)))}{\sin((\pi)\sin(\theta))}, \quad -\pi \leq \theta < \pi. \tag{7.5}$$

A polar plot of this $|H(\theta)|$ again displayed from 0° to 360° for $N = 4$ is shown in Fig. 7.3(a) and for $N = 8$ is shown in Fig. 7.3(b). We note, in these two figures, in addition to the desired high gains in the 0° and 180° directions, there are also two undesired equal high gains with large angular spreads at 90° and 270°. These two additional large gains may cause considerable interference to the desired source signal from unwanted sources in these directions. The spatial Nyquist criterion requires the inter-sensor spacing d of a uniform linear array to be less than or equal to $\lambda/2$ to avoid grating lobes (also called spatial aliasing). This phenomenon is analogous to spectral aliasing due to the periodicity in the frequency-domain created by sampling in time. Thus, a uniform linear array is most commonly operated at the $d = \lambda/2$ condition.

The minimum-variance distortionless response (MVDR) method [6] was an early practical beamformer that provided a computationally attractive solution for constraining the array response of a linear array of the desired source DOA angle θ_1 to a fixed value of unity value, while minimizing the response to all other interference sources. From the output of an Nth order beamformer, denote its $N \times N$ autocorrelation matrix as R, the array weight vector as $W = [w_1, \ldots, w_N]^T$, and the steering vector as $C = [1, \exp(-i\theta), \ldots, \exp(-i(N-1)\theta)]^T$. Then the MVDR solution (see Exercise 4) satisfies

$$\min\{W^{HRW}\}, \quad \text{subject to } W^C(\theta_1) = 1, \tag{7.6}$$

and is given by

$$W = R^{-1}C(\theta_1)(C^H(\theta_1)R^{-1}C(\theta_1))^{-1}. \tag{7.7}$$

Consider an $N = 20$ uniform linear array MVDR beamformer, with a spatial angle constrained to a unit amplitude single tone source of $f = 900$ Hz at DOA of $\theta_1 = 0°$. It is subjected to a broadband white Gaussian interferer of variance $\sigma^2 = 2$ at DOA of $\theta_2 = 40°$, in the presence of additive white Gaussian noise of unit variance. This beamformer has an input *SINR* $= -3.7$ dB and an output *SINR* $= 10.0$ dB resulting in a gain of *SINR* $= 13.7$ dB. In Fig. 7.4, this MVDR beamformer achieved a unity gain (i.e., 0 dB) at the known desired angle of $\theta_1 = 0°$ as constrained and placed a null of about -52 dB. If we had used the sample correlation matrix, the MVDR beamformer's performance would be degraded with a null of only – 23 dB. Many other forms of

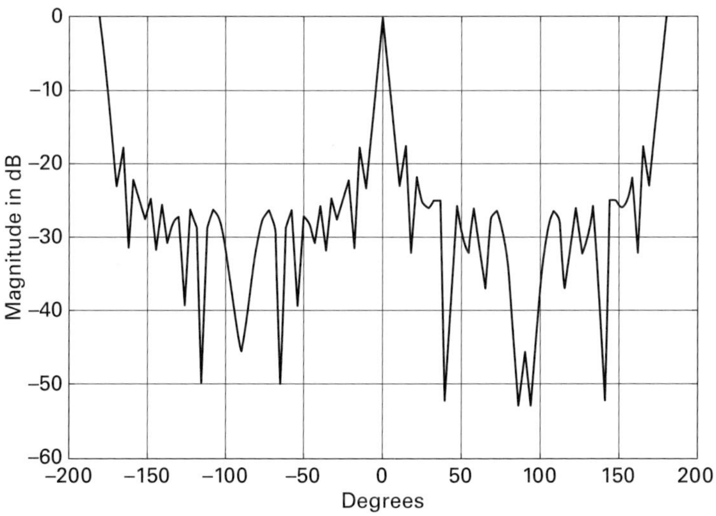

Figure 7.4 $N = 20$ uniform linear array MVDR beamformer response (dB) using theoretical correlation matrix vs θ

mean-square error criteria-based linear beamforming arrays have been developed [7] and [8], but will not be discussed here.

7.3 Wideband Beamformer

Source localization using sensor arrays is of interest to sensor, radar, navigation, geophysics, seismic, and acoustic systems. Localization, the determination of the location of an object, whether that of a source emitting some energy (e.g., RF, acoustic, ultrasonic, optical, IR, seismic, thermal) of interest, or a sensor node in the SN, is a major issue of interest. Often, in many applications, the localization of the source is of primary interest, while the localization of the sensor node is needed to support some other functions. Various techniques in DOA estimation for narrowband sources in the far-field have been summarized in [9]. Locating wideband sources in the near-field may involve range and DOA estimations. Other related works include: [10], [11], [12], and [13].

In this section, we will consider three different types of beamformers in some detail. First, in trilateration, consider that all the range-sensing nodes and the source are situated on the x–y plane. Range-sensing nodes A, B, and C are located at (x_A, y_A), (x_B, y_B), and (x_C, y_C) as shown in Fig. 7.5. If node A estimates a source at range d_A and node B estimates that source at range d_B, then the intersection of these two circles yields two possible ambiguous source locations marked by S and S'. Similarly, if node C estimates a range of d_C, then the intersection between node B and node C yields two possible ambiguous source locations S and S'', while the intersection of node C and node A yields two possible ambiguous source locations of S and S'''. However, the true location of the source is at the intersection of all three circles at S. Of course, in the presence of noisy estimations on d_A, d_B, d_C, the three circles will not intersect at a single point to yield a unique location S. In practice, denote the source at (x, y), then d_A, d_B, d_C satisfy the equations of

$$\begin{aligned}(x - x_A)^2 + (y - y_A)^2 &= d_A^2,\\ (x - x_B)^2 + (y - y_B)^2 &= d_B^2,\\ (x - x_C)^2 + (y - y_C)^2 &= d_C^2.\end{aligned} \tag{7.8}$$

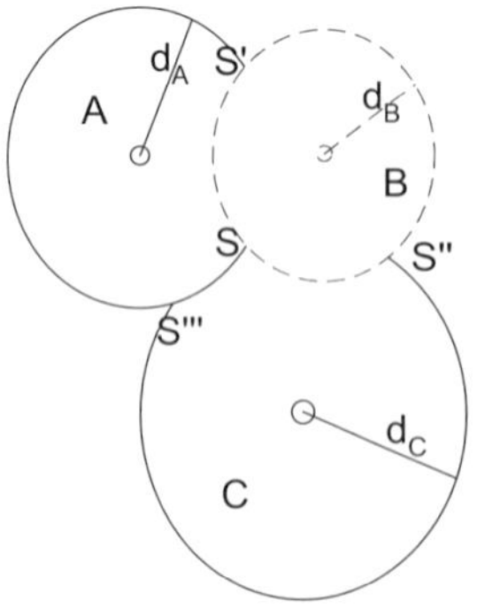

Figure 7.5 Trilateration

The solution $[x, y]^T$ of (7.8) can be written as the least-squares (LS) solution of

$$\begin{bmatrix} x \\ y \end{bmatrix} = \begin{bmatrix} 2(x_A - x_C) & 2(y_A - y_C) \\ 2(x_B - x_C) & 2(y_B - y_C) \end{bmatrix}^{-1} \begin{bmatrix} x_A^2 - x_C^2 - y_A^2 - y_C^2 + d_C^2 - d_A^2 \\ x_B^2 - x_C^2 - y_B^2 - y_C^2 + d_C^2 - d_B^2 \end{bmatrix}. \tag{7.9}$$

In multi-lateration, N range-sensing nodes with known locations (x_1, y_1), (x_2, y_2), ..., (x_N, y_N) are used instead of three nodes in trilateration. Their distances from the unknown source are denoted by $d_1, d_2, \ldots, d_N$. Then corresponding to (7.8), we have

$$\begin{aligned} &(x - x_1)^2 + (y - y_1)^2 = d_1^2, \\ &(x - x_2)^2 + (y - y_2)^2 = d_2^2, \\ &\vdots \\ &(x - x_N)^2 + (y - y_N)^2 = d_N^2. \end{aligned} \tag{7.10}$$

The LS solution of $\mathbf{AX} = \mathbf{B}$ corresponding to (7.9) is now given by

$$\mathbf{X} = (\mathbf{A}^T\mathbf{A})^{-1}\mathbf{A}^T\mathbf{B}, \tag{7.11}$$

where

$$\begin{aligned} \mathbf{A} &= \begin{bmatrix} 2(x_1 - x_N) & 2(y_1 - y_N) \\ \vdots & \vdots \\ 2(x_{N-1} - x_N) & 2(y_{N-1} - y_N) \end{bmatrix}, \\ \mathbf{X} &= \begin{bmatrix} x \\ y \end{bmatrix}, \\ \mathbf{B} &= \begin{bmatrix} x_1^2 - x_N^2 + y_1^2 - y_N^2 + d_N^2 - d_1^2 \\ \vdots \\ x_{N-1}^2 - x_N^2 + y_{N-1}^2 - y_N^2 + d_N^2 - d_{N-1}^2 \end{bmatrix}. \end{aligned} \tag{7.12}$$

Second, another beamformer uses the energy-based criterion to perform source localization. One of the key requirements of the time-difference of arrival (TDOA)-based and many other source localization algorithms is the synchronization among the sensor nodes. However, accurate synchronization requires extra energy consumption and communication bandwidth, which can be costly for wireless sensor network applications. More recently, energy-based approaches have been proposed for acoustic source localization to relax such constraints.

Let P be the number of sensors and M be the number of acoustic sources in the sensor field. Then the received signal measured on the pth sensor over the time interval n can be expressed as

$$x_p(n) = s_p(n) + v_p(n), \tag{7.13}$$

where

$$s_p(n) = \gamma_p \sum_{m=1}^{M} \frac{a^{(m)}(n - t_p^{(m)})}{\|\boldsymbol{\rho}^{(m)}(n - t_p^{(m)}) - \mathbf{r}_p\|} \tag{7.14}$$

is the signal intensity, $\nu_p(n)$ is the background noise modeled as a zero-mean additive white Gaussian noise with variance ς_p^2, $t_p^{(m)}$ is the propagation delay from the mth source to the pth sensor, and $a^{(m)}(n - t_p^{(m)})$ represents the intensity of the mth acoustic source measured 1 meter away from the source and is modeled as a random variable uncorrelated to other sources. The notations $\boldsymbol{\rho}^{(m)}$, $\mathbf{r}_p$, and γ_p denote the position vector of the mth source, position vector of the pth sensor, and the sensor gain factor of the pth sensor. $s_p(n)$ and $\nu_p(n)$ are assumed to be uncorrelated, and as a result the acoustic energy received at the pth sensor can be represented as

$$\mathrm{E}\left[s_p^2(n)\right] = \gamma_p^2 \sum_{m=1}^{M} \frac{\mathrm{E}\left[a^{(m)}(n - t_p^{(m)})\right]}{\|\boldsymbol{\rho}^{(m)}(n - t_p^{(m)}) - \mathbf{r}_p\|^2}, \tag{7.15}$$

$$= g_p \sum_{m=1}^{M} \frac{\mathcal{E}_s^{(m)}(n - t_p^{(m)})}{\|\boldsymbol{\rho}^{(m)}(n - t_p^{(m)}) - \mathbf{r}_p\|^2}, \tag{7.16}$$

where $g_p = \gamma_p^2$, and $\mathcal{E}_s^{(m)}(n - t_p^{(m)}) = \mathrm{E}\left[a^{(m)}(n - t_p^{(m)})\right]$. In practice, the ensemble average is implemented using time average over a window $\left[t - \frac{N}{2F_s}, t + \frac{N}{2F_s}\right]$, where N is the number of samples and F_s is the sampling frequency. Under some practical assumptions [15], the received energy at the pth sensor $\mathcal{E}_p(t)$ can then be modeled as

$$\mathcal{E}_p(t) = g_p \sum_{m=1}^{M} \frac{\mathcal{E}_s^{(m)}(n)}{\left[d_p^{(m)}(t)\right]^2} + \epsilon_p(t), \tag{7.17}$$

where $d_p^{(m)}(t) = \|\boldsymbol{\rho}^{(m)}(t) - \mathbf{r}_p\|$. Note that ν_p^2 has a χ^2 distribution with mean ς_p^2 and variance $2\varsigma_p^4/N$. When N is sufficiently large ($N \gg 30$), $\epsilon_p(t)$ can be approximated by a normal distribution, $\mathcal{N}(\mu_p, \sigma_p^2)$, where $\mu_p = \varsigma_p^2$ and $\sigma_p^2 = 2\varsigma_p^4/N$. Using (7.17) and omitting the time index t for brevity, the energy-based source localization problem can be formulated using the maximum-likelihood criterion

$$\hat{\boldsymbol{\psi}} = \arg\min_{\boldsymbol{\psi}} \ell(\boldsymbol{\psi}), \tag{7.18}$$

where

$$\boldsymbol{\psi} = \left[\boldsymbol{\rho}^T, \mathcal{E}_s{}^T\right]^T, \tag{7.19}$$

$$\ell(\boldsymbol{\psi}) = \|\mathbf{Z} - \mathbf{G}\mathbf{D}\boldsymbol{\mathcal{E}}_s\|^2, \tag{7.20}$$

and

$$\mathbf{Z} = \left[\frac{(\mathcal{E}_1 - \mu_1)}{\sigma_1}, \cdots, \frac{(\mathcal{E}_P - \mu_P)}{\sigma_P} \right]^T, \tag{7.21}$$

$$\mathbf{G} = \text{diag}\left[\frac{g_1}{\sigma_1}, \cdots, \frac{g_P}{\sigma_P} \right], \tag{7.22}$$

$$\mathbf{D} = \begin{bmatrix} \frac{1}{|d_1^{(1)}|^2} & \frac{1}{|d_1^{(2)}|^2} & \cdots & \frac{1}{|d_1^{(M)}|^2} \\ \frac{1}{|d_2^{(1)}|^2} & \frac{1}{|d_2^{(2)}|^2} & \cdots & \frac{1}{|d_2^{(M)}|^2} \\ \vdots & \vdots & \ddots & \vdots \\ \frac{1}{|d_P^{(1)}|^2} & \frac{1}{|d_P^{(2)}|^2} & \cdots & \frac{1}{|d_P^{(M)}|^2} \end{bmatrix}, \tag{7.23}$$

$$\boldsymbol{\mathcal{E}}_s = \left[\mathcal{E}_s^{(1)}, \cdots, \mathcal{E}_s^{(M)} \right]^T, \tag{7.24}$$

$$\boldsymbol{\rho} = \left[\boldsymbol{\rho}^{(1)T}, \cdots, \boldsymbol{\rho}^{(M)T} \right]^T. \tag{7.25}$$

More details on this approach can be found in [14] and [15].

The third wideband beamformer problem is based on the approximate maximum likelihood (AML) method [16]. The maximum likelihood (ML) method is a well-known statistical estimation method. In the AML approach, we consider a sensor array of N omni-directional sensors which is impinged by M sources $s^{(m)}(t), m = 1, \ldots, M$, from M distinct directions in the far-field. The data collected by the nth sensor are given by

$$x^{(n)}(t) = \sum_{m=1}^{M} s^{(m)}(t - \tau^{(n)}(\theta^{(m)}, \phi^{(m)})) + n^{(n)}(t),$$

in which $\tau^{(n)}(\theta^{(m)}, \phi^{(m)})$ is the time-delay of the mth source to the nth sensor relative to the centroid of the array, $\tau^{(n)} = [x^{(n)}, y^{(n)}, z^{(n)}]^T$ is the location of the nth sensor, and $\theta^{(m)}$ and $\phi^{(m)}$ are the azimuth and elevation of the mth source, and $n^{(n)}$ is the noise. If all the data are separated into H successive snap-shots, the array signal model of the hth frame, after a size K DFT/FFT operation yields the kth frequency bin output of $\mathbf{x}_k(h) = \mathbf{D}_k(\Theta)\mathbf{s}_k(h) + \mathbf{w}_k(h)$, where $\mathbf{x}_k(h) = [x_k^{(0)}(h), \ldots, x_k^{(N-1)}(h)]^T$, $\mathbf{D}_k(\Theta) = [\mathbf{d}_k(\theta^{(1)}, \phi^{(1)}), \ldots, \mathbf{d}_k(\theta^{(M)}, \phi^{(M)})]$ is the steering matrix, $\mathbf{d}_k(\theta^{(m)}, \phi^{(m)}) = [e(^{-2\pi j F_s k \tau^{(0)}(\theta^{(m)}, \phi^{(m)})}/K), \ldots, e(^{-2\pi j F_s k \tau^{(N-1)}(\theta^{(m)}, \phi^{(m)})}/K)]^T$ is the mth steering vector, the DOAs of the sources are denoted by $\Theta = [\theta^{(1)}, \phi^{(1)}, \ldots, \theta^{(M)}, \phi^{(M)}]^T$, the frequency component of the sources is given by $s_k(h) = [s_k^{(1)}(h), \ldots, s_k^{(M)}(h)]^T$, the noise spectrum $\mathbf{w}_k(h)$ is modeled as a zero-mean i.i.d. Gaussian vector, and F_s is the sampling rate of the array.

Then the general 3D AML DOA estimation can be obtained by solving the following maximization problem

$$\max_{\Theta} J(\Theta) = \max_{\Theta} \sum_{k=k}^{k_{K_0}} \sum_{h=1}^{H} \| P_k(\Theta) \mathbf{x}_k(h) \|^2, \tag{7.26}$$

where $P_k(\Theta) = \mathbf{D}_k(\Theta)\mathbf{D}_k^+(\Theta)$ is the orthogonal projection, K_0 is the number of spectral terms, and $\mathbf{D}_k^+(\Theta) = \left(\mathbf{D}_k^H(\Theta)\,\mathbf{D}_k(\Theta)\right)^{-1}\mathbf{D}_k^H(\Theta)$ is the pseudo-inverse.

For the multiple-source case, the above parameter estimation problem becomes a challenging multi-dimensional search problem. The alternating projection (AP) approach, originally proposed by [17], breaks the multi-dimensional parameter search into a sequence of single-source parameter search problems. For the two-source problem,

Step 1: Estimate the DOA of the strongest source on a single-source grid search by using

$$\theta_{s_1}^{(0)} = \arg\max_{\theta_{s_1}} \left\{ J(\theta_{s_1}) \right\}.$$

Step 2: Estimate the location of the second source on a single-source grid under a two-source model, while using the first source location estimate from Step 1 constraint. Then perform

$$\theta_{s_2}^{(0)} = \arg\max_{\theta_{s_2}} \left\{ J(\theta_{s_1}^{(0)}, \theta_{s_2}) \right\}.$$

Repeat Step 1 and Step 2, until convergence. While we, as well as [17], do not seem to have a proof on the convergence of the AP algorithm, all of our practical experiences with two sources yield convergence of both estimated source angles within two to three iterations. An example using the above algorithms will be provided in Section 7.4.2.

Next, we propose a new eigensystem-based fast 3D wide band AML DOA estimator with significant reduction in processing complexity caused by multiple snap-shots of the data in each processing block. Define the $N \times H$ matrix of $\mathbf{X}_n = [\mathbf{x}_k(1), \ldots, \mathbf{x}_k(H)]$. Then the $N \times N$ cross-sample covariance is given by $\mathbf{C}_k = \mathbf{X}_k{}^* \mathbf{X}_k^H$. Next, consider an eigenvalue decomposition of $\mathbf{C}_k$ given in terms of the Matlab notation of $[\mathbf{V}_k, \mathbf{D}_k] = \mathrm{eig}(\mathbf{C}_k)$, where both the eigenvector and eigenvalue matrices are of dimension $N \times N$. Assume the eigenvalues are arranged in increasing order, then we can replace our prior maximization metric in (7.26) by

$$\max_{\Theta} J(\Theta) = \max_{\Theta} \sum_{k=k_1}^{k_{K_0}} \| P_k(\Theta)\,\mathbf{x}_k(h) \|^2. \tag{7.27}$$

In practice, since the number of sensors N (typically may have values of 4 to 8), while the number of snap-shots H (typically may have values of 20 to 1,000), by eliminating the sum of H term in the maximization metric of (7.27) significantly reduces the processing complexity by a factor of H. Eigendecomposition methodology has been used in [18], [19], and in the MUSIC algorithm [20] but will not be considered here.

7.4 Array System Performance Analysis by CRB Method, Simulations, and Field Measurements

7.4.1 CRB Method

The Cramér–Rao Bound (CRB) is most often used as a theoretical lower bound for any unbiased estimator [21]. First, we assume the source signals are known and the unknown parameters are denoted by $\Theta = [\theta_1, \ldots, \theta_M]$. In this subsection, we derive the CRB for the single-source case from the signal model. We can construct the Fisher Information Matrix (FIM) from [27] using the signal model defined by $F = 2\Re[H^H R_\Psi^{-1} H] = \frac{2}{N\sigma^2}\Re[H^H H]$. Note that $H = \frac{\partial G}{\partial \phi^T}$ and $\phi = [\phi, e]^T$. For the single source, vector G can be simplified as

$$G_{(NP)/2x1} = [e^{-2\pi t_{c1}/N} S_1(\omega_1), \ldots, e^{-2\pi t_{cp}/N} S_1(\omega_1), \ldots, e^{-2\pi t_{c1}/N} S_1(\omega_{N/2}), \ldots, e^{-2\pi t_{cp}/N} S_1(\omega_{N/2})]^T,$$

$R_\xi = E[\xi\xi^H] = (N\sigma^2) I_{NP/2}$, where σ^2 denotes the variance of the zero-mean i.i.d. Gaussian noise.

In this case,

$$F = \alpha T, \tag{7.28}$$

$$\alpha_{1\times 1} = \frac{2}{N\sigma^2}\sum_{k=1}^{N/2}\left(\frac{2\pi k|S_1(\omega_k)|}{N}\right)^2, \tag{7.29}$$

$$T = \sum_{p=1}^{P}\begin{bmatrix} \left(\frac{\partial t_{cp}}{\partial \psi}\right)^2 & \frac{\partial t_{cp}}{\partial \psi} \times \frac{\partial t_{cp}}{\partial e} \\ \frac{\partial t_{cp}}{\partial e} \times \frac{\partial t_{cp}}{\partial \psi} & \left(\frac{\partial t_{cp}}{\partial e}\right)^2 \end{bmatrix}. \tag{7.30}$$

The first and second diagonal entries of the inverse Fisher Information Matrix indicate the variances in the DOA estimation corresponding to azimuth and elevation angles of source 1, respectively. The third and fourth diagonal entries of the inverse FIM indicate the variances of DOA estimations corresponding to azimuth and elevation angles of source 2, respectively.

Next, consider the CRB performance of the 3D AML algorithm by simulations of two sources. The first source comes from the far-field of a Mexican Antthrush bird source with an azimuth of $\theta_1 = 160°$ and elevation of $\phi_1 = 20°$. Let the second source come from the far-field of a Dusky Antbird source with an azimuth of $\theta_1 = 30°$ and an elevation of $\phi_1 = 50°$. An isotropic 3D array of eight omni-directional sensors are located at the corners of a cube with side length of 5 centimeters. We assume an SNR of 20 dB. The CRBs of the first and second source are given in Fig. 7.6 and Fig. 7.7 respectively.

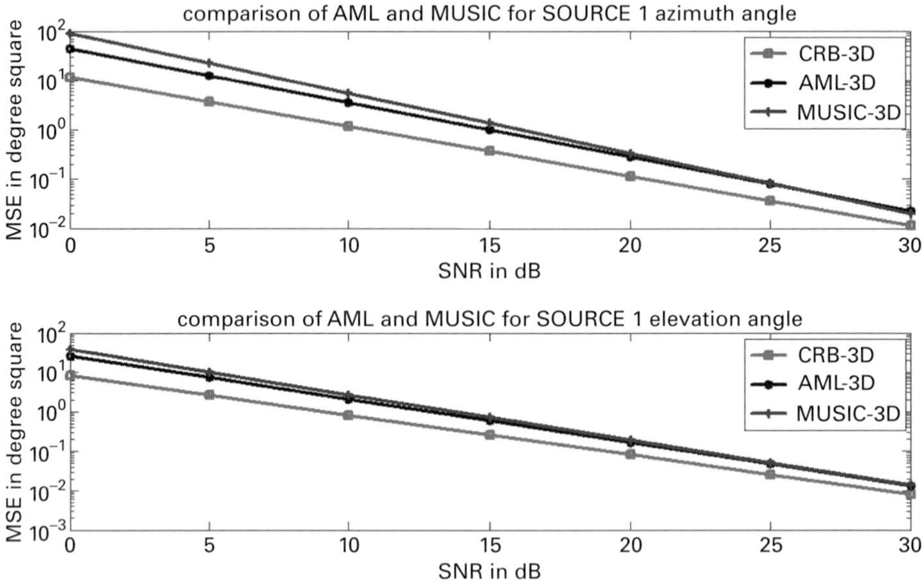

Figure 7.6 Performance comparison of the 3D AML and MUSIC for the first source

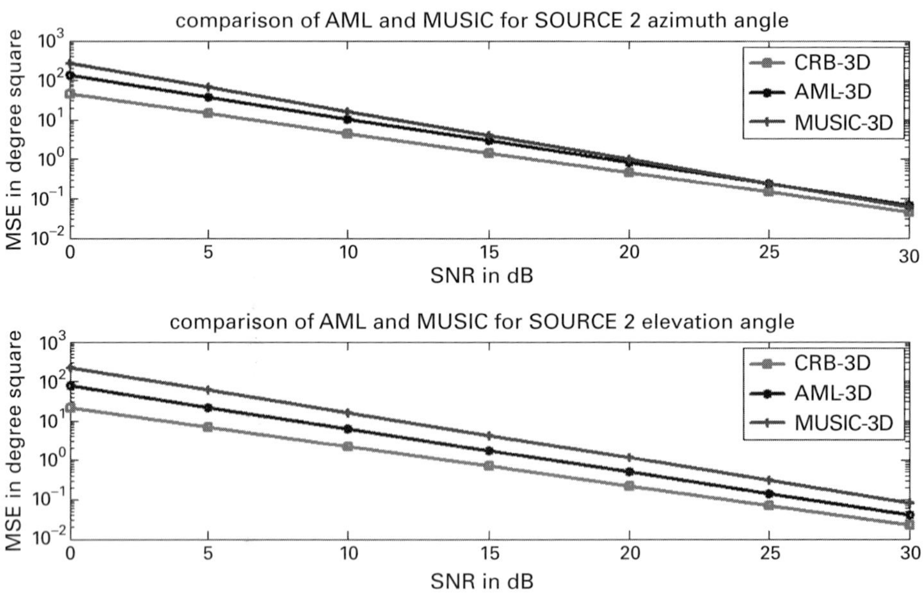

Figure 7.7 Performance comparison of the 3D AML and MUSIC for the second source

7.4.2 AML Simulation Results

Consider the simulation of a cubic array of side length of 5 cm centered at the origin. The locations of the sensors in spherical coordinates are given by

$$\varsigma = \begin{bmatrix} 0.5\sqrt{3} & 5\pi & -\tan^{-1}\left(\frac{1}{\sqrt{2}}\right) \\ 0.5\sqrt{3} & 7\pi & -\tan^{-1}\left(\frac{1}{\sqrt{2}}\right) \\ 0.5\sqrt{3} & \pi & -\tan^{-1}\left(\frac{1}{\sqrt{2}}\right) \\ 0.5\sqrt{3} & 3\pi & -\tan^{-1}\left(\frac{1}{\sqrt{2}}\right) \\ 0.5\sqrt{3} & 5\pi & \tan^{-1}\left(\frac{1}{\sqrt{2}}\right) \\ 0.5\sqrt{3} & 7\pi & \tan^{-1}\left(\frac{1}{\sqrt{2}}\right) \\ 0.5\sqrt{3} & \pi & \tan^{-1}\left(\frac{1}{\sqrt{2}}\right) \\ 0.5\sqrt{3} & 3\pi & \tan^{-1}\left(\frac{1}{\sqrt{2}}\right) \end{bmatrix}^T. \tag{7.31}$$

To eliminate the impact of the array geometry, we consider isotropic arrays. An array is said to be isotropic if it has a constant mean-square angular error (MSAE) for all azimuth and all elevation angles. [22] proved the following theorem, which gives the necessary and sufficient conditions for an array to be isotropic.

THEOREM 7.1 *Suppose that a P element 3D array centered at the origin is represented by the array geometry matrix B as*

$$B_{3\times 3} = \sum_{p=1}^{P} l_p l_p^T, \; l_p = r_p \times \frac{f_s}{v}, \tag{7.32}$$

where l_p is the normalized location of the pth sensor in the Cartesian coordinate system and f_s is the sampling frequency. Then the array is isotropic if and only if $B = kI_{3\times 3}$ where I is the identity matrix.

It can be shown that the FIM for a single signal source and an isotropic array is given by

$$F = \alpha \begin{bmatrix} k\cos^2(e) & 0 \\ 0 & k \end{bmatrix}. \tag{7.33}$$

Then the CRB for an isotropic array becomes $\text{CRB}(\theta) = \frac{1}{\alpha k \cos^2 e}$, $\text{CRB}(e) = \frac{1}{\alpha k}$. Now, if we remove the scaling of the CRB of the azimuth angle, then the CRB of the elevation angle and the CRB of the azimuth angle become equal, and as a result, the MSAE becomes constant for all azimuth and elevation angles ($\text{MSAE}(B) = \cos^2(e)\,\text{CRB}(\psi)$ $+ \text{CRB}(e) = \frac{2}{\alpha k}$).

Table 7.1 3D AML for two far-field sources

Source	True DOA	Est. DOA
Mexican Antthrush	Azimuth: 160°	160.5°
Mexican Antthrush	Elevation: 20°	20.25°
Dusky Antbird	Azimuth: 30°	30.2°
Dusky Antbird	Elevation: 50°	50.25°

Sources: Mexican Antthrush and Dusky Antbird (SNR: 20 dB).

For two signal sources and an isotropic array, the FIM is given by

$$F = \alpha \begin{bmatrix} k\cos^2(e_1) & 0 & 0 & 0 \\ 0 & k & 0 & 0 \\ 0 & 0 & k\cos^2(e_2) & 0 \\ 0 & 0 & 0 & k \end{bmatrix}. \tag{7.34}$$

Although the Fisher information matrix for an isotropic array is diagonal, the global likelihood function is not necessarily decoupled. Thus, a number of computationally costly 2D grid searches are required to find the global maximum for the 3D AML problem.

More details on CRB [27] and for the CRB of the AML method can be found in [23], [24], [25], [26], and [28].

7.5 Some Recent Array Processing Methodologies

7.5.1 Robust Beamforming Method

Many engineering designs are based on the use of some optimized criteria. However, when some parameters in these designs are perturbed (due to imperfect tolerance issues), the performance of these designs can degrade significantly. Recently, various robust optimization methods (such as the semi-definite programming (SDP) method) [29] have been proposed for beamforming array designs [30] and [31]. For example, consider an array beamforming problem with sidelobe constraints. The goal of the sidelobe level minimization problem is to minimize the gains in all other directions, while having a fixed unity gain in some desired target direction. The constrained least-squares version of this problem can be solved using the practical Matlab cvx SDP program. We can formulate this problem as

$$\min ||A_j\, w||_2^2, \text{ subject to } A_{tar}\, w = 1,$$

where

$$\begin{aligned} A_{tar} &= \begin{bmatrix} e^{-i\gamma_1(\theta_{tar})} & e^{-i\gamma_2(\theta_{tar})} & \dots & e^{-i\gamma_n(\theta_{tar})} \end{bmatrix}, \\ A_j &= \begin{bmatrix} e^{-i\gamma_1(\theta_{tar})} & e^{-i\gamma_2(\theta_{tar})} & \dots & e^{-i\gamma_n(\theta_{tar})} \end{bmatrix}, \\ \gamma_n &= x_n \cos\left(\theta_j\right) + y_n \sin\left(\theta_j\right). \end{aligned}$$

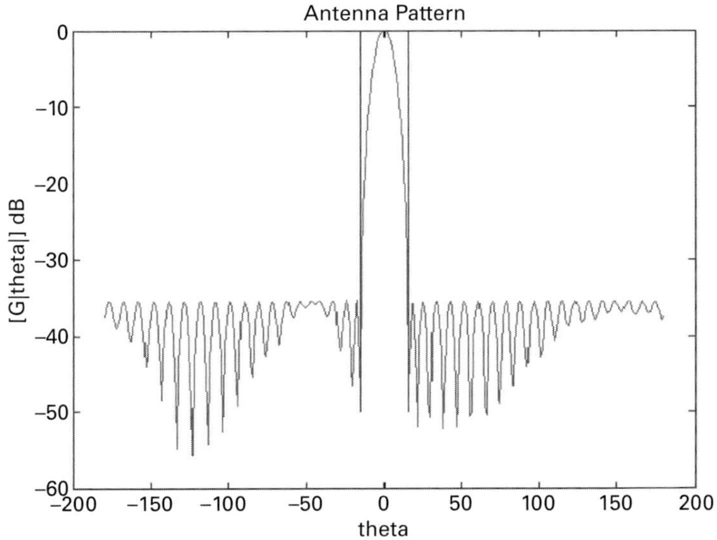

Figure 7.8 Antenna pattern under original ideal condition

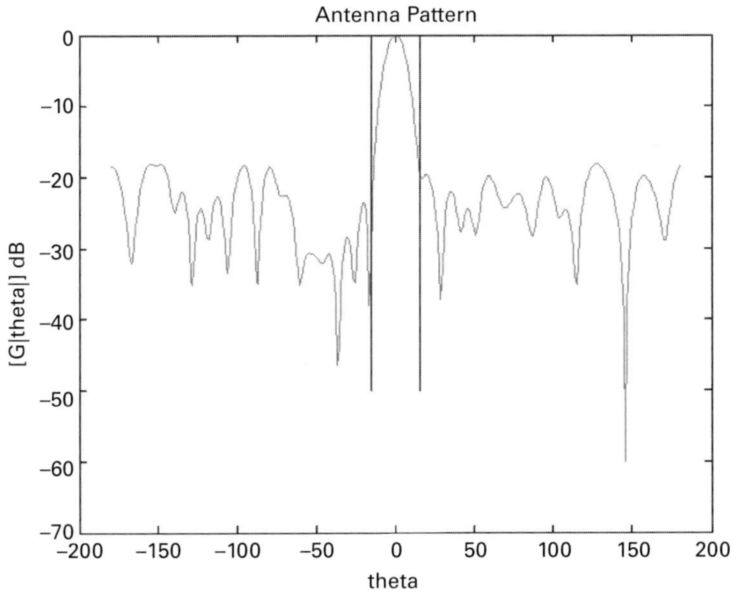

Figure 7.9 Antenna pattern with perturbed array weights

The array response under the ideal constrained least-squares solution is such that it has a sidelobe peak magnitude of about -35 dB. Now, suppose the array weights are modified with a 10% increment, then the array response is given by which now has a sidelobe peak magnitude of about -18 dB, which may not be acceptable. But upon using

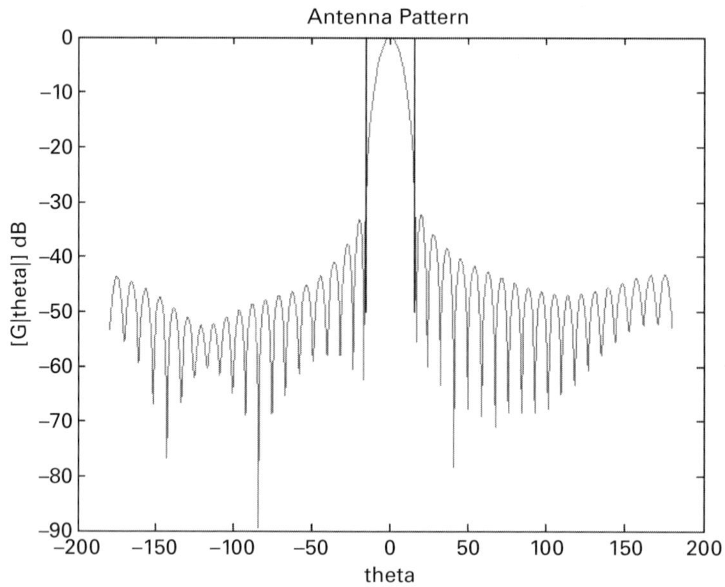

Figure 7.10 Antenna pattern obtained using a SCOP optimization

a robust second-order cone program (SCOP) optimization, we obtain the following array response where the sidelobe peak magnitude is now at about −32 dB, which is better than the previous −18 dB case. Additional robust array processing information can be found in [32].

7.5.2 Random Finite Set Method for Array Design

In modeling various theoretical and practical sensor networks, the number of targets (or sources) can randomly change (including no target) and the number and kinds of sensors can also change randomly. Present methods first detect/estimate these discrete random quantities and then perform optimum detection/estimation/localization/tracking, leading to overall sub-optimum solutions. Random set theory (RST) is a mathematical tool that generalizes the conventional probability theory to a new probability theory over finite sets whose randomness lies both in the number of their elements and in the values they take on. RST was originated by [33]. Mahler [34] introduced the use of a simplified version of RST, called random finite set theory (RFST) (where the number of the possible random sets is finite) to estimation and tracking problems. RFST formulation can jointly and systematically handle the randomness in the discrete number of targets, discrete number of sensors, and continuous-valued channel/measurement/processing noises in many general sensor systems. Thus, in the problems where the entities to be estimated (active users, propagation paths, active neighbors, number of targets, number of sensors) can be thought of as elements of a finite random set, RFST provides a fairly natural approach, as it unifies in a single step more than two steps that would be taken separately without it: viz, detection of the number of parameters to be estimated, and estimation of the values they take on. A central point in RFST is the generation of "densities" which are not the usual Radon–Nikodým derivatives of probability measures, but rather "set derivatives" of non-additive "belief functions." Alternatively, these densities, which capture what is known about measurement state space, user state space, and user dynamics, can be derived in a rather straightforward way from the system model by using an RFST toolbox, and using essentially in the same ways as ordinary densities in "classical" detection/estimation theory. Thus, RFST provides a tool which is mathematically sharp and yet matched to engineering intuition. Next, we consider two simple tracking examples to illustrate some aspects of RFST.

From Fig. 7.11 and Fig. 7.12, we observe that using the RST tracker, from time instants 1–11, there is one target. Then from time instants 12–15, RST declares no target and from time instants 16–32, it declares again one target. The RST tracker has

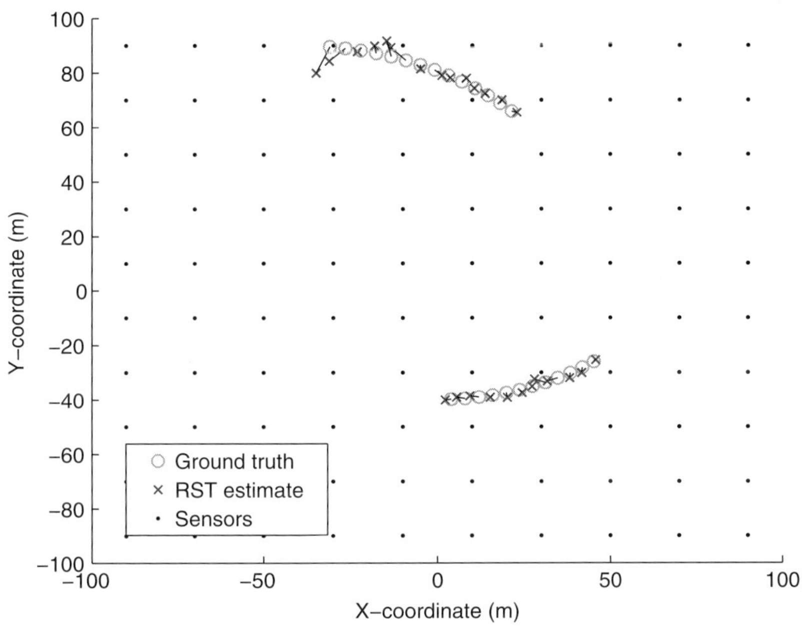

Figure 7.11 Two separate tracks over two time instants

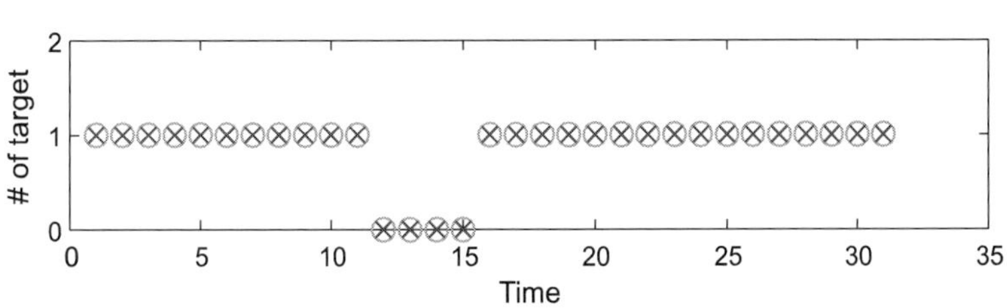

Figure 7.12 Declaration of number of tracks over two time instants

no information whether the second target is the continuation of the first target. More information on RST can be found in [35] and on RST tracking in [26] and [28].

7.5.3 Large Arrays

In recent years, large arrays with many sensors have been used for long distance deep-space astronomical applications and for wideband space–time multi-target detection and tracking of aerospace applications. For example, RF phase array antennae based on beamforming have been used in avionic, aerospace, and communication systems for many years. However, almost all of these systems are narrowband in the sense that the ratio of the highest frequency to the lowest frequency is close to one. On the other hand, beamforming in acoustic and seismic processing is wideband. However, in recent years, there is increasing interest in using beamforming in wideband radar systems. Wideband digital beamformer (DBF) can use commercially available technologies at the L, S, and X band frequencies. Furthermore, the DSP portion of the DBF can use subband analysis filter methodology implemented by modern FPGA hardware. The subband approach is known to be one of the most efficient ways (in terms of multiplications) to implement a WDBF using narrowband channelization of the wideband signal. In this approach, one exploits the usage of decimating polyphase filtering in conjunction with IDFT to perform the analysis filter bank for the subbanding purpose. Similarly, the reverse operation of DFT followed by interpolating polyphase filtering performs the synthesis filter. We note each of the beamformers operating at the K subbands at the outputs of the analysis filter is a narrowband beamformer. Thus, all the known narrowband beamforming methodologies, such as that of MVDR beamforming, are applicable. More details on this approach can be found in [36] and [37].

7.6 Conclusion

Section 7.1 introduced the concepts of beamforming for narrowband waveforms, including SNR, mainbeam, sidelobes, and grating lobes. Section 7.2 considered some early array processing concepts, including the simple time-delay method and the minimum-variance distortionless response array. In Section 7.3, beamforming for wideband waveforms was considered. Various DOA estimation and source localization methods for acoustic and seismic signals were summarized. Details on the AML beamforming and alternating projections for multiple sources were discussed. Section 7.4 considered array system performance analysis based on the Cramér–Rao bounding (CRB) methods for DOA and source localization. Section 7.5 introduced some recent methods for array processing. Section 7.5.1 dealt with the concept of robust array design using semi-definite programming optimization method. In Section 7.5.2, multi-source estimation and tracking based on random set theory were introduced. Finally, in Section 7.5.3 we briefly summarized the implementation of a radar wideband beamformer using the subband approach. An extensive list of references dealing with various beamforming methods is provided.

7.7 References

[1] J.J. Faran and R. Hills, "The Application of Correlation Techniques to Acoustic Receiving System," *Tech. Mem.* 28, Acoustical Research Laboratory, Harvard University, Nov. 1, 1953.

[2] H.T. Freis and C.B.A. Feldman, "A Multiple Unit Steerable Antenna for Short-Wave Reception," *Proc. IRE*, 1937, vol. 25, pp. 841–847.

[3] N. Fourikis, *Advanced Array Systems, Applications, and RF Technologies*, Academic Press, 2000.

[4] G.J. Pottie and W.J. Kaiser, "Wireless Integrated Network Sensors," *Comm. of the ACM*, May 2000, vol. 43, pp. 51–58.

[5] J.C. Chen, L. Yip, J. Elson, H. Wang, D. Maniezzo, R.E. Hudson, K. Yao, and D. Estrin, "Coherent Acoustic Array Processing and Localization on Wireless Sensor Networks," *Proceedings of the IEEE*, Aug. 2003, vol. 91, pp. 1154–1162.

[6] J. Capon, "High-Resolution Frequency-Wavenumber Spectrum Analysis," *Proceedings of the IEEE*, 1969, vol. 57, pp. 1408–1418.

[7] H.L. Van Trees, *Optimum Array Processing*, J. Wiley, 2002.

[8] G.C. Carter, ed., *Coherence and Time Delay Estimation: An Applied Tutorial for Research, Development, Test, and Evaluation Engineers*, IEEE Press, 1993.

[9] J. Krim and M. Viberg, "Two Decades of Array Signal Processing Research: The Parametric Approach," *IEEE Signal Processing Magazine*, Jul. 1996, vol. 13, pp. 67–94.

[10] S. Bancroft, "Algebraic Solution of the GPS Equation," *IEEE Trans. Aerospace Electronic Systems*, 1985, pp. 56–59.

[11] H.C. Schau and A.Z. Robinson, "Passive Source Localization Employing Intersecting Spherical Surfaces from Time-of-Arrival Differences," *IEEE Trans. on Acoustics, Speech, and Signal Processing*, Aug. 1987, vol. ASSP-35, no. 8, pp. 1223–1225.

[12] J.O. Smith and J.S. Abel, "Closed-Form Least-Squares Source Location Estimation from Range-Difference Measurements," *IEEE Trans. on Acoustics, Speech, and Signal Processing*, Dec. 1987, vol. ASSP-35, no. 12, pp. 1661–1669.

[13] Y.T. Chan and K.C. Ho, "A Simple and Efficient Estimator for Hyperbolic Location," *IEEE Trans. on Signal Processing*, Aug. 1994, vol. 42, no. 8, pp. 1905–1915.

[14] D. Li and Y. Hu, "Energy-based Collaborative Source Localization Using Acoustic Microphone Array," *EURASIP J. Applied Signal Processing*, 2003, pp. 321–337.

[15] X. Sheng and Y. Hu, "Maximum Likelihood Multiple Source Localization Using Acoustic Energy Measurements," *IEEE Trans. on Signal Processing*, 2005, pp. 44–53.

[16] J.C. Chen, R.E. Hudson, and K. Yao, "Maximum-Likelihood Source Localization and Unknown Sensor Location Estimation for Wideband Signals in the Near-Field," *IEEE Trans. on Signal Processing*, Aug. 2002, vol. 50, no. 8, pp. 1843–1854.

[17] I. Ziskind and M. Wax, "Maximum Likelihood Localization of Multiple Sources by Alternating Projection," *IEEE Trans. on Acoustics, Speech, and Signal Processing*, 1988, pp. 1553–1560.

[18] V.F. Pisarenko, "The Retrieval of Harmonics from a Covariance Function," *Geophys. J. Royal Astron. Soc.*, 1973.

[19] R.A. Hudson and K. Yao, "A New Eigenvector-based 3D Wideband Acoustic DOA Estimator," *Proc. IEEE Inter. Symposium on Phase Array System & Technology*, 2016, pp. 1447-7/16.

[20] R.O. Schmidt, "Multiple Emitter Location and Signal Parameter Estimation," *IEEE Trans. on Antennas and Propagation*, Mar. 1986, vol. AP-34, no. 3, pp. 276–280.

[21] C.R. Rao, *Linear Statistical Inference and Its Applications*, J. Wiley, 1973.

[22] U. Baysal and R.L. Moses, "On the Geometry of Isotropic Arrays," *IEEE Trans. on Signal Processing*, 2003, pp. 1469–1478.

[23] Joe C. Chen, "Wideband Source Localization Using Passive Sensor Array," Thesis of Doctor of Electrical Engineering, UCLA, 2002.

[24] S. Asgari, "Far Field Source Localization in Sensor Networks," Thesis of Doctor of Electrical Engineering, UCLA, 2008.

[25] L. Yip, "Array Signal Processing for Source DOA Estimation and Source Localization in a Distributed Sensor Network," Thesis of Doctor of Electrical Engineering, UCLA, 2004.

[26] A.M. Ali, "Distributed Acoustic Localization and Tracking Design and Analysis," Thesis of Doctor of Electrical Engineering, UCLA, 2010.

[27] S.M. Kay, *Fundamentals of Statistical Signal Processing: Estimation Theory*, Englewood Cliffs, NJ: Prentice-Hall, 1993.

[28] J.Y. Lee, "Inference in Sensor–Actor–Evaluator Networks," Thesis of Doctor of Electrical Engineering, UCLA, 2012.

[29] S. Boyd and L. Vandenberghe, *Convex Optimization*, Cambridge University Press, 2004.

[30] S. Lebret and S. Boyd, "Antenna Array Pattern Synthesis Via Convex Optimization," *IEEE Trans. on Signal Processing*, March 1997, vol. 45, pp. 5256–5532.

[31] S.A. Vorobyov, A.B. Gershman, Z.Q. Luo, "Robust Adaptive Beamforming Using Worst-Case Performance Optimization: A Solution to the Signal Mismatch Problem," *IEEE Trans. on Signal Processing*, Feb. 2003, vol. 51, pp. 313–324.

[32] J. Li and P. Stoica, *Robust Adaptive Beamforming*, J. Wiley, 2006.

[33] G. Matheron, *Random Sets and Integral Geometry*, J. Wiley, 1975.

[34] R.P.S. Maher, *Statistical Multisource-Multitarget Information Fusion*, Artech House, 2007.

[35] H.T. Nguyen, *An Introduction to Random Sets*, Chapman and Hall/CRC, 2006.

[36] D.R. Martinez, T.J. Moeller, and K. Teitelbaum, "Application of Reconfigurable Computing to a High Performance Front-End Radar Signal Processor," *J. VLSI Signal Processing*, May–June 2001, vol. 28, nos. 1–2, pp. 65–83.

[37] J. Litva and T.K.-Y. Lo, *Digital Beamforming in Wireless Communications*, Artech House, 1996.

7.8 Exercises

1. Consider a linear array with $N = 5$ in Fig. 7.2. Except now, instead of having the desired uniform spacing of $d = \lambda/2$ separating all the sensors, the actual inter-sensor spacing (due to random misalignment) is $d = (\lambda/2) + \epsilon$, where ϵ is a uniformly distributed r.v. on $(-\lambda/10, \lambda/10)$. Plot the magnitude of this array response $|H(\theta)|$ versus θ as well as the polar plots (similar to Fig. 7.3), assuming all these random misalignments are uncorrelated using five Monte Carlo simulations of each ϵ. Clearly, some responses may be quite good (similar to the $\epsilon = 0$ case), while others may be quite different from the uniform case.
2. Consider a linear array with $d = \lambda$, where the beamformer transfer function (7.17) holds. What is the magnitude of the transfer function for $\theta = 0$? Are the grating lobe magnitudes for $N = 4$ and $N = 8$ consistent with those in Fig. 7.3?

3. Derive the MVDR array weight W of (7.19).
4. Duplicate the $N = 20$ uniform MVDR array of Fig. 7.5. using theoretical correlation method. Repeat this problem to achieve the results of Fig. 7.4 using the sample correlation method. Note: Your results may not be identical to that of Fig. 7.4.
5. Plot the frequency responses using FFT (with some zero paddings) to show that h_0 is a low-pass filter and h_1 is a high-pass filter for the following cases:

   a. The Haar wavelet.
   b. The Shannon wavelet.
   c. The Daubechies wavelet of order 2.

8 Introduction to Compressive Sampling

8.1 Introduction

Compressive sampling (cs) is a relatively new and active research topic in the last few years. The series of tutorial papers (e.g. [1]) on cs has motivated intense interest in this topic. Compressive sampling represents a modern approach to the notion that the number of samples needed to faithfully represent a certain type of speech, image, on data signal that is "sparse" in some sense is much lower than the traditionally accepted Nyquist condition. It turns out that many practically encountered signals are moderately to very sparse in some sense. For those classes of signals, cs can lead to a reduced number of samples at the original sensing level as well as significant compression for storage and transmission. In this chapter, we will only deal with some basic concepts and examples illustrating some elementary aspects of cs. A full understanding of cs is mathematically intricate and involved and is presently still evolving. However, the actual cs algorithms for various applications can be stated quite simply and can be readily implemented using the public-domain cvx optimization package utilizing Matlab. In Section 8.2, we first review the classical "Shannon sampling theorem" and related issues for a bandlimited signal. In Section 8.3, we show use of a cs methodology for solving the sparse solution of an under-determined system of equations.

8.2 Classical Shannon Sampling Theorem and Related Issues

Consider a continuous-time signal $x(t),\ -\infty < t < \infty$, whose Fourier transform $X(f)$, is supported only on $(-B, B)$. In other words, $X(f)$ is identically zero outside of that frequency band. Then if a countably infinite number of samples of $x(t)$ are taken at the Nyquist sampling instants of $\{t_n = n/(2B),\ -\infty < n < \infty\}$, then the collection of sampled values, $\{x(t_n),\ -\infty < n < \infty\}$, can interpolate the original continuous-time signal $x(t)$ exactly at any time instant $-\infty < t < \infty$. Furthermore, the Shannon sampling theorem reconstruction formula is given simply by

$$x(t) = \sum_{n=-\infty}^{\infty} x\left(\frac{n}{2B}\right) \frac{\sin\left(2\pi B(t - \frac{n}{2B})\right)}{2\pi B(t - \frac{n}{2B})}, \quad -\infty < t < \infty. \tag{8.1}$$

The expansion in (8.1) can be derived by performing a Fourier series expansion of $X(f) = \mathcal{F}\{x(t)\}$, where $\mathcal{F}$ denotes the Fourier transform, over the interval $(-B, B)$.

Then the coefficients of this Fourier series turn out to be proportional to the sampled values of $\{x(t_n), -\infty < n < \infty\}$. Then after applying the inverse Fourier transform, $\mathcal{F}^{-1}$, on the Fourier series $X(f), -B < f < B$, we obtain the expression of (8.1).

Even though (8.1) is commonly known as the Shannon sampling theorem [2], this interpolation formula was known as the cardinal series expansion of Whittaker [3] in 1915, and in fact this expansion was even known to Cauchy [4] in 1841. The expansion given by (8.1) has many interesting properties. If we only use a finite number of all the sampled values in $\{x(t_n), -\infty < n < \infty\}$, then there is an error between the original continuous-time signal $x(t)$ and the expansion of (8.1), with only a finite number of terms. Various tight error bounds can be found in [5] and [6].

Since $x(t)$, $t \in \Re$, where t is on the real-line is a special case of

$$x(s) = \int_{-B}^{B} X(f)e^{i2\pi fs}df, \; s = t + i\sigma \in \mathbf{C}, \tag{8.2}$$

where s is a complex variable in the complex plane $\mathbf{C}$. However, (8.2) shows that $x(s)$, including the special case of $s = t$, is an entire function of type B [7], and thus is an analytic function in the theory of complex variables. In other words, a bandlimited signal $x(t)$, having the inverse Fourier transform definition of (8.2), being an analytic function possesses many interesting properties, including the fact that it is infinitely differentiable, and has a valid Taylor series expansion at any real-valued sampling point t_0. In other words, by having all the derivative values $\{f^p(t_0), \; p = 0, 1, \ldots\}$ of $f(t)|_{t_0}$, the Taylor series can yield the value of $f(t)$ at any other time instant t, regardless of how far t is from t_0. Unfortunately, such expansions are not *stable* [8] in the sense that small perturbations in the sampled values may lead to arbitrarily large deviations in the reconstructed interpolatory functions.

These early uniform and non-uniform sampling expansions all assumed the finite energy signals are only bandlimited, but otherwise have no other restrictions. In fact, all these reconstruction interpolatory functions only involve linear combinations of either the sampled signal values or sampled derivatives of $x(t)$. However, in addition to bandlimitedness, when we impose the additional constraint that there is sparseness in the signals (either in the time-domain or in the transformed domain), various non-linear recovery cs schemes (which involves some optimization operations) can reconstruct the original signals that do not need the classical Nyquist sampling constraints. Some tutorial introduction to non-linear sampling reconstruction appeared in [9] and more detailed discussions appeared in [10].

8.3 Compressive Sampling for Solving Under-Determined System of Equations

Consider a linear system of equations given by

$$\mathbf{Ax} = \mathbf{b}, \tag{8.3}$$

where $\mathbf{A}$ is an $m \times n$ matrix, $\mathbf{x}$ is an $n \times 1$ vector, and $\mathbf{b}$ is an $m \times 1$ vector. In (8.3), the matrix $\mathbf{A}$ and the vector $\mathbf{b}$ are assumed to be known, and $\mathbf{x}$ is the unknown vector to be solved. If $m = n$ and $\mathbf{A}$ is a non-singular matrix, then clearly $\hat{\mathbf{x}} = \mathbf{A}^{-1}\mathbf{b}$ yields the desired solution. Of course, we assume $\mathbf{A}$ is fairly well behaved, such that the condition number, cond($\mathbf{A}$), is not too large, and thus $\mathbf{A}^{-1}$ is numerically stable. In practice, if $\mathbf{A}$ is not singular, but has a moderately large value of cond($\mathbf{A}$) number, then various methods (i.e., the QR method or the SVD method) can be used to solve the $m = n$ linear system of equations. In the over-determined system case, when $m > n$, implying there are more equations than unknowns (i.e., m equations with n unknowns with $m > n$), the classical "normal equation" method can be used to solve $\hat{\mathbf{x}} = (\mathbf{A}^{\mathbf{T}}\mathbf{A})^{-1}\mathbf{A}^{\mathbf{T}}\mathbf{b}$. The normal equation method assumes $\mathbf{A}^{\mathbf{T}}\mathbf{A}$ is non-singular or equivalently all the columns of $\mathbf{A}$ are linearly independent. In practice, the QR method or the SVD method may also be used to solve for the unknown solution in the over-determined system case.

Now, consider (8.3) for the under-determined system of equations, where $m < n$, implying that there are more unknowns than equations. Then there is no unique solution for $\hat{\mathbf{x}}$ of (8.1). Consider a simple example of an under-determined system of equations given by Example 8.1.

Example 8.1 Consider the case of $m = 1$ and $n = 3$, with $\mathbf{A} = [2, 4, 6]$ and $\mathbf{b} = [6]$. By inspection, one possible solution for $\hat{\mathbf{x}}$ is $\hat{\mathbf{x}}_1 = [0, 0, 1]^T$, since $[2, 4, 6] * [0, 0, 1]^T = [0 + 0 + 6] = [6]$. Another solution for $\hat{\mathbf{x}}$ is $\hat{\mathbf{x}}_2 = [0.5, 0.5, 0.5]^T$, since $[2, 4, 6] * [0.5, 0.5, 0.5]^T = [1 + 2 + 3] = [6]$. Clearly, there are an uncountable number of possible $\hat{\mathbf{x}}$ solutions of (8.3).

The classical normal equation approach for solving an under-determined system of equations is given by

$$\hat{\mathbf{x}}_{NE} = \mathbf{A}^{\mathbf{T}} * (\mathbf{A} * \mathbf{A}^{\mathbf{T}})^{-1} * \mathbf{b}, \tag{8.4}$$

assuming the $n \times n$ matrix $\mathbf{A} * \mathbf{A}^{\mathbf{T}}$ is not singular. When the $\hat{\mathbf{x}}_{NE}$ solution of (8.4) exists, it has the minimum ℓ_2 norm among all possible solutions of (8.3). Now, apply the $\hat{\mathbf{x}}_{NE}$ solution of (8.4) to the Example 8.1 problem.

Example 8.2 (8.1 Continued)

$$\begin{aligned}\hat{\mathbf{x}}_{NE} &= \mathbf{A}^{\mathbf{T}} * (\mathbf{A} * \mathbf{A}^{\mathbf{T}})^{-1} * \mathbf{b} = [2, 4, 6]^T * [4 + 16 + 36]^{-1} * [6] \\ &= [2, 4, 6]^T * [6/56] = [0.2143, 0.4286, 0.6429]^T.\end{aligned} \tag{8.5}$$

Next, consider the ℓ_2 norms (i.e., Matlab ℓ_2 norm is given by norm($\cdot$,2)) of $\hat{\mathbf{x}}_{NE}$ of (8.5), $\hat{\mathbf{x}}_1 = [0, 0, 1]^T$, and $\hat{\mathbf{x}}_2 = [0.5, 0.5, 0.5]^T$. Explicit evaluations show norm$(\hat{\mathbf{x}}_{NE}, 2) = 0.8018$, norm$(\hat{\mathbf{x}}_1, 2) = 1$, and norm$(\hat{\mathbf{x}}_2, 2) = 0.8660$. This shows $\hat{\mathbf{x}}_{NE}$ has the minimum ℓ_2 norm, among these three $\hat{\mathbf{x}}$ solutions.

For an under-determined system of equations, compressive sampling theory is able to provide a method to solve for the "most sparse" $\hat{\mathbf{x}}$ solution of (8.3) given by the following lemma.

Table 8.1 Comparisons of the ℓ_p, $p = 1, 2, 4, 6$ norms of vectors $\{\mathbf{a}_1, \mathbf{a}_2, \mathbf{a}_3, \mathbf{a}_4\}$

	$\mathbf{a}_1 = [1,0,0,0]$	$\mathbf{a}_2 = [0.5, \sqrt{3}/2, 0, 0]$	$\mathbf{a}_3 = [0.5, 0.5, \sqrt{1/2}, 0]$	$\mathbf{a}_4 = [0.5, 0.5, 0.5, 0.5]$
$\ell_1(\cdot)$	1	1.3660	1.7071	2
$\ell_2(\cdot)$	1	1	1	1
$\ell_4(\cdot)$	1	0.8891	0.7828	0.7071
$\ell_6(\cdot)$	1	0.8713	0.7339	0.6300

Lemma 8.1 *Consider an under-determined system of equations of (8.3), with $m < n$. Among all the $\hat{\mathbf{x}}$ solutions of (8.3), the "most sparse" $\hat{\mathbf{x}}$ solution is given by the solution to*

$$\min ||\mathbf{x}||_{\ell_1}, \text{ subject to } \mathbf{Ax} = \mathbf{b}. \tag{8.6}$$

The lemma as stated above is not very precise, since some quite sophisticated technical details need to be imposed on the conditions of the matrix $\mathbf{A}$, and some restrictions need to be imposed on how small m can be relative to n, etc. We note, that the minimization in (8.6) must use the ℓ_1 norm. If the minimization uses ℓ_p, $p \neq 1$, then this lemma is not valid.

The use of ℓ_1 norm for characterizing the "sparseness" of a vector can be seen from the simple demonstration of the following four vectors, $\{\mathbf{a}_1, \mathbf{a}_2, \mathbf{a}_3, \mathbf{a}_4\}$, all of length 4 and also of unit energy (i.e., $(\ell_2)^2 = 1$) in Table 8.1.

Three of the components of vector $\mathbf{a}_1$ are zero, two of the components of vector $\mathbf{a}_2$ are zero, and only one of the components of vector $\mathbf{a}_3$ is zero, while no components of the vector $\mathbf{a}_4$ is zero. Thus, $\mathbf{a}_1$ is more sparse than $\mathbf{a}_2$, $\mathbf{a}_2$ is more sparse than $\mathbf{a}_3$, and $\mathbf{a}_3$ is more sparse than $\mathbf{a}_4$. By using the ℓ_1 norm, we note that $\ell_1(\mathbf{a}_1) < \ell_1(\mathbf{a}_2) < \ell_1(\mathbf{a}_3) < \ell_1(\mathbf{a}_4)$. This observation gives a hint on why we used the minimization under the ℓ_1 norm, rather than some other ℓ_p norm in (8.6).

Now, we want to demonstrate the use of this lemma (without worrying about these technical details) to find a solution for Example 8.1.

Example 8.3 (8.1 Continued) Consider the use of the modern convex optimization program cvx (inside Matlab) to perform the minimization of the ℓ_1 norm for this problem. The ch12ex1.m Matlab program is given by:

```
%%ch12ex1.m
cvx_setup␣A=[2␣4␣6];␣b=[6];
cvx_begin
variable␣x(3);
minimize(norm(x,1));
subject␣to␣A*x␣==␣b;
cvx_end
x
%%
```

The result of this program is given by:

```
Status:␣Solved␣Optimal␣value␣(cvx_optval):␣+1
x␣=

0.0000
0.0000
1.0000
---
```

Thus, ch12ex1.m found the desired "most sparse" solution as given by $\hat{\mathbf{x}}_1 = [0, 0, 1]^T$. Indeed, this solution has two zeros among the three possible solutions of $\hat{\mathbf{x}}$ of (8.3). Of course, it is not possible to have more zeros (i.e., three zeros) for a possible solution of $\hat{x}$.

Since the dimensions of m and n in the example in Example 8.4 are very small, it is reasonable to ask whether the algorithm in the lemma is applicable to the larger dimensions of m and n. Now, we consider Example 8.4 (taken from an example of Candes [9], with $m = 256$ and $n = 512$).

Example 8.4 The Matlab program of

```
␣candes_lec5_l1.m
```

is given below:

```
%%candes_lec5_l1.m
n␣=␣512;␣%␣Size␣of␣the␣signal
m␣=␣256;␣%␣Number␣of␣samples
randn('state',2010)%␣This␣initial␣condition␣for␣randn␣was
added␣by␣K.␣Yao,
%␣so␣we␣can␣duplicate␣the␣outcomes␣of␣this␣program
A␣=␣randn(m,n);␣%␣Gaussian␣matrix
S␣=␣round(m/3);␣%␣Oversample␣by␣a␣factor␣3␣only!
support␣=␣randsample(n,S);
x0␣=␣zeros(n,1);␣x0(support)␣=␣randn(S,1);
b␣=␣A*x0;
%%␣Solve␣l1
cvx_begin
variable␣x(n);
minimize(norm(x,1));
subject␣to␣A*x␣==␣b;
cvx_end
%%%
```

At this point, we will not comment on the use of the Gaussian random variable to generate the matrix **A** (these are very technical details to be worried about later).

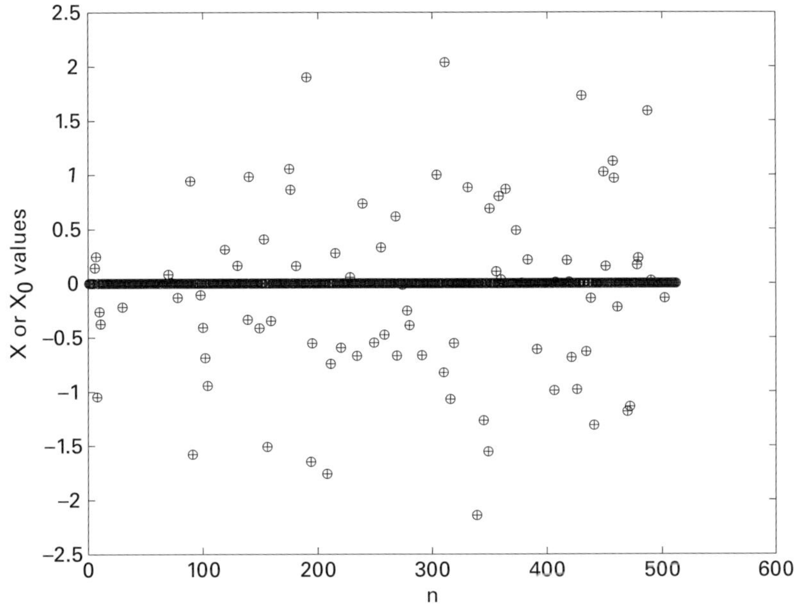

Figure 8.1 Plot of true $\mathbf{x}_0$ solution ("o") and cs est. sparse $\mathbf{x}$ solution ("+")(using ℓ_1 norm) vs n

We insert the line of "randn('state',2010)," so we can duplicate the outcomes of this program; otherwise, each time we run this program, we will obtain different results. With the initial condition of "randn('state',2010)," we first obtained the true $\mathbf{x}_0$ as a 512×1 vector with 427 zeros and only 85 non-zero values. Thus, we can consider $\mathbf{x}_0$ to be a quite "sparse" vector. With this $\mathbf{x}_0$ and the already generated $\mathbf{A}$ matrix, we construct $\mathbf{b} = \mathbf{A} * \mathbf{x}_0$. Then the cvx program is used to find the desired sparse solution of $\hat{\mathbf{x}}$. Next, we want to ask how the $\hat{\mathbf{x}}$ solution obtained by this program (based on the lemma) compares to the true solution of $\mathbf{x}_0$? In Fig. 8.1, we plot the cs algorithm (using the proper minimization based on the ℓ_1 criterion in (8.6) computed non-zero-valued $\hat{\mathbf{x}}$ with the symbol of "+", while the non-zero-valued true $\mathbf{x}_0$ have the symbol of "o"). We note, all the "+" symbols are on top of the "o" symbols, and all the 427 zero values of both solutions form a thick horizontal line at the ordinate value of 0. Thus, Example 8.4 shows the algorithm in the lemma is able to yield the sparse solution of an under-determined system of equations. In Fig. 8.2, we plot the cs algorithm (using the minimization based on the ℓ_2 criterion in (8.6) computed non-zero-valued $\hat{\mathbf{x}}$ with the symbol of "+", while the non-zero-valued true $\mathbf{x}_0$ have the symbol of "o"). Unlike Fig. 8.1, none of the "+" is on top of the "o" and in fact none of the 512 values of this $\hat{\mathbf{x}}$ has a zero value, indicating it is not sparse. We also note, that if we used the classical normal equation solution of (8.4), the resulting $\mathbf{x}_{NE}$ is not sparse; in fact, it also does not have a single zero value.

Thus, far, we have shown the $\hat{\mathbf{x}}$ solution of (8.6) in the lemma worked fine for finding the sparse solution in Example 8.4. However, the dimensions of Example 8.4 are trivially small and the matrix $\mathbf{A}$ in Example 8.3 was chosen as the realization of a random

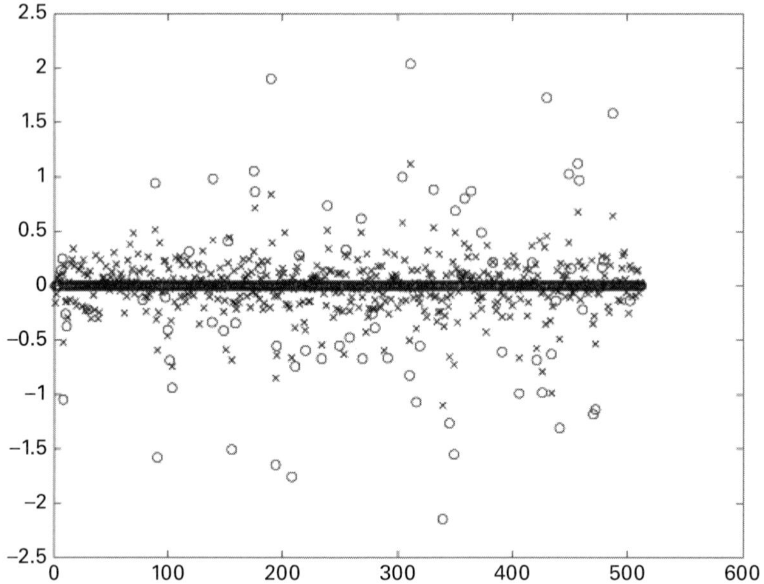

Figure 8.2 Plot of true $\mathbf{x}_0$ solution ("o") and cs est. sparse $\mathbf{x}$ solution ("x")(using ℓ_2 norm) vs n

Gaussian matrix. A more general version of the lemma is now given by the following theorem.

THEOREM 8.1 *Consider an under-determined system of equations given in (8.7),*

$$\mathbf{A}\mathbf{x} = \mathbf{b}, \tag{8.7}$$

where $\mathbf{A}$ *is a known* $m \times n$ *matrix,* $\mathbf{b}$ *is a known* $m \times 1$ *vector, and* $\mathbf{x}$ *is the* $n \times 1$ *unknown solution we seek with* $m \leq n$*. Suppose* $\mathbf{x}$ *is known to be sparse with no more than* S *number of non-zero values. Furthermore,* m *cannot be smaller than the right-hand side of (8.8) in*

$$m \geq S \log\left(\frac{n}{S}\right). \tag{8.8}$$

Define an $m \times m$ *matrix* Φ*, where all of its elements* $\Phi(i, j)|_{i,j=1,\ldots,m}$*, are taken to be realizations of independent and identically distributed (iid)* $N(0, 1/m)$ *rv. Then define the new known* $m \times n$ *matrix* $\tilde{\mathbf{A}} = \Phi * \mathbf{A}$ *and the new known* $m \times 1$ *vector by* $\tilde{\mathbf{b}} = \Phi * \mathbf{b}$*. Thus among all the* $\hat{\mathbf{x}}$ *solutions of (8.3), the "most sparse"* $\hat{\mathbf{x}}$ *solution is given by*

$$\min||\mathbf{x}||_{\ell_1}, \text{ subject to } \tilde{\mathbf{A}}\mathbf{x} = \tilde{\mathbf{b}}. \tag{8.9}$$

Next, consider the application of this theorem to the solution of a linear system of equations encountered in the "Super Efficient Monte Carlo Methodology Based on

Novel Use of Correlated Chaotic Sequences" problem [12]. In this problem, we are given a function $B(x)$ defined on $(-1, 1)$.

$$B(x) = \frac{e^{-x^2/2}}{\sqrt{2\pi}\rho(x)}, \rho(x) = \frac{1}{\pi\sqrt{1-x^2}}, -1 < x < 1. \tag{8.10}$$

Since the family of Chebyshev polynomials of the first kind, $\{T_k(x), k = 0, 1, \ldots\}$, consists of a complete orthogonal set of functions over the interval of $(-1, 1)$, with respect to the weight function of $\rho(x)$ defined in (8.10), we can expand $B(x)$ with respect to these Chebyshev polynomials as follows

$$B(x) = \sum_{k=0}^{L} c_k T_k(x), \ -1 < x < 1, \tag{8.11}$$

where the coefficient c_k is defined by

$$c_k = \int_{-1}^{1} B(x)T_k(x)\rho(x)dx, k = 0, 1, \ldots, L. \tag{8.12}$$

In theory, the number of terms L in (8.11)–(8.12) needs to be infinite, but in practice L must be a finite positive integer. Since the coefficient c_k in (8.12) does not have an explicit closed-form expression, it must be evaluated numerically. First, we note that since $B(x)$ is an even function and the Chebyshev polynomial $T_k(x)$ is an odd function for all odd integral values of k over the interval $(-1, 1)$, then all $c_k = 0$ for all odd integral values of k. Thus, we only need to evaluate c_k for all even integral values of k. However, using Mathematica's NIntegrate function, it appears the evaluation of c_k for $k > 20$ may have numerical stability problems. Table 8.2 shows the valid values of c_k versus k for $k = 2, 4, \ldots, 20$.

Using the cs methodology to solve sparse solutions of a linear system of equations as given in the theorem was proposed by [11]. As a preliminary step, consider the generation of the sequence $\{x_{i+1} = T_2(x_i), i = 1, \ldots, n-1\}$ using the Chebyshev polynomial $T_2(x)$, where we set $L = n - 1$. Then evaluate $B(x)$ at these sequence values as follows

Table 8.2 Chebyshev coefficients c_k versus k for $k = 2, 4, \ldots, 20$ evaluated using Mathematica NIntegrate

$k = 0$	2	4	6	8	10
$c_k = 0.6827$	– 0.570	– 0.0177	– 0.0269	– 0.0146	– 0.00948
12	14	16	18	20	
$c_k = -0.00662$	– 0.00489	– 0.00375	– 0.00297	– 0.00241	

$$B(x_i) = \sum_{k=0}^{n-1} c_k T_k(x_i) = [1,\ T_1(x_i),\ T_2(x_i),\ \cdots, T_{n-1}(x_i)] \begin{bmatrix} c_0 \\ c_1 \\ \vdots \\ c_{n-1} \end{bmatrix}, i = 1, 2, \ldots, n. \tag{8.13}$$

Next, form an $n \times 1$ vector $\mathbf{B}$ from the $B(x_i)$ values in (8.13)

$$\mathbf{B} \triangleq \begin{bmatrix} B(x_1) \\ B(x_2) \\ \vdots \\ B(x_n) \end{bmatrix} = \begin{bmatrix} 1 & T_1(x_1) & T_2(x_1) & \cdots & T_{n-1}(x_1) \\ 1 & T_1(x_2) & T_2(x_2) & \cdots & T_{n-1}(x_2) \\ \cdots & \cdots & \cdots & \cdots & \cdots \\ 1 & T_1(x_n) & T_2(x_n) & \cdots & T_{n-1}(x_n) \end{bmatrix} \begin{bmatrix} c_0 \\ c_1 \\ \vdots \\ c_{n-1} \end{bmatrix} \triangleq \mathbf{Ac}. \tag{8.14}$$

Equation (8.14) is now a linear system of equations with $m = n$, where the $n \times n$ matrix $\mathbf{A}$ is known and the $n \times 1$ vector $\mathbf{B}$ is also known, but the $n \times 1$ vector $\mathbf{b}$ is the unknown variable to be solved. For our problem, suppose we take $n = 1{,}000$. Clearly, it is difficult to solve this problem for $\mathbf{b}$ as it stands with $m = n = 1{,}000$. However, we know $\mathbf{b}$ is moderately sparse, owing to the nature of about half of its components (i.e., $c_k = 0$, for all odd integral k values).

Thus, consider the use of the cs methodology of the theorem. Consider a new $m' \times n$ random matrix Φ, where the elements of $\Phi(i, j)|_{i=1,\ldots,m';\, j=1,\ldots,n}$, are taken to be realizations of iid $N(0, 1/n)$ rv. Furthermore (for technical reasons to be mentioned later), we want to make all the rows of Φ orthonormalized. Thus, we define

$$\Phi = \text{ortho}(\Phi')'. \tag{8.15}$$

Multiply on the left of (8.14) by the new Φ (the one obtained from (8.15)) to obtain

$$\tilde{\mathbf{b}} \triangleq \Phi\mathbf{B} = \Phi\mathbf{Ac} \triangleq \tilde{\mathbf{A}}\mathbf{c}, \tag{8.16}$$

where $\tilde{\mathbf{b}}$ is a known $m' \times 1$ vector, $\tilde{\mathbf{A}}$ is a known $m' \times n$ matrix, and $\mathbf{c}$ is the $n \times 1$ unknown vector to be determined. By comparing (8.16) to (8.9), we note m' and the desired $\mathbf{c}$ of (8.16) correspond to m and the solution of the minimization of $\mathbf{x}$ under the ℓ_1 norm of (8.9).

As stated earlier, we take $m = n = L + 1 = 1{,}000$ and $m' = 500$. Results using the theorem are tabulated in Table 8.3.

We note that most of the c_k values in Table 8.3 compare well with those in Table 8.2 for low values up to about $k = 20$. The cs method seems to be able to provide reliable values of c_k for k up to about 40, while the direct NIntegrate numerical method was not able to do so. It is interesting to check whether the condition of (8.8) is satisfied. With $m' = 500$, $n = 1{,}000$, and $S = 500$, we do have $500 = m' \geq S \log(n/S) = 500 * \log(1{,}000/500) = 500 * \log(2) = 500 * 0.693 = 347$. We should note that even if this $\tilde{\mathbf{c}}$ solution is not very sparse, since only about half of its coefficients have zero values, the cs methodology of the theorem still seems to work fairly well in terms of its ability to yield the desired $\mathbf{c}$ solution.

Table 8.3 Chebyshev coefficients c_k versus k for $k = 2, 4, \ldots, 40$ evaluated using (8.9) of the theorem

$k = 0$	2	4	6	8	10
$c_k = 0.6827$	-0.570	-0.0178	-0.0271	-0.0143	-0.00960
$k = 12$	14	16	18	20	
$c_k = -0.0065$	-0.0047	-0.0039	-0.0029	-0.0022	
$k = 22$	24	26	28	30	
$c_k = -0.0019$	-0.0013	-0.0011	-0.0013	-0.0008	
$k = 32$	34	36	38	30	
$c_k = -0.0004$	-0.0007	-0.0004	-0.0004	-0.0005	

The reason that we need to perform the orthonormalization of all the rows in (8.15) is due to the rather technical issue of wanting to have "low coherence" between the Φ and the original system matrix $\mathbf{A}$ in the matrix $\tilde{\mathbf{A}} = \Phi\mathbf{A}$ in (8.16), which is used in the minimization in the theorem. Specifically, the "coherence" of these two matrices denoted by $\mu(\Phi, \mathbf{A})$ is a measure of the largest correlation between any two elements of Φ and $\mathbf{A}$. It appears that there is very low coherence between a random matrix Φ and the most "human" generated matrix $\mathbf{A}$. It is claimed that

$$\mu(\Phi, \mathbf{A}) \approx \sqrt{2\log(n)}. \tag{8.17}$$

For the above problem with $n = 1{,}000$, its coherence based on (8.17) is only 3.7 . More technical detailed analysis can show the probability that the sparse solution obtained from the ℓ_1 optimization in the theorem is related to some function of the inverse of the square of the coherence of $\mu(\Phi, \mathbf{A})$. The smaller the coherence, the higher the probability the sparse solution will be found by the solution of the theorem.

8.4 Conclusion

In Section 8.2, we first reviewed the classical "Shannon sampling theorem" and related issues for linear reconstruction of samples of the bandlimited signal. In Section 8.3, we showed the use of a cs methodology for solving the sparse solution of an underdetermined linear system of equations.

8.5 References

A series of tutorial papers in [1] introduced various issues of cs problems. Classical sampling theorems have appeared in [3] and [2], and even earlier in [4]. Sampling truncation errors and stability issues and related expansions have appeared in [5], [6], and [8]. The bandlimited property of these sampling expansions are all treated in the

Fourier transform work of [7]. The use of cs in Section 8.3 for solving the underdetermined system of equations was considered in [9]. A simple example exploiting the sparseness of the solution, using the cs approach demonstrated its benefits compared to the standard orthonormal series expansion approach in [11] and [12].

[1] R.G. Baraniuk *et al*, "Sensing, Sampling, and Compression," *IEEE Signal Processing Magazine*, March 2008.

[2] C.E. Shannon, "A Mathematical Theory of Communication," *Bell System Technical Journal*, 1948, vol. 27, no. 3, pp. 379–423.

[3] E.T. Whittaker, "On the Functions which are Represented by the Expansion of the Interpolation Theory," *Proceedings of the Royal Society*, Edinburgh, 1914–1915.

[4] A.L. Cauchy, "Memoroire sur Diverses Formulaes de Analyze," *Comptre Rendu*, Paris, 1841.

[5] K. Yao, "On Truncation Error Bounds for Sampling Representation of Bandlimited Signal," *IEEE Trans. on Aerospace and Electronic Systems*, 1966, pp. 644–647.

[6] K. Yao, "Applications of Reproducing Kernal Hilbert Spaces – Bandlimited Signal Models," *Information and Control*, 1967, pp. 429–444.

[7] R.E.A. Paley and N. Wiener, *Fourier Transforms in the Complex Domain*, American Mathematical Society, 1934.

[8] K. Yao and J.B. Thomas, "On Some Stability and Interpolatory Properties of Nonuniform Sampling Expansions," *IEEE Trans. on Circuit Theory*, 1967, pp. 404–408.

[9] E. Candes, "Compressive Sampling and Frontiers in Signal Processing," Short Course Lecture.

[10] Y.C. Elder, *Sampling Theory: Beyond Bandlimited Systems*, Cambridge University Press, 2015. Notes, IMA, Lecture 5, June 2007.

[11] J.Y. Lee, "Using CS Methodology to Numerically Evaluate Coefficients of the Chebyshev Polynomial Expansion of the B(x) Function," Private Communications, May 2009.

[12] K. Yao, "Super Efficient Monte Carlo Simulations," *Simulations Technologies in Networking and Communications*, CRC Press, 2010, Ch. 3, pp. 69–92.

8.6 Exercises

1. Duplicate the results in Table 8.1.
2. Duplicate the results in Example 8.3. You can use the Matlab program of Candes given in Example 8.3.
3. Consider the Chebyshev coefficient problem considered in (8.10)–(8.16). Duplicate the results in Table 8.3.

9 Chaotic Signal Analysis and Processing

9.1 Introduction

Most electrical engineers have been trained in and exposed to deterministic and random linear mathematics and system theory. Linear mathematics and associated system theory are quite well understood and have proven to be useful in engineering. In recent years, tremendous progress has been made in the basic understanding and application of chaos. Chaos has provided many intellectually interesting and intuitive but not obvious explanations of diverse physical phenomena and theoretical models. While chaos theory was originated by mathematicians at the beginning of the twentieth century in the study of non-linear dynamical systems, its modern developments were made essentially by theoretical physicists. In the last thirty years, a large number of highly technical as well as tutorial papers and books have appeared on chaos. The extremely readable non-technically oriented book by J. Gleick [1], *Chaos: The Amazing Science of the Unpredictable*, was on the top selling list for many months after it was published in 1988. Quoting from Gleick's book, we have "The most passionate advocates of this new science go so far as to say that twentieth-century science will be remembered for just three things: relativity, quantum mechanics, and chaos. Chaos, they contend, has become the century's third great revolution in the physical sciences. Like the first two revolutions, chaos cuts away at the tenets of Newton's physics. As one physicist put it: "Relativity eliminated the Newtonian illusion of absolute space and time; quantum mechanics eliminated the Newtonian dread of a controllable measurement process; and chaos eliminates the Laplacian fantasy of deterministic predictability." "Of the three, the revolution in chaos applies to the universe we see and touch, to objects at human scale."

In 1963, E. Lorenz, a theoretical meteorologist at MIT, observed the extreme sensitivity of the initial condition on the discretized solution of his set of non-linear differential equations on a digital computer [2]. Later the parameters of these equations were known to yield chaotic trajectories with unpredictable behaviors. Since these equations were used in modeling some long-range weather forecasts, this phenomenon led to the simplistic and dramatic "Butterfly Effect" of chaotic behavior. It claims that if a butterfly in the Amazon flips its wings once, six month later the weather in North America will be fine, but if it flips its wing twice, the weather will be terrible. While this claim is clearly exaggerated, we will study in Section 9.2 the extreme sensitivity of the solutions with respect to initial conditions and their bifurcation dependency on the parameter values of the non-linear dynamical system. Since then, chaotic behavior has been observed in turbulence, biological population growth, modeling of cardiac fibrillation, electronic non-linearity, etc. More recently, chaotic properties of observed data with chaotic features have been used for their identifications. Engineering systems have been designed to incorporate chaotic behaviors such as in spread spectrum-like communication systems in Section 9.3. We study various basic techniques using limited observed data from the state of a non-linear dynamical system to determine its chaoticness and extract some basic system parameters. In Section 9.4, we consider the use of chaos in the generation of a specific class of chaotic sequences for applications in "super-efficient" Monte Carlo simulations.

9.2 Chaos and Non-linear Dynamical Systems

A non-linear dynamical system (NLDS) is a system in which the time evolution of the equation is non-linear. Consider Newton's law of motion, in which a particle of mass m subject to a force $F_x(x,t)$ in the x direction is governed by

$$F_x(x,t) = ma = m\frac{d^2x}{dt^2}.$$

If the mass is connected to an ideal spring obeying Hooke's law with a spring constant k, then

$$F_x(x,t) = -kx.$$

By combining these two equations, we obtain

$$\frac{d^2x}{dt^2} = -\frac{k}{m}x.$$

This dynamical system is linear in the state variables $\{x, d^2x/dt^2\}$ and the system is known to yield the harmonic oscillation solution with an angular frequency of $\omega = (k/m)^{0.5}$. For a linear system, if $g(x,t)$ and $h(x,t)$ are two solutions of the system, then superposition holds and $cg(x,t) + dh(x,t)$ is also a solution of the system. Then the study of solutions of a linear system is more tractable. On the other hand, if we have a non-linear spring with

$$F_x(x,t) = ax^2,$$

then

$$\frac{d^2x}{dt^2} = \frac{a}{m}x^2,$$

is an NLDS in $\{x, d^2x/dt^2\}$. The key observation is that in the original linear system, if the spring constant k is changed slightly, the overall behavior of the system remains the same, with the angular frequency changed only slightly. Similarly, a slight change of the initial condition leads also to a slight change of the solution. However, for an NLDS, a slight change of a system parameter can lead to dramatic change in the solution. For one parameter value, the solution may be periodic, while for a slightly different value, the solution can be aperiodic or even chaotic. Similarly, a slight change of initial condition can lead to a dramatically different solution later in time.

9.2.1 Bifurcation and Chaos in the Logistic Map NLDS

The logistic NLDS is a model that was used by R.M. May in 1976 [3] to demonstrate the bifurcation of the solution of a dynamical system. By *bifurcation*, we mean a sudden change in the nature of the solutions into two regions: one region when the parameter is below some threshold value and another region when the parameter is slightly above the threshold. As we shall see, the evolution of the bifurcation phenomena moves from regular to less regular to chaotic behaviors.

Consider a biological population growth model of a species. Let N_i denote the species population in some region at time i. The unit of time may be a year, or month, or minute depending on the time it takes to have a new generation. For mammals, a time unit of a year may make sense, while for bacteria, the time unit may be a minute. It is reasonable to assume that $N_i = AN_{i-1}$, $i = 1, 2, \ldots$, where A is a positive constant. This means the population at time $i = 1$ is proportional to its parental population at time $i = 0$. If the constant of proportionality $A > 1$, then the population will explode exponentially with time, while if $A < 1$, then the species will become extinct. Of course, assuming the above model is valid for all values of N_i is probably not very realistic. As the population explodes, many other factors will come in, such as the food supply may become limited. One possible modification of the model is to assume

$$N_i = AN_{i-1} - BN_{i-1}^2, \; i = 1, 2, \ldots \tag{9.1}$$

Here we assume B is also a positive constant much smaller than A. For small values of N_{i-1}, the BN_{i-1}^2 effect in the equation is insignificant, but for large population values, the quadratic term in N_{i-1} helps to reduce the population at time i. Then upon defining $C = A/B$, $N_{i+1} > 0$ follows from $N_i \leq C$. Upon defining a normalized variable $x_i = N_i/C$, (9.1) becomes

$$x_i = Ax_{i-1}(1 - x_{i-1}) = f_A(x_{i-1}), \; i = 1, 2, \ldots \tag{9.2}$$

Equation (9.2) is clearly a set of non-linear difference equations, called the logistic equations.

We want to study the behavior of the sequence of values of $\{x_i, i = 1, 2, \ldots\}$, called the trajectories, in (9.2). Any function $f_A(\cdot)$, such that $x_i = f_A(x_{i-1}), i = 1, 2, \ldots$ is called an iterated map function, if the domain and the range of $f_A(\cdot)$ are identical. Then the function $f_A(\cdot)$ in the logistic equations of (9.2) becomes a logistic iterated map function if the parameter A is restricted to $0 \leq A \leq 4$. In this case, the domain and range of $f_A(\cdot)$ are given by [0, 1]. Fig. 9.1 shows the plot of $f_A(x)$ vs x for different values of A. In particular, we note that the diagonal line intersects the logistic map function for all values of A at the origin but intersects at non-origin values only for $A > 1$.

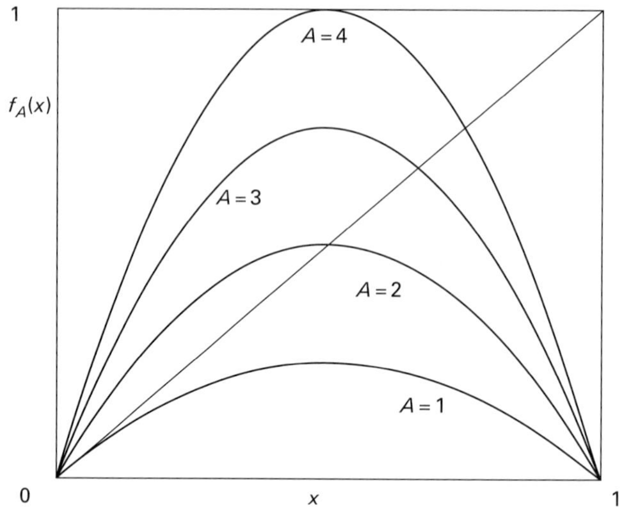

Figure 9.1 Plot of $f_A(x)$ vs x for the logistic map function and the diagonal line with $y = x$

The value x_A^* is called a *fixed point* of an iterated map function $f_A(\cdot)$ if

$$x_A^* = f_A(x_A^*). \tag{9.3}$$

From (9.2), we find there are two fixed points. One obvious fixed point is $x_A^* = 0$. The second fixed point is $x_A^* = 1 - (1/A)$. We note, for $A < 1$, the $x_A^* = 1 - (1/A) > 1$ is outside the domain of the logistic map function of (9.2), and thus is not a valid fixed point of (9.2). However, for $A > 1$, $x_A^* = 1 - (1/A)$ is indeed an interior point of [0, 1] as verified in Fig. 9.1.

For $A < 1$ and all initial values of x_0 in $0 \leq x_0 \leq 1$, the logistic map function iterates to the final value of $x = 0$. This can be seen in the example of Fig. 9.2, where we consider $A = 0.6$ and $x_0 = 0.7$. An *attractor* is a point at which a set of trajectories converge as the number of iterations goes to infinity. The set of trajectories associated with an attractor is called a *basin of attraction*. For the logistic map function with $A < 1$, $x = 0$ is an attractor with its basin of attraction given by the entire domain [0, 1] of the function.

For the case of $1 < A < 3$ with initial values x_0 in [0, 1], all the trajectories converge to the second fixed point $x_A^* = 1-(1/A)$. This second fixed point is the attractor with its basin of attraction given by [0, 1]. The first fixed point $x_A^* = 0$ is then a *repelling fixed point.* As an example, consider the case of $A = 1.5$ with $x_0 = 0.1$. Then the trajectories converge to the fixed point at $x_A^* = 1 - (1/1.5) = 0.3$ as shown in Fig. 9.3.

So far, the results are simple with no surprises. Now, consider $3 < A$. Consider the example of $A = 3.1$ and iterating from $x_0 = 0.25$. As shown in Table 9.1, the trajectories do not converge to a single point, but oscillate between the two values of $x = 0.558$ and $x = 0.764$. In the original population problem, this case implies in the steady state, the population is high one year and low another year. Thus, at $A = 3$ a *period-2 bifurcation* occurs. At $A = 3.44948\ldots$ a *period-4 bifurcation* occurs with values at a steady state

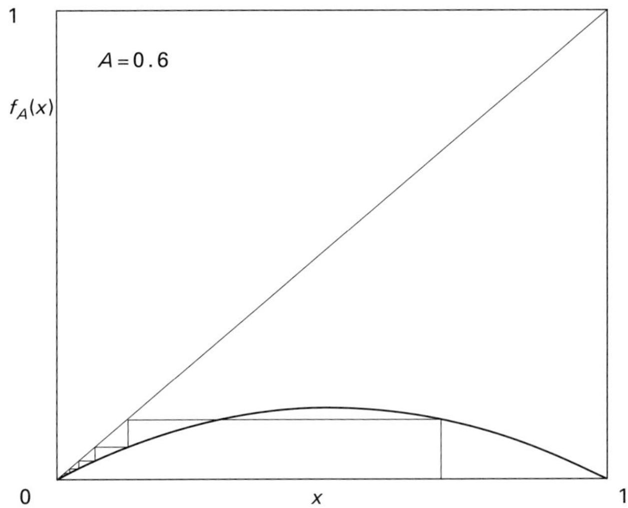

Figure 9.2 Iterations for the logistic map function with $A = 0.6$ and initial value given by $x_0 = 0.7$

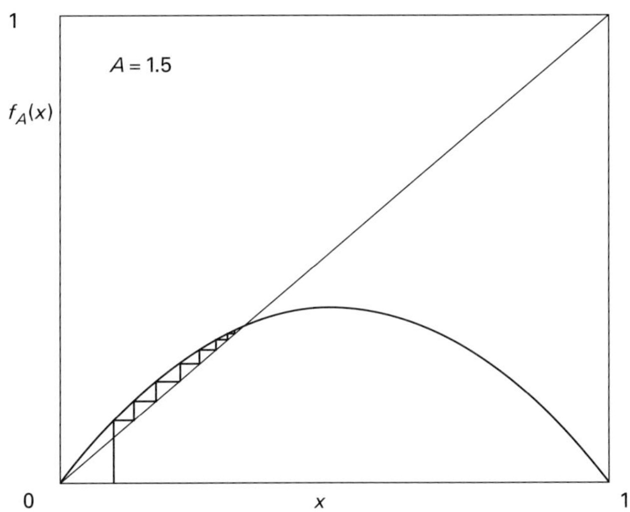

Figure 9.3 Iterations for the logistic map function with $A = 1.5$ and initial value given by $x_0 = 0.1$

oscillating among $x \in \{0.852, 0.433, 0.847, \text{and } 0.447\}$. As A increases, the logistic map function yields *period-8 bifurcation*, followed by *period-16 bifurcation*, etc. For $3.5699\cdots < A$, the trajectory values do not repeat, and the trajectories are chaotic. Fig. 9.4 is a bifurcation diagram for the logistic map function showing the values of x in which doubling and chaotic phenomena occur as a function of A.

Table 9.1 Trajectories for the logistic map function with $A = 3.1$ and initial value of $x_0 = 0.25$

n	x_n
0	0.250
1	0.581
2	0.754
3	0.574
4	0.758
5	0.569
6	0.760
7	0.565
8	0.762
9	0.562
10	0.763
11	0.561
12	0.764
13	0.559
14	0.764
15	0.559
16	0.764
17	0.559
18	0.764
19	0.558
20	0.764

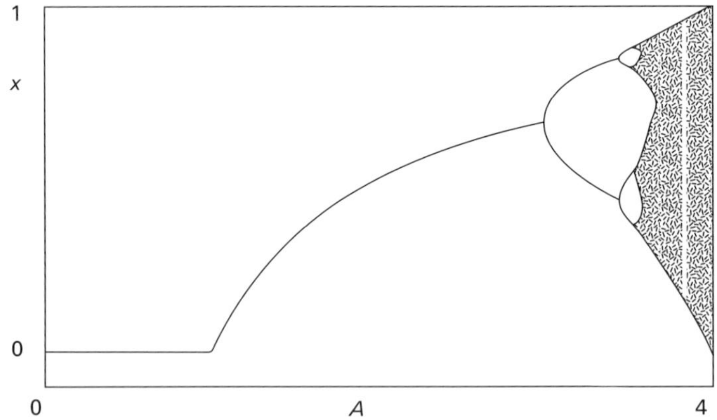

Figure 9.4 Bifurcation diagram for the logistic map function

In order to verify the extreme sensitivity of the chaotic trajectories, consider $A = 3.99$ with initial values at $x_0 = 0.400$, $x_0 = 0.401$, and $x_0 = 0.4005$, as shown in Table 9.2. We see after ten iterations that there is no resemblance among the x_i values starting from the three initial x_0 values. From this logistic map function modeling a biological population growth problem, it becomes clear that there need not be significant

Table 9.2 Trajectories for the logistic map function with $A = 3.99$ and initial value of $x_0 = 0.400$, 0.4010, and 0.4005

n	x_n	x_n	x_n
0	0.4000	0.4010	0.4005
1	0.9576	0.9584	0.9580
2	0.1620	0.1591	0.1605
3	0.5417	0.5338	0.5377
4	0.9906	0.9929	0.9918
5	0.0373	0.0280	0.0324
6	0.1432	0.1085	0.1250
7	0.4894	0.3860	0.4365
8	0.9970	0.9456	0.9814
9	0.0117	0.2051	0.0727
10	0.0462	0.6506	0.0269
11	0.1758	0.9070	0.7847
12	0.5782	0.3366	0.6740
13	0.9731	0.8910	0.8767
14	0.1046	0.3874	0.4312
15	0.3736	0.9470	0.9786

environmental changes for the population of a species to vary greatly. This type of conclusion cannot be obtained using conventional techniques, but can be obtained using chaotic NLDS analysis.

9.2.2 Bifurcation and Chaos in the Lorenz NLDS Differential Equations

Lorenz [2] studied a set of non-linear differential equations that characterize a simplified convection fluid model. His equations belong to the class of Navier–Stokes equations for fluid and have been studied previously by many other people. There are three variables denoted by $X(t)$, $Y(t)$, and $Z(t)$. These variables are a function of time, but they are not spatial variables. $X(t)$ describes the time-dependence of the components of fluid flow velocity. $Y(t)$ is proportional to the temperature difference between the rising and falling parts of the fluid, while $Z(t)$ is proportional to the deviation from temperature linearity. The Lorenz equations are given by

$$\dot{X} = p(Y - X), \quad \dot{Y} = -XZ + rX - Y, \quad \dot{Z} = XY - bZ, \tag{9.4}$$

where $\dot{X} = dX(t)/dt$ is the derivative of X with respect to time, and similarly for $\dot{Y}$ and $\dot{Z}$. We omit the time variable t for simplicity of notation. Here p, r, b are parameters; all have quite specific physical interpretations but will not be considered here. While the equations in (9.4) appear quite simple, they can have quite complex behaviors involving bifurcation and chaos. For small values of $r < 1$, Rayleigh predicted that the trajectories should converge to the non-convective state of $X = Y = Z = 0$. Similarly, we can have $Z - Y$, $Y - X$, and other planes of the state space. A point in the state space such that all of the time derivatives of the state space are zero is called a *fixed point*. The

Lorenz equations have three fixed points. One is at $X = Y = Z = 0$. The others are at $X = Y = \pm(b(r-1)^{0.5})$ with $Z = r-1$. Clearly, the last two fixed points are of interest only for $r > 1$. For $r < 1$, all trajectories regardless of initial conditions converge to the origin. Thus, the $X - Y - Z$ space is the basin of attraction of the attractor at the origin. For $r > 1$, the origin is a repelling fixed point and all trajectories move away from it. The other two fixed points are attracting fixed points, if r is not too large. Thus, $r = 1$ is a bifurcation of the Lorenz equations [2].

Nothing too dramatic happens until $r > 13.93$. Then there are still small regions about the two non-origin fixed points that each have basins of attraction. However, trajectories outside of these small regions are repelled from its fixed point to the other fixed point. Here is a case in which there is no external time-varying forcing term in the Lorenz equations, and yet the three variables have complex time behaviors. This is an example of chaotic behavior. This behavior is a chaotic manifestation of an NLDS. The importance of the divergence of trajectories from two slightly different initial conditions is that even though the outcomes of the NLDS are *deterministic,* they are *unpredictable.* We also call the unpredictable future of a chaotic dynamical system *indeterminable* even though it is deterministic.

9.3 Analysis of Chaos in State Space

The use of *state space* to characterize the behavior of a dynamical system was proposed by the French mathematician H. Poincaré in the late 1800s. Similarly, this concept was called the *phase space* in the study of statistical mechanics by the American physicist J.W. Gibbs. Consider a dynamical system as characterized by a set of three first-order differential equations in terms of the state variables $u, v,$ and w. Then the derivatives of $u, v,$ and w with respect to time t are given by

$$\dot{u} = \frac{du}{dt} = f(u,v,w); \;\; \dot{v} = \frac{dv}{dt} = g(u,v,w); \;\; \dot{w} = \frac{dw}{dt} = h(u,v,w), \qquad (9.5)$$

where $f, g,$ and h depend on the $u, v,$ and w, and possibly on some parameters, but not on their derivatives nor on time t. The system of equations in (9.5) is called *autonomous.* We note, the Lorenz system of equations is such an autonomous system.

If a system of equations has an external time-dependent forcing term, then the system is called *non-autonomous.* For example, the two-state non-autonomous system

$$\dot{u} = f(u,v,t), \;\; \dot{v} = g(u,v,t),$$

can be converted into a three-state autonomous system by introducing the new variable $w = t$. Then

$$\dot{u} = f(u,v,t), \;\; \dot{v} = g(u,v,t), \;\; \dot{w} = \frac{dt}{dt} = 1.$$

Now, we want to consider the chaotic behavior of a continuous-time NLDS. First, we want to show heuristically that chaotic behavior can occur only when the system has at least three state variables. For a chaotic NLDS, the trajectories must satisfy all three conditions of:

a. bounded trajectories;
b. exponential divergence of nearby trajectories;
c. no intersection of different trajectories.

The first condition is required so that the solution of the NLDS is bounded. The second condition requires that the trajectories be arbitrarily different from each other with increasing time. The last condition states that two trajectories cannot meet, otherwise the deterministic solution of the system will be the same from that point on. With these three constraining conditions, it is not possible to draw trajectories on a two-dimensional paper which do not intersect. However, in a three-dimensional space, it is possible for two trajectories starting at the same point to differ significantly from each other by weaving about each other and still staying within a bounded region.

In order to characterize the exponential divergence of the trajectories of a chaotic NLDS, we need to consider the *Lyapunov exponent* of the system. At the simplest level of understanding, suppose two nearby trajectories of a chaotic system are separated by a distance of d_0 at time $t = 0$. Then the trajectories at time t are separated by

$$d(t) = d_0 e^{\lambda t}, \tag{9.6}$$

where λ is a positive number and is called the Lyapunov exponent of the chaotic system.

The Lyapunov exponent can be evaluated from the system of differential equations as follows. First, consider a one-dimensional state space model given by $\dot{x}(t) = f(x)$. Use a Taylor series expansion of $f(x)$ about the point x_0 close to x. Then

$$f(x) = f(x_0) + \frac{df(x)}{dx}|_{x_0}(x - x_0) + \cdots \tag{9.7}$$

Denote the distance between the two trajectories and their derivatives by

$$d(t) = x(t) - x_0; \quad \dot{d}(t) = \dot{x}(t) - \dot{x}_0 = f(x) - f(x_0) = \frac{df(x)}{dx}|_{x_0}(x - x_0), \tag{9.8}$$

where only the first derivative term in the Taylor series expansion is kept. Then the derivative of the expression in (9.6) yields

$$\dot{d}(t) = \lambda d_0 e^{\lambda t} = \lambda d(t) = \frac{df(x)}{dx}|_{x_0}(x - x_0). \tag{9.9}$$

This shows that the Lyapunov exponent is given directly as

$$\lambda = \frac{df(x)}{dx}|_{x_0}. \tag{9.10}$$

In general, the derivative of $f(\cdot)$ with respect to x varies with x. Thus, we may want to evaluate df/dx at different values of x as well as consider its average value for an interval of x.

The autonomous system with three state variables can be denoted by

$$\dot{x}_1 = f_1(x_1, x_2, x_3), \; \dot{x}_2 = f_2(x_1, x_2, x_3), \; \dot{x}_3 = f_3(x_1, x_2, x_3). \tag{9.11}$$

Then the three Lyapunov exponents are given by

$$\lambda_1 = \frac{\partial f_1}{dx_1}, \quad \lambda_2 = \frac{\partial f_2}{dx_2}, \quad \lambda_3 = \frac{\partial f_3}{dx_3}, \tag{9.12}$$

where the partial derivatives are evaluated at the state space point of interest. An NLDS is said to be chaotic if it has at least one positive average Lyapunov exponent.

9.3.1 Poincaré Section

While the discussion in this subsection on the Poincaré section is applicable to arbitrary finite-dimensional state space, for simplicity of notation and the discussion, we will only consider the three-dimensional case. The Poincaré section is generated from a two-dimensional plane on which the locations of the trajectories are traversed. The plane is oriented in such a way that all the trajectories are not parallel or nearly parallel to the plane. The fixed points of the Poincaré section are those points that satisfy

$$\dot{x}_1^* = F_1(x_1^*, x_2^*), \ \dot{x}_2^* = F_2(x_1^*, x_2^*), \tag{9.13}$$

where $F_1(\cdot,\cdot)$ and $F_2(\cdot,\cdot)$ are functions induced from (9.11). Then each fixed point on the Poincaré section corresponds to a limit cycle in the full three-dimensional state space.

9.4 Characterizing Chaos from Time-Series Data

From the last section, we realize that for a continuous-time the NLDS must have three or more variables in order to exhibit chaotic behavior. This means on a two-dimensional Poincaré plane, there is no way of fully characterizing the chaotic behavior. Indeed, for a large dimensional NLDS, it is very difficult (if not impossible) to obtain data from all the state variables when studying the NLDS. In this section, we show that by appropriately using one-dimensional time-series data obtained from the NLDS, it is possible to characterize the chaotic behavior and parameters of the system.

In general, from observed time-series data, we are interested in characterizing whether the data are deterministic or random. If deterministic, are they chaotic? If they are chaotic, how many state variables are needed to model the dynamics of the system? In some sense, if the NLDS is chaotic, we hope the dimension of the system is low. If it turns out that the system is chaotic but has a high dimension, the behavior of the chaotic data and random data is difficult if not distinguishable.

9.4.1 Lyapunov Exponent from a One-Dimensional Time-Series

Consider an observable state variable $x(t)$ in an NLDS sampled every τ second. Starting at time t_0, the nth sample will be given by $t_n = t_0 + n\tau$. Then we denote the observed time-series $x(t_0), x(t_1), x(t_2), \ldots$ by $x_0, x_1, x_2, \ldots$ For a chaotic system, the divergence of nearby trajectories can be quantified in the following manner. Suppose we initially pick x_i and then search for another x value denoted by x_j that is close to x_i. We denote the absolute value of the difference of their values as d_0. Similarly, we repeat this for x_{j+1} from x_{i+1} and so forth. We can state this formally as

$$d_0 = |x_j - x_i|,\ d_1 = |x_{j+1} - x_{i+1}|,\ d_2 = |x_{j+2} - x_{i+2}|\ \ldots,\ d_n = |x_{j+n} - x_{i+n}|. \tag{9.14}$$

If the difference grows exponentially, then we have

$$d_n = d_0 e^{\lambda n}, \tag{9.15}$$

which can be solved for λ as

$$\lambda = \frac{1}{n} \ln\left(\frac{d_n}{d_0}\right). \tag{9.16}$$

In practice, (9.16) is used operationally as the definition of the Lyapunov exponent. If λ is positive, then the NLDS is chaotic. Thus, the Lyapunov exponent measures the exponential divergence of two nearby points. While the concept defining the Lyapunov exponent is simple, some comments are in order.

9.4.2 Fractal Dimensions for a Chaotic NLDS

A fundamental property of a chaotic attractor is its dimension. This is a basic issue in geometry and is related to the issue of how the trajectories are distributed in space. The dimension of an object at first seems to be a simple concept. A point is of dimension zero, a line or a curve is of dimension one, a surface is of dimension two, and a volume is of dimension three. Hypervolumes of higher dimensions are also possible. In any case, all these dimensions are integral valued. However, Mandelbrot [4] showed that the dimension of a chaotic attractor is not an integer and such geometric objects are called *fractals*. Indeed, a chaotic attractor with a non-integral dimension is also called a *strange attractor*. It turns out there are several different ways of measuring the dimension of an object.

Capacity Dimension

First, consider a one-dimensional line of length L. Consider $N(\epsilon)$ number of boxes of dimension one needed to cover this line segment. As shown in Fig. 9.5a, if $\epsilon = L$, then $N = 1$. If $\epsilon = L/2^n$, we need $N = 2^n$ such boxes. In general, for a line segment of length L, the number of required boxes of length ϵ is given by

$$N(\epsilon) = L \times \frac{1}{\epsilon}. \tag{9.17}$$

Similarly, for a two-dimensional square with a side of length L. As shown in Fig. 9.5b, in general if $\epsilon = L/2^n$, then we need $N = 2^{2n}$ such boxes. That is, for a square of length L on each side, the number of required boxes of length ϵ on each side is given by

$$N(\epsilon) = L^2 \times \frac{1}{\epsilon^2}. \tag{9.18}$$

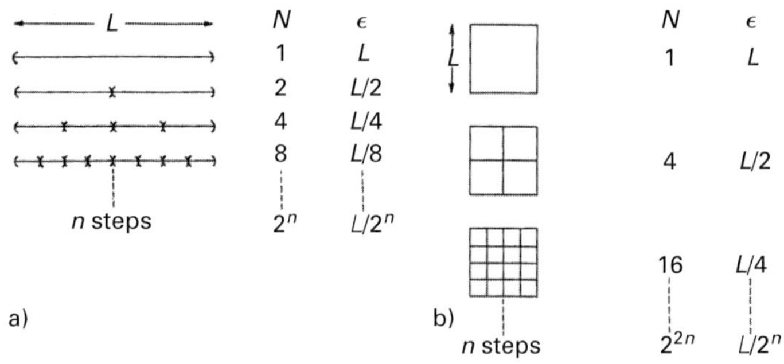

Figure 9.5 Box covering method for capacity dimension evaluations: (a) for a one-dimensional line segment; (b) for a two-dimensional box

In general, for a box of length L on each side in dimension d, the number of required boxes of length ϵ on each side is given by

$$N(\epsilon) = L^d \times \frac{1}{\epsilon^d}. \tag{9.19}$$

Taking the logarithm for the dimension, we solve for the dimension d and obtain

$$d = \frac{\log(N(\epsilon))}{\log(L) + \log(1/\epsilon)}. \tag{9.20}$$

In the limit of small ϵ, the term involving $\log(L)$ in the denominator of (9.20) becomes negligible. The *capacity dimension* is defined by

$$d_c = \lim_{\epsilon \to 0} \frac{\log(N(\epsilon))}{\log(1/\epsilon)}. \tag{9.21}$$

While the capacity dimension d_c of (9.21) yields an integral value of 1 for a line segment, 2 for a two-dimensional box, etc., it can yield non-integral values for less regular behaving geometric figures. Consider the *Cantor set* of Fig. 9.6. It starts at iteration $n = 1$ with an $N = 1$ segment of length $\epsilon = 1$. Then at the next iteration of $n = 2$, the middle third of the segment is removed, resulting in $N = 2$ segments each with a segment length of $\epsilon = 1/3$. At iteration n, there are $N = 2^n$ segments each with length of $\epsilon = 1/3^n$. Then the capacity dimension according to (9.21) is given by $d_c = \lim_{\epsilon \to 0}[\log(N(\epsilon))/\log(1/\epsilon)] = \lim_{\epsilon \to 0}[\log(2^n)/\log(3^n)] = \log(2)/\log(3) = 0.6309\ldots$ which is definitely not an integer. The fact that the Cantor set has a non-integral capacity dimension should not be too surprising since in the limit this is a set with an uncountable number of points but has "no length." We can see this since the length at each iteration n is given by $L = (2/3)^{n-1}$. In the limit as n goes to infinity, the length L goes to zero.

In the Koch curve, as seen in Fig. 9.7, we start at iteration $n = 1$ with the $N = 1$ segment of length $\epsilon = 1$. Then at the next iteration of $n = 2$, the middle third of the segment is doubled, resulting in $N = 4$ segments with each segment of length $\epsilon = 1/3$. At iteration n, there are $N = 4^n$ segments with each length of $\epsilon = 1/3^n$. Then the capacity dimension according to (9.21) is given by $d_c = \lim_{\epsilon \to 0}[\log(N(\epsilon))/\log(1/\epsilon)] =$

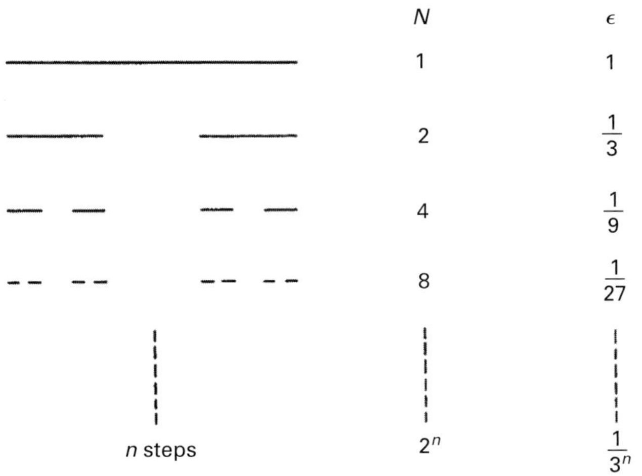

Figure 9.6 Capacity dimension of the Cantor set

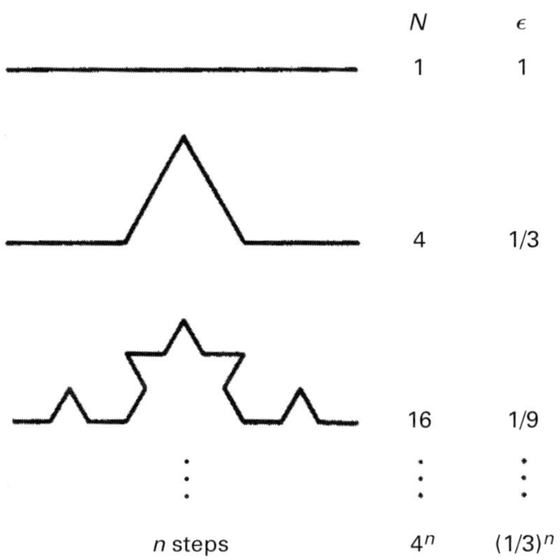

Figure 9.7 Capacity dimension of the Koch curve

$\lim_{\epsilon\to 0}[\log(4^n)/\log(3^n)] = \log(4)/\log(3) = 2 \times (\log(2)/\log(3)) = 1.2618\ldots$, which is definitely not an integer. We note, this capacity dimension is twice that of the Cantor set. It is in this spirit that Mandelbrot showed that the length of the coast line of England can be made arbitrarily large with an arbitrarily fine segmental approximation. All of these objects are fractals and all possess the *self-similar* property that a small section of the object, when suitably magnified, is idential to the original object. It is claimed that many physical objects (e.g., clouds, rock formations, etc.) as well as biological objects (e.g., blood vessels, nerves, etc.) all possess these self-similar properties (at least up to some quantum effect level).

9.4.3 Pseudo-Random Sequence Generation by NLDS

Many engineering and scientific problems have random behaviors and thus are modeled by different random processes. However, when we want to do a simulation of these problems on a digital computer, we need to find some deterministic algorithms for generating these "random sequences." Since these sequences are really deterministic sequences but have almost all the desired properties of random sequences, they are often called "pseudo-noise" (PN) sequences. One possible approach for the generation of PN sequences is based on algebraic theory of finite fields, leading to the shift-register (SR) sequences including the m-sequences, Kasami sequences, and Gold sequences. These sequences are used in spread-spectrum communication systems as well as in cryptosystems. However, another approach is to use NLDS to generate chaotic sequences for PN applications.

Consider a map $\tau(\cdot)$ defined on a finite interval $I \rightarrow I$, where $I \in \Re$. Then we can define the sequence $\{x_i, i = 0, \ldots\}$ generated from an iterated map function by

$$x_i = \tau(x_{i-1}),\ i = 1, \ldots, \tag{9.22}$$

where x_0 is called the "seed" of the map. We have seen the logistic map of (9.2) as an example of such a map that can be used to generate a chaotic sequence for $A \geq 3.5699\ldots$ In general, if the map of (9.22) has some desired properties (to be specified later), then the sequence generated by the map can also be chaotic and its values are not predictable. For a given map, each such generated chaotic sequence is fully deterministic and is completely specified by x_0. However, even if we are given a large number of these sequences, what general conclusions can we say about this map? Clearly, the study of all the sequences (i.e., trajectories in the state space) is not fruitful.

Invariant Measure of a Chaotic Map

Instead of studying all the individual trajectories, it turns out that we may want to consider how often the trajectories fall within some region of the state space. Suppose we partition the interval I into N subintervals and consider a chaotic sequence of length M. Let m_n denote the number of the sequence values that fall into subinterval n. Then we use the relative frequency concept of probability $p_n = m_n/M, n = 1, \ldots, N$, to measure how often the sequence falls into the nth subinterval. For many chaotic NLDS, the set of $\{p_n, n = 1, \ldots, N\}$ does not depend on the initial x_0, as long as the length of the sequence M is taken to be sufficiently long. In this case, this set of $\{p_n, n = 1, \ldots, N\}$ is called an "invariant measure." The term "invariant" denotes the fact the probabilities do not depend on the initial x_0. In order to understand and derive this result, we need to consider the Frobenius–Perron (FP) method needed in the construction of the invariant measure.

Frobenius–Perron Method

Consider a map $\tau(\cdot)$ defined on a finite interval $I \rightarrow I$, where $I = [0, 1]$. For this map, for every $x \in I$, it has two pre-images, y and z, such that $x = \tau(y) = \tau(z)$ or equivalently, $\tau^{-1}(x) \in \{y, z\}$. Then $p(\cdot)$ is the invariant measure of the map $\tau(\cdot)$ if $p(\cdot)$ satisfies

$$p(x) = p(\tau^{-1}(x)), \text{ for every } x \in I, \tag{9.23}$$

and $\int_I p(x)dx = 1$. In general, for an arbitrary map $\tau(\cdot)$, finding its invariant measure is not obvious. From probability theory on the transformation of a random variable, the probability over $[x, x + dx]$ equals the probability over $[y, y + dy]$ plus the probability over $[z, z + dz]$. Thus,

$$p(x)|dx| = p(y)|dy| + p(z)|dz|. \tag{9.24}$$

Since $dx = d\tau(y) = \tau'(y)dy$ and $dx = d\tau(z) = \tau'(z)dz$, then (9.24) becomes

$$p(x)|dx| = p(y)\frac{|dx|}{|\tau'(y)|} + p(z)\frac{|dx|}{|\tau'(z)|},$$

or equivalently

$$p(x) = \frac{p(y)}{|\tau'(y)|} + \frac{p(z)}{|\tau'(z)|}. \tag{9.25}$$

Equation (9.25) is called the Frobenius–Perron equation for finding the invariant measure of the map $\tau(\cdot)$. As stated above, the solution of (9.25) for an arbitrary $\tau(\cdot)$ is difficult. We will consider the solution of the Bernoulli shift map.

Bernoulli Shift Map

One of the simplest non-linear iterated maps defined on $I = [0, 1]$ is the Bernoulli shift map defined by

$$x_i = 2x_{i-1} \bmod(1), \; i = 1, \ldots, \tag{9.26}$$

for any $x_0 \in I$, as shown in Fig. 9.8. This map states to take any number in the unit interval, multiply it by two and keep only the remainder that is still less than one. Every number x_0 in the unit interval has a binary representation of the form $x_0 = 0.a_1a_2a_3\ldots$, where each $a_n, n = 1, 2, \ldots,$ takes only the binary values of $\{0, 1\}$ to the right of the

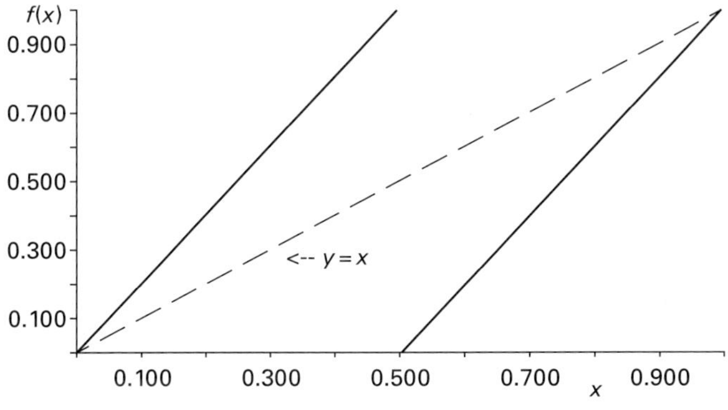

Figure 9.8 Bernoulli shift map

"binary point" in (9.26). For example, if $x_0 = 0.101000\ldots$ (which is equivalent to the decimal number 0.625), then after multiplication by 2, $x_1 = 1.0100$, $\text{mod}(1) = 0.0100$ (which in decimal, becomes $1.25 \,\text{mod}(1) = 0.25$). Thus, each iteration of this map is equivalent to a shift to the left by one place.

The interesting question is what are the trajectories of the states of the Bernoulli shift map? If x_0 is a rational number represented with only a finite number of 1s, then after a finite number of iterations, x_i will be zero and all future iterated values remain at zero. If x_0 is a rational number represented with repeating binary numbers (and thus also repeating decimals), then the iterated value after any number of left shifts remains a rational number with repeating binary values. Thus, the trajectories are periodic. On the other hand, if x_0 is an irrational number (with no repeating binary numbers), then after any number of left shifts, it remains an irrational number (with no repeating binary numbers). Thus, the trajectories are non-periodic. For every rational number in I, there is another irrational number arbitrarily close to it. Similarly, for every irrational number in I, there is another rational number arbitrarily close to it. Clearly, there are an infinite number of periodic trajectories and also an infinite number of non-periodic trajectories. However, if x_0 is chosen randomly in I, since there are many more irrational points than rational points in I, then it is much more likely that the trajectories will be non-periodic than periodic.

Now, consider the "random" behavior of the Bernoulli shift map. Suppose we flip a coin repeatedly and record the sequence of heads and tails as "HTHTTT..." We can replace an "H" by "1" and a "T" by "0." Then the sequence of the flipping of a coin becomes "101000...," which has the same form as the binary representation of an irrational number in I to the right of the binary point. Since the outcomes of the flipping of a coin can be considered to be "random," then so too are the outcomes of the Bernoulli shift map.

Now, use the FP method to determine the invariant measure of the Bernoulli shift map. We note the Bernoulli shift map of (9.26) can be rewritten as

$$\tau(x) = \begin{cases} 2x, & 0 \le x \le 0.5, \\ 2x - 1, & 0.5 < x \le 1. \end{cases} \tag{9.27}$$

Thus, $\tau'(y) = 2$, $y = x/2$, $0 \le y \le 0.5$, and $\tau'(z) = 2$, $z = (x+1)/2$, $0.5 < z \le 1$. Then (9.25) for this case becomes

$$2p(x) = p(x/2) + p\big((x+1)/2\big). \tag{9.28}$$

But the only solution to the above equation is $p(x) = 1$, $0 \le x \le 1$. Thus, the invariant measure of the Bernoulli shift map is the uniform density over the interval $[0, 1]$. That is, in the steady state, the trajectories take values equally between zero and one, regardless of the initial starting x_0. This also implies that the Bernoulli shift map can be used for pseudo-random generation of a uniform r.v. over the interval $[0, 1]$.

9.6 CDMA System Models

We consider a CDMA system with K users and spreading factor N. The received signal is

$$r(t) = \sum_{k=1}^{K} s^{(k)}(t - \tau^{(k)}) + n(t), \tag{9.33}$$

where $s^{(k)}(t) = \mathrm{Re}\{b^{(k)}(t)a^{(k)}(t)\exp(\omega_{ct} + \phi_0^{(k)})\}$ is the transmitted signal from the kth user, $n(t)$ is the white Gaussian noise with two-sided spectral density $N_0/2$, $b^{(k)}(t)$ and $a^{(k)}(t)$ are the data and spreading signal, respectively, ω_c is the carrier frequency, and $\phi_0^{(k)}$ is the phase. The data signal $b^{(k)}(t)$ is a sequence of ± 1 rectangular pulses with a duration of T_b. The spreading sequence has a period N, and is composed of rectangular pulses with duration T_c and chip amplitude $a_j^{(k)}$ such that $\sum_{j=0}^{N-1} a_j^{(k)2} = 1$. We assume that $N = T_b/T_c$, and $\tau^{(k)}$ is the time delay.

Because we are concerned with relative phase shifts modulo 2π and relative time delays modulo T_b, there is no loss of generality in assuming $\phi^{(i)} = \phi_0^{(i)} - \omega_c \tau^{(i)} = 0$, $\tau^{(i)} = 0$ for the desired receiver i, and $T_c = 1$ and considering only $0 \leq \phi^{(k)} < 2\pi$ (assumed to be uniformly distributed in $[0, 2\pi)$), and $0 \leq \tau^{(k)} < N$ for $k \neq i$.

The output of the single-user matched filter at the ith receiver is

$$\begin{aligned} Z_i &= 2\int_0^{T_b} r(t)a_i(t)\cos(\omega_{ct})dt \\ &= b_0^{(i)} + \sum_{k\neq i}^{K} I^{(k)}(b^{(k)}, \phi^{(k)}, \tau^{(k)}) + \sum_{l=0}^{N-1} a_l^{(i)}\eta_l, \end{aligned} \tag{9.34}$$

where η_l is the equivalent Gaussian noise in the lth chip with $E[\eta_l] = 0$ and $E[\eta_i \eta_j] = N_0\delta(i - j)$, and $\sum_{k\neq i}^{K} I^{(k)}(b^{(k)}, \phi^{(k)}, \tau^{(k)})$ is the MAI. The channel SNR is given by $\mathrm{SNR} = 1/N_0$.

The interference term $I^{(k)}(b^{(k)}, \phi^{(k)}, \tau^{(k)})$ due to the k-user has been found [10] to be

$$I^{(k)}(b^{(k)}, \phi^{(k)}, \tau^{(k)}) = (b_{-1}^{(k)} R_{k,i}(\tau^{(k)}) + b_0^{(k)} \hat{R}_{k,i}(\tau^{(k)}))\cos(\phi^{(k)}), \tag{9.35}$$

where $R_{k,i}(\tau)$ and $\hat{R}_{k,i}(\tau)$ for $0 \leq l \leq \tau < l + 1 \leq N$, are given by

$$\begin{aligned} R_{k,i}(\tau) &= C_{k,i}(l - N) + [C_{k,i}(l + 1 - N) - C_{k,i}(l - N)](\tau - l), \\ \hat{R}_{k,i}(\tau) &= C_{k,i}(l) + [C_{k,i}(l + 1) - C_{k,i}(l)](\tau - l), \end{aligned} \tag{9.36}$$

where $C_{k,i}(l)$ is the partial cross-correlation between the kth and the ith user and is defined by

$$C_{k,i}(l) \equiv \begin{cases} \sum_{j=0}^{N-l-1} a_j^{(k)} a_{j+l}^{(i)}, & 0 \leq l \leq N-1 \\ \sum_{j=0}^{N+l-1} a_{j-l}^{(k)} a_j^{(i)}, & 1-N \leq l \leq -1 \\ 0, & |l| \geq N. \end{cases} \tag{9.37}$$

If $\tau^{(k)}$ for $k \neq i$ is a discrete random variable uniformly distributed in $\{0, 1, 2, \ldots, N-1\}$, the system is called the chip-synchronous CDMA (CS-CDMA) system, whereas the system is called the asynchronous CDMA (A-CDMA) system if $\tau^{(k)}$ is a continuous random variable uniformly distributed in $[0, N)$.

9.7 Derivation of Optimal Sequences

9.7.1 Asynchronous CDMA

The overall interference variance for the ith user from all other users can be computed [10] as

$$\begin{aligned} \sigma_A^2(i) &\equiv E_{\phi^{(k)}, b_{-1}^{(k)}, b_0^{(k)}, \tau^{(k)}} \left\{ \left[\sum_{k \neq i}^{K} I^{(k)}(b^{(k)}, \phi^{(k)}, \tau^{(k)}) \right]^2 \right\} \\ &= \frac{1}{6N} \sum_{k \neq i}^{K} \sum_{l=1-N}^{N-1} [2C_{k,i}^2(l) + C_{k,i}(l) C_{k,i}(l+1)]. \end{aligned} \tag{9.38}$$

With the SGA assumption, the error probability of the ith user is given by

$$P_e(i) = Q\left(\sqrt{\frac{1}{\sigma_A^2(i) + N_0}}\right), \tag{9.39}$$

where $Q(\cdot)$ is the complementary standard Gaussian probability distribution. The evaluation of the error probability of an A-CDMA system based on the moment space bounding technique and its relationship to the SGA assumption were discussed in [11].

Using the following identity

$$\sum_{l=1-N}^{N-1} C_{x,y}(l) C_{x,y}(l+n) = \sum_{l=1-N}^{N-1} C_{x,x}(l) C_{y,y}(l+n) \tag{9.40}$$

given in [12] and the trivial identity $C_x(l) \equiv C_{x,x}(l) = C_{x,x}(-l)$, (9.38) can be simplified to

$$\begin{aligned}\sigma_A^2(i) &= \frac{1}{6N}\sum_{k\neq i}^{K}\Bigg[2C_k(0)C_i(0) + 4\sum_{l=1}^{N-1} C_k(l)C_i(l) \\ &\quad + \sum_{l=1-N}^{N-1} C_k(l)C_i(l+1)\Bigg] \\ &= \frac{1}{6N}\sum_{k\neq i}^{K}\Bigg[2C_k(0)C_i(0) + 4\sum_{l=1}^{N-1} C_k(l)C_i(l) \\ &\quad + \sum_{l=0}^{N-1} C_k(l)C_i(l+1) + C_k(l+1)C_i(l)\Bigg].\end{aligned} \tag{9.41}$$

The importance of (9.41) lies in the fact that $\sigma_A^2(i)$ can be computed from the autocorrelation functions alone; the cross-correlation functions are not needed. Moreover, the bit error probability is independent of the distribution of the spreading sequences due to the SGA assumption.

Because the Q-function is convex for positive arguments, the lower bound of average error probability of all users in the system can be attained by assigning the same interference variance to every user. This can be done with $C_i(l) = C_k(l) = C(l)$ for all i, k, l. With the normalization $C_i(0) = 1$, we minimize the average error probability by minimizing the interference power

$$\sigma_A^2 = \frac{K-1}{6N}\left[2 + 4\sum_{l=1}^{N-1} C^2(l) + 2\sum_{l=0}^{N-1} C(l)C(l+1)\right]. \tag{9.42}$$

This is a positive quadratic form in $C(l)$ whose unique minimum is achieved when $\partial\sigma^2/\partial C(l) = 0$ for $l = 1, 2, \ldots, N-1$, that is, when

$$4C(l) + C(l+1) + C(l-1) = 0, \qquad \forall l = 1, 2, \ldots, N-1. \tag{9.43}$$

The solution to (9.43) is given by

$$C_k(l) = (-1)^l \frac{r^{l-N} - r^{N-l}}{r^{-N} - r^N}, \quad l = 0, 1, 2, \ldots, N-1, \forall k, \tag{9.44}$$

where $r = 2 - \sqrt{3}$. Substituting (9.44) into (9.42), we obtain the minimum interference power [9] as

$$\sigma_{A-opt}^2 = \frac{\sqrt{3}(K-1)}{6N}\frac{r^{-2N} - r^{2N}}{r^{-2N} + r^{2N} - 2}. \tag{9.45}$$

The ensemble-averaged partial autocorrelation function obtained in [8] when the spreading sequences are assumed to be stationary and independent random processes is identical to the deterministic constant partial autocorrelation vector derived here. Note that when $l \ll N$, $C_k(l) \approx (-r)^l$ which decays exponentially with alternative sign. Moreover, the minimum interference variance is given by $\sigma^2_{A-opt} = \sqrt{3}(K-1)/6N$ as N is large, which increases by 15% the number of users achieved with white sequences, that is, $(K-1)/3N$.

9.7.2 Chip-Synchronous CDMA

The variance of MAI for a CS-CDMA system with $\tau^{(k)}$ uniformly distributed in $\{0, 1, 2, \ldots, N-1\}$ for $k \neq i$ is computed in a similar way to (9.38) and given by

$$\begin{aligned}\sigma^2_C(i) &= \frac{1}{2N}\sum_{k\neq i}^{K}\sum_{l=0}^{N-1}[C^2_{k,i}(l-N) + C^2_{k,i}(l)] \\ &= \frac{1}{2N}\sum_{k\neq i}^{K}\sum_{l=1-N}^{N-1} C^2_{k,i}(l) \\ &= \frac{1}{2N}\sum_{k\neq i}^{K}\left[C_k(0)C_i(0) + 2\sum_{l=1-N}^{N-1} C_k(l)C_i(l)\right],\end{aligned} \tag{9.46}$$

where the last equality is due to the identity in (9.40). Similarly, the solution that minimizes (9.46) is given by

$$C_k(l) = \delta(l), \qquad \forall k. \tag{9.47}$$

That is, the optimal sequences for the CS-CDMA system are the random white sequences and the corresponding minimum MAI power is

$$\sigma^2_{C-opt} = \frac{K-1}{2N}, \tag{9.48}$$

which decreases by 73% and 50% with the number of users achieved with optimal sequences and white sequences for an A-CDMA system, respectively.

9.8 Ergodic Dynamical Systems

9.8.1 Ergodic Theory

The second-order time-averaged statistic of spreading sequences is needed for sequence design and performance analysis. For a spreading sequence generated by a deterministic dynamical system, the performance can be computed analytically or numerically by using the Birkhoff individual ergodic theory, which is restated as follows.

THEOREM 9.1 (4.2.4 in [13]) *Let $(X, \mathcal{A}, \mu)$ be a finite measure space and $S : X \mapsto X$ be a measure-preserving and ergodic transformation. Then, for any integrable f, the average of f along the sequence generated by S, that is, $\{S^{(k)}(x)\}_{k=0}^{\infty}$ for any given* *"initial" $x \in X$, is equal almost everywhere to the average of f over the space X; that is,*

$$\lim_{n\to\infty} \frac{1}{n} \sum_{k=0}^{n-1} f(S^{(k)}(x)) = \frac{1}{\mu(X)} \int_X f(x)\mu(dx) \qquad a.e. \tag{9.49}$$

The use of this theorem allows us to evaluate the autocorrelation function of a sequence generated by any measure-preserving ergodic transformation. As an example of measure-preserving and ergodic transformation, consider the tent map $S(x)$ defined by

$$S(x) \equiv 1 - 2|x|, \qquad |x| \le 1. \tag{9.50}$$

The uniform measure is invariant for this transformation; the autocorrelation function of the sequence generated by the tent map $S(x)$ can be evaluated by

$$\begin{aligned} < C(l) > &\equiv < \frac{1}{n} \sum_{j=1}^{n} S(x_j)S(x_{j+l}) > \\ &= \frac{1}{2} \int_{-1}^{1} xS^{(l)}(x)dx \\ &= \frac{1}{3}\delta(l), \end{aligned} \tag{9.51}$$

where $< >$ denotes the ensemble average with respect to the initial condition x_0 under the corresponding invariant measure.

Another ergodic transformation, the nth degree Chebyshev polynomials, defined by $T_n(x) \equiv \cos(n \arccos(x))$ over the interval $[-1, 1]$, have been considered as a sequence generator for the synchronous CDMA system [14]. Examples of Chebyshev polynomials are given by

$$T_0(x) = 1, \quad T_1(x) = x, \quad T_2(x) = 2x^2 - 1, \quad T_3(x) = 4x^3 - 3x, \ldots \tag{9.52}$$

Adler and Rivlin [15] have shown that Chebyshev polynomials of degree $n \ge 2$ are mixing and thus ergodic, and their invariant measure is given by $\rho(x)dx = dx/(\pi\sqrt{1-x^2})$. Furthermore, by investigating the asymptotical stability of the Frobenius–Perron operator corresponding to Chebyshev transformations, Chebyshev polynomials are shown to be exact, and thus mixing and ergodic transformations [13].

The Chebyshev polynomials have the orthogonality

$$\int_{-1}^{1} T_i(x)T_j(x)\rho(x)dx = \delta_{i,j}\frac{1+\delta_{i,0}}{2}, \tag{9.53}$$

where $\delta_{i,j}$ is the Kronecker delta function such that

$$\delta_{i,j} = \begin{cases} 1, & i = j, \\ 0, & i \ne j. \end{cases} \tag{9.54}$$

The autocorrelation functions for sequences generated by these Chebyshev polynomials are given by

$$
\begin{aligned}
< C(l) > &\equiv < \frac{1}{n}\sum_{j=1}^{n} T_p(x_j)T_p(x_{j+l}) > \\
&= \int_{-1}^{1} T_p(x)T_{p^{l+1}}(x)\rho(x)dx \\
&= \frac{1}{2}\delta(l). \qquad (9.55)
\end{aligned}
$$

9.8.2 Dynamical Systems with Lebesgue Spectrum

One class of ergodic dynamical systems with special properties are those with Lebesgue spectrum [16]. These systems, denoted as $\phi(x)$, not only have an ergodic invariant measure, but are also associated with a special set of orthonormal basis functions $\{f_{\lambda,j}(x)\}$ for the Hilbert space L_2. This orthonormal basis can be split up into classes and written as $\{f_{\lambda,j}(x) : \lambda \in \Lambda, j \in F\}$, where λ labels the classes and j labels the functions within each class. The cardinality of Λ can be proven to be uniquely determined and is called the *multiplicity of the Lebesgue spectrum.* If Λ is (countably) infinite, we speak of the (countably) infinite Lebesgue spectrum. If Λ has only one element, the Lebesgue spectrum is called simple. The important property that these particular basis functions $f_{\lambda,j}$ have is

$$
f_{\lambda,j} \circ \phi = f_{\lambda,j+1}, \qquad \forall \lambda \in \Lambda, j \in F. \qquad (9.56)
$$

That is, all the other basis functions in the same class can be generated from one of the basis functions by using compositions with powers of the dynamical system $\phi(x)$. Furthermore, because the basis functions are orthogonal, every function is orthogonal both to every other function in the same class, and to every function in other classes.

An example of chaotic dynamical systems with a Lebesgue spectrum is the Bernoulli shift map $\phi(x)$ with invariant measure density $\rho(x) = 1$, as considered in [16], [17], and defined by

$$
\phi(x) = \begin{cases} 2x, & 0 \le x \le 1/2, \\ 2x - 1, & 1/2 < x \le 1. \end{cases} \qquad (9.57)
$$

The associated basis functions for L_2 space are Walsh functions and are defined by

$$
\begin{aligned}
w_1(x) &= 1, \\
w_{k+1}(x) &= \prod_{i=0}^{r-1} \text{sgn}\{\sin^{k_i}(2^{i+1}\pi x)\}, \qquad k = 1, 2, \ldots, \qquad (9.58)
\end{aligned}
$$

where the values of k_i, either 0 or 1, are the binary digits of k, that is, $k = \sum_{i=0}^{r-1} k_i 2^i$. Thus, this generator can produce random white binary sequences. A two-dimensional dynamical system with a Lebesgue spectrum can also be exhibited [16], [17].

Another class of chaotic (ergodic) dynamical systems with a Lebesgue spectrum are the Chebyshev polynomial maps as considered above. In particular, we consider the pth degree Chebyshev polynomial map, that is, $\phi(x) = T_p(x)$, where $p \geq 2$ is prime. The associated basis functions for $L_2([-1, 1])$ are also Chebyshev polynomials $\{T_i(x)\}_{i=0}^{\infty}$. Then, the $f_{\lambda,j}(x)$ can be defined by

$$f_{\lambda,j}(x) = T_{\lambda \cdot p^j}(x), \qquad \forall \lambda \in \Lambda, j \in F, \tag{9.59}$$

where $\Lambda = \{n | n$ is non-negative integer and relative prime to $p\}$, and F is the set of non-negative integers. To see this, we consider the composition of ϕ with one of the basis functions:

$$\begin{aligned} f_{\lambda,j} \circ \phi(x) &= T_{\lambda \cdot p^j} \circ T_p(x) \\ &= T_{\lambda \cdot p^{j+1}}(x) \\ &= f_{\lambda,j+1}(x). \end{aligned} \tag{9.60}$$

Note that the basis function $f_{0,j}(x) = T_0(x) = 1$ constitutes its own class and the basis function we used in (9.55) is the particular case when $\lambda = 1$.

9.9 Chaotic Optimal Spreading Sequences Design

9.9.1 Sequences Construction for CS-CDMA Systems

The previous section shows some chaotic dynamical systems that can produce independently identically distributed (i.i.d.) sequences. Therefore, the system performance of CS-CDMA and A-CDMA systems using these sequences is identical to the random white sequences with the interference power $\sigma^2 = (K-1)/2N$ and $\sigma^2 = (K-1)/3N$, respectively.

9.9.2 Sequences Construction for A-CDMA Systems

Let us consider a polynomial function $G(x)$ in the Hilbert space $L_2([-1, 1])$ with the form

$$G(x) \equiv \sum_{j=1}^{N} d_j T_{p^j}(x) = \sum_{j=1}^{N} (-r)^j T_{p^j}(x), \qquad x \in [-1, 1], \tag{9.61}$$

where $p \geq 2$. Thus, the coefficients of the Chebyshev expansion of $G(x)$ are given by

$$d_0 \equiv 0, \quad d_j = (-r)^j \quad \text{for} \quad 1 \leq j \leq N. \tag{9.62}$$

By using ergodic theory, the average of G^2 along the sequence generated by the Chebyshev transformation $T_p(\cdot)$ is given by

$$\begin{aligned} < C(0) > &\equiv < \frac{1}{n}\sum_{i=1}^{n} G^2(x_i) > \\ &= \int_{-1}^{1} G^2(x)\rho(x)dx \\ &= \frac{1}{2}\frac{r^2(1-r^{2N})}{1-r^2} \\ &\equiv A, \end{aligned} \tag{9.63}$$

and the normalized autocorrelation function of such a sequence can be evaluated by

$$\begin{aligned} < C(l) > /A &\equiv \frac{1}{A} < \frac{1}{n}\sum_{i=1}^{n} G(x_i)G(x_{i+l}) > \\ &= \frac{1}{A}\int_{-1}^{1} G(x)G(T_{p^l}(x))\rho(x)dx \\ &= \frac{1}{2A}\sum_{m=l}^{N} d_m d_{m+l} \\ &= \frac{r^2}{2A}\frac{(-r)^l(1-r^{2(N-l)})}{1-r^2} \\ &= (-1)^l\frac{r^{l-N}-r^{N-l}}{r^{-N}-r^{N}}. \end{aligned} \tag{9.64}$$

Thus, with the condition $r = 2-\sqrt{3}$, the output sequences $\{y_1, y_2, \ldots, y_N\}$ generated by

$$\begin{aligned} y_j &= \frac{1}{\sqrt{A}}G(x_j), \\ x_{j+1} &= T_p(x_j), \end{aligned} \tag{9.65}$$

are the optimal spreading sequences for A-CDMA systems. Because of the property $T_{p^{j+1}}(x) = T_p \circ T_{p^j}(x)$, the function $G(x)$ in (9.61) actually is a non-causal finite impulse response (FIR) filter fed by the input sequence generated by Chebyshev polynomial map $T_p(x)$. This FIR filter can be easily implemented to produce the output sequence $\{y_j\}$ [9].

When the spreading factor N is large, an alternative practical design is given as follows. Because the autocorrelation function of a Chebyshev sequence is a Kronecker delta function, we can design the optimal spreading sequences by passing these Chebyshev sequences through an infinite impulse response (IIR) low-pass filter with a single pole at $(-r)$. That is,

$$y_{j+1} = -ry_j + \sqrt{2(1-r^2)}\,x_j,$$
$$x_{j+1} = T_p(x_j). \tag{9.66}$$

The output sequence $\{y_1, y_2, \ldots, y_N\}$ of the filter will have an exponential autocorrelation function $(-r)^l$. Then each sequence generated by the same Chebyshev map, but starting from different initial condition or generated by a different-degree Chebyshev map, is assigned to a different user. The sequences designed in (9.66) are used for simulation of asynchronous CDMA systems considered in the next section.

9.10 Performance Comparisons of CDMA Systems

First, we simulate an A-CDMA communication system using optimal and white spreading sequences generated by Gaussian, uniform, Chebyshev, and binary random-number generators. The simulation result in Fig. 9.10 show that the error probability is independent of the distribution of the spreading sequences, and that the optimal sequences are better than random white sequences, which justifies our design. Particularly, the performances using optimal Chebyshev sequences and Gaussian sequences are similar,

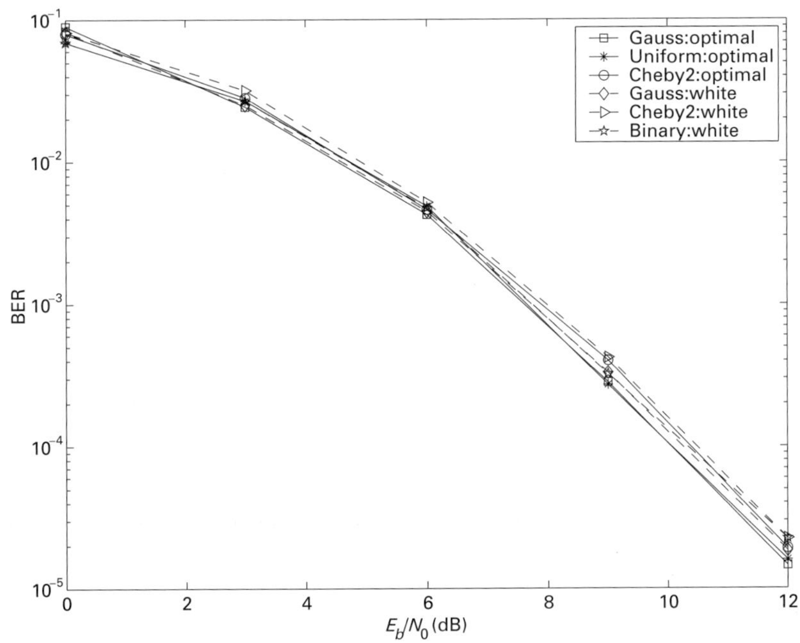

Figure 9.10 Comparison of error probabilities of an asynchronous CDMA system in AWGN channel using optimal and white sequences generated by Gaussian, uniform, second-degree Chebyshev, and binary random-number generators for sequence length of 31 with $K = 3$

which is consistent with our design of chaotic spreading sequences using ergodic theory. Moreover, the performance difference between optimal and random white sequences becomes more distinct when the number of users becomes larger.

Similar results can be obtained for $K = 5, 7, 10$. In order to understand the behavior of the sequences, we also performed simulations for different numbers of users, as shown in Fig. 9.11. These simulation results show that the optimal sequences are better than random white sequences by about 15% in terms of allowable number of users, which is consistent with the analytical expression of (9.45). We also observe that when the number of users is smaller, simulation results do not quite match with analytical results obtained under the SGA condition. This confirms the well-known fact that the Gaussian approximation is not valid when the user number is small.

As compared to PWAM maps, the tent map and Bernoulli shift map have finite-precision computational difficulty because each iteration of these two maps will shift out one bit of the current value. Unlike PWAM maps, Chebyshev polynomials considered here are expected to be more robust against the finite precision problem. We evaluate the performance loss of A-CDMA systems when a second-degree Chebyshev polynomial

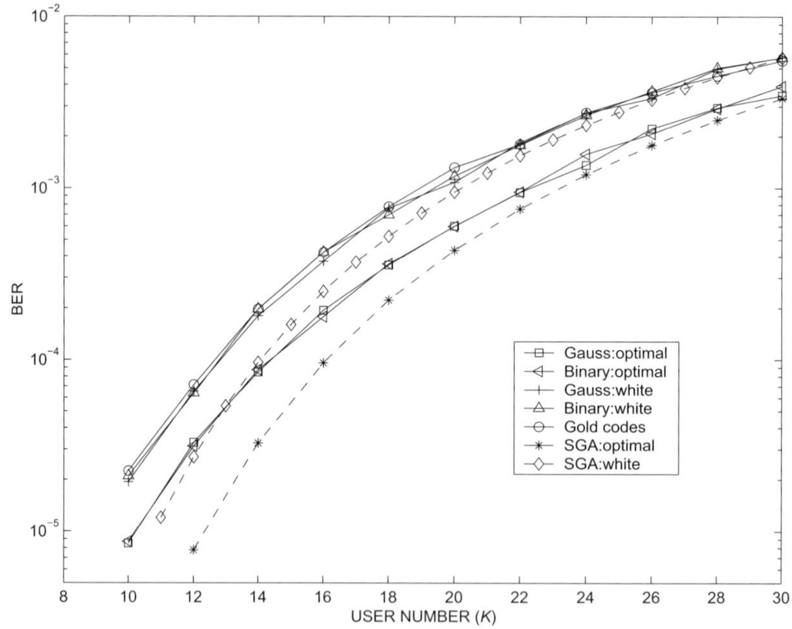

Figure 9.11 Comparison of error probabilities of an asynchronous CDMA system in AWGN channel using optimal and white sequences generated by Gaussian and binary random-number generators and Gold codes for sequence length of 63 and for different number of users (channel SNR = 25 dB)

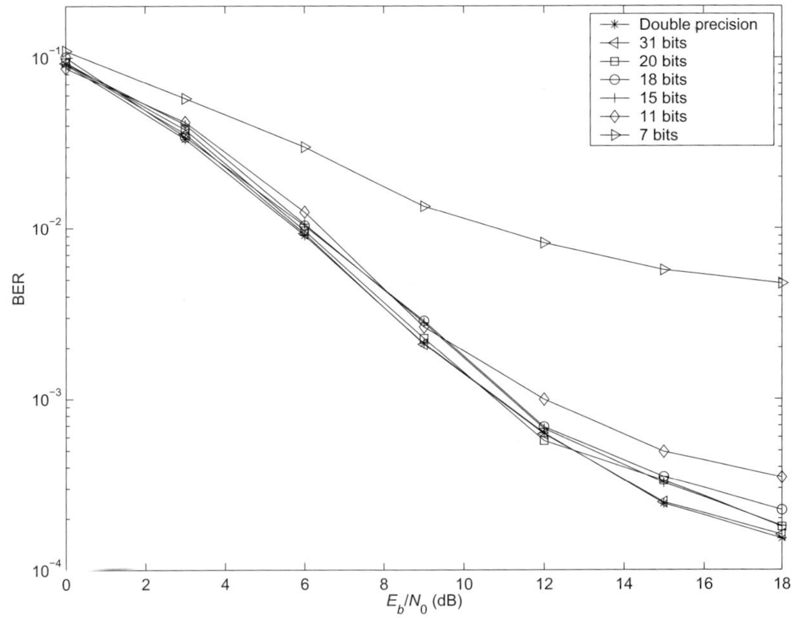

Figure 9.12 Performance of asynchronous CDMA system in AWGN channel using various finite-precision optimal sequences generated by the second-degree Chebyshev polynomial map with $N = 31$ and $K = 7$

generator using finite-precision computation is employed. The simulation results are shown in Fig. 9.12. We observe from these simulation results that an A-CDMA system has only slight performance loss by using more than 15–20 bits for sequence generation compared to double precision (52 bits).

9.11 Construction of Optimal Spreading Sequences from Gold Codes

As is well known, Gold codes of length $N = 2^n - 1$ are a family of optimal binary sequences that attain the Sidelnikov bound on the maximum $\theta_{\max}$ of periodic non-zerolag autocorrelation peak θ_a and cross-correlation peak θ_c for any set of N or more binary sequences of period N when n is odd. When n is even, the $\theta_{\max}$ for Gold codes is larger than the Sidelnikov bound by a factor of approximately $\sqrt{2}$ [18]. In other words, Gold codes mimic purely random binary sequences, excluding some "bad" sequences, and are expected to have similar performance parameters as purely random binary sequences when employed in an asynchronous CDMA system.

Neglecting the small values of autocorrelation and cross-correlation on a set of Gold codes, we can design the optimal spreading sequences by passing these Gold codes through an infinite impulse response (IIR) low-pass filter with a single pole at $(-r)$ defined in the previous section. That is,

$$\begin{aligned} y_0 &= x_0, \\ y_{j+1} &= -ry_j + \sqrt{2(1-r^2)}\,x_{j+1}, \end{aligned} \tag{9.67}$$

where $r = 2 - \sqrt{3}$ and $\{x_j\}_{j=0}^{N-1} \in \{-1, 1\}$ is a Gold code. The output sequence $\{y_0, y_1, \ldots, y_{N-1}\}$ of the filter will have an nearly exponential autocorrelation function $(-r)^l$.

First, we simulate an asynchronous CDMA communication system for the spreading factors $N = 31$ using Gold codes and purely random binary sequences. Then we perform the simulation of an asynchronous CDMA system using the proposed optimal spreading sequences. These simulation results show that the proposed optimal spreading sequences are better than Gold codes and purely random binary sequences by about 15% in terms of allowable number of users, which is consistent with the analytical expression. We also observe that when the number of users is smaller, simulation results do not quite match with analytical results obtained under the SGA condition. This confirms the well-known fact that the Gaussian approximation is not valid when the user number is small. These simulation results are omitted here but can be found in [19].

9.12 Conclusions on Chaotic CDMA System

We proposed a new design methodology for optimal spread-spectrum sequences for asynchronous and chip-synchronous CDMA systems with respect to minimum error probability under the SGA condition. Without any assumption on spreading sequences, the optimal partial autocorrelation function of the spreading sequences is derived. Moreover, based on the ergodic theory of dynamical systems, a simple method to construct such sequences using Chebyshev polynomials, as well as analytical performance expression are provided. Using the ergodic theory of dynamical systems, a method to construct and to analyze such sequences based on ergodic transformations is shown. Our method generalizes some previous approaches proposed in [8], [14], [20]. Under the SGA condition, an asynchronous CDMA system using the optimal spreading sequences allows 15% more users than when random white sequences are employed, and 73% more users than when chip-synchronous systems are employed. Simulation results also show that system performances using this family of chaotic Chebyshev spreading sequences are superior and similar to Gold codes when employed in asynchronous and chip-synchronous CDMA systems, respectively. Moreover, the Chebyshev polynomial map generator is robust against the finite-precision computational problem in terms of asynchronous CDMA system performance, as shown in the simulation results. The proposed optimal spreading sequences still perform better than Gold codes in an asynchronous CDMA system over frequency non-selective fading channels [26], [9], [19]. Then some details on the generation of these optimum chaotic CDMA codes upon transformation of Gold codes were given. The acquisition time of these optimum chaotic CDMA codes were also shown to be competitive to previously known CDMA codes. Extensive simulations were given to verify the performance of these chaotic CDMA codes (details are omitted here).

9.13 Super-Efficient Chaos-Based Monte Carlo Simulation

Ulam and von Neumann [21] first formulated the Monte Carlo (MC) simulation methodology as one using random sequences to evaluate high-dimensional integrals. Since then, MC simulation methods have been widely used to solve complex engineering and scientific problems. Unlike other deterministic methods, MC methods use statistical sampling to produce approximate solutions. A fundamental question of implementing MC simulation is how to generate random samples. It turned out that the generation of truly random sequences in a controlled manner is a non-trivial problem. Fortunately, in many applications it suffices to use pseudo-random (PR) sequences. A PR sequence can be generated deterministically by some transformations, and it appears to be random from the statistical point of view. As the processed sample size N grows, the uncertainty of the MC solution is reduced. It is well known that the variance of the approximation error decreases as $1/N$. However, for computationally intensive simulations, MC methods may take an extremely large number of samples to obtain a solution with acceptable tolerance. In this chapter, we describe the novel super-efficient (SE) Monte Carlo simulation method, originated by Umeno, which produces a solution whose variance of the approximation error decreases as fast as $1/N^2$. In order to achieve this SE result, one needs to consider the generation of the PR sequences using a particular class of chaotic dynamical systems. The greatest distinction between conventional and chaotic MC simulation is that the chaotic sequence has correlation between samples. For conventional MC simulation, good PR number generators produce near independent and identically distributed (iid) samples. Correlation between samples is generally considered to be a bad thing, because it may decrease the convergence rate of the simulation. However, if we select the chaotic mapping carefully, the correlation between samples may actually improve the convergence rate for certain integrands. In this section, we show how an appropriate choice of the correlation affects the variance of the approximation error. Let N denote the number of samples and denote the variance of the approximation error of the chaotic MC simulation. It turns out that the convergence rate has two contributors, one decaying as $1/N$ and the other as $1/N^2$. Eventually, the convergence rate will be dominated by $1/N$, which suggests that the chaotic MC simulation has the same performance as the standard MC simulation. However, if the dynamical system introduces the right amount of negative correlation such that the $1/N$ part becomes zero, the convergence rate becomes $1/N^2$, which is a huge improvement over the conventional MC simulation. Based on the Lebesgue spectrum representation of the integrand, a necessary and sufficient condition (NSC) for SE is given. Various examples based on the use of the Chebyshev polynomials are given. When the integrand does not meet the NSC, we describe an approximation SE (ASE) Monte Carlo simulation method that is applicable to a wider class of problems than the original SE method, and yields a convergence rate as fast as $1/N^\alpha$ for $1 \le \alpha \le 2$. The SE and ASE methods can also be generalized to two- and higher-dimensional MC simulation problems.

9.13.1 Introduction to MC Simulation

Ulam and von Neumann first formulated the Monte Carlo (MC) simulation methodology as one using random sequences to evaluate high-dimensional integrals. Since then, MC simulations have been used in many applications to evaluate the performance of various systems that are not analytically tractable. The simplest, yet most important, form of MC simulation is used to approximate the integral $I = \int_\Omega A(x)dx$, where the integrand $A(x)$ is defined on the domain $\Omega = [a,b]$ for some real number $a < b$. To do this, we first choose a probability density function (pdf) $\rho(x) \neq 0$ in Ω, and define the function $B(x) = A(x)/\rho(x)$. Then the above integral I is approximated by calculating the N-sample average of $(1/N)\sum_{i=1}^{N} B(X_i) \approx \mathbf{E}[B(X_j)] = I$, $j = 1,2,\ldots,N$, where N is the sample size, X_i are the independent identically distributed (iid) random samples whose common pdf is $\rho(x)$, and $\mathbf{E}[\cdot]$ denotes the expectation operator with respect to $\rho(x)$. By the strong law of large numbers, the above summation converges almost surely to I if the random samples are independent. Furthermore, the variance of the approximation decreases at rate $1/N$, given by $\mathrm{var}\left[(1/N)\sum_{i=1}^{N} B(X_i)\right] = (1/N)\mathrm{var}[B(X_j)]$, $j = 1,2,\ldots,N$. Note that the above expression holds regardless of the dimension of the domain Ω of the integrand $A(x)$, which makes Monte Carlo simulation suitable for performing multi-dimensional integrations.

Umeno's Super-Efficient Monte Carlo (SEMC) algorithm [20] is a variation of an MC based on a certain class of chaotic sequences, and exhibits a superior rate of convergence. Umeno and Yao's approximate SEMC [22] removed some restrictions from the original method to make the concept of super-efficiency applicable to more general situations. In the following sections, we review the pseudo-random number generation used in conventional MC simulation, and describe the concept of chaotic sequences and chaotic MC simulation. The correlation between samples of the chaotic sequence gives rise to the super-efficient convergence rate, which makes chaotic MC simulation super-efficient. We illustrate how to generate chaotic sequences from the practical point of view, and how to apply super-efficient simulation methods to a wide class of integrands using the notion of approximate SEMC.

9.14 Pseudo-Random Number and Chaotic Sequence

A fundamental question of implementing Monte Carlo simulation is how to generate random samples. It turns out that the generation of truly random sequences in a controlled manner is a non-trivial problem. Fortunately, in many applications it suffices to use pseudo-random (PR) sequences. A PR sequence is generated deterministically by some transformations, and it appears to be random from the statistical point of view. For example, the sequence of linear congruential PR numbers (LCPRN) $\{x_0, x_1, \ldots,\}$ is produced by the recursion $x_{n+1} = (ax_n + c) \mod m$, where $0 < a < m$, $0 \leq c < m$, and $0 \leq x_0 < m$ is the seed of the sequence [1]. When the parameters a, c, m, and x_0 are properly selected, the linear congruential recursion can produce a sequence of period m. LCG is one of the oldest and popular algorithms for generating PR sequences due to its simplicity and well-understood properties. Although an LCPRN sequence passes many randomness tests, LCPRN has some serious defects. Most notably, it exhibits correlation between successive samples. The Mersenne twister algorithm is a better choice for generating high-quality PR numbers for reducing this correlation. For example, Matlab uses the Mersenne twister algorithm as the default uniform random number generator starting from its Version 7.4 in 2007.

We can think of the process of generating the PR sequence as applying a deterministic transformation on some state variable repeatedly. More precisely, let Ω denote the collection of all possible states of the PR generator and $T : \Omega \to \Omega$ denote the transformation. We select a seed or initial state $x_0 \in \Omega$ and generate the sequence $\{x_0, x_1, \ldots\}$ by $x_{i+1} = T(x_i)$, $i = 0, 1, \ldots$ The output of the PR generator can be written as $y_i = g(x_i)$ for some suitable output function g.

Another way of generating PR sequences is through *dynamical systems*. Formally, a measure-preserving dynamical system is the quadruple $(\Omega, \mathcal{A}, \mu, T)$, where Ω is the state space, $\mathcal{A}$ is the σ-algebra on Ω, μ is a probability measure on $\mathcal{A}$, and T is a mapping from Ω to itself such that $\mu(T^{-1}(E)) = \mu(E)$ for all measurable $E \in \mathcal{A}$. A mapping T that satisfies the above expression is called a measure-preserving transformation. The initial state x_0 of a dynamical system at time 0 is a point in the domain Ω, and the evolution of the state is governed by a mapping T such that $x_{i+1} = T(x_i)$, $i = 0, 1, \ldots$ The sequence $\{x_0, x_1, \ldots\}$ with the seed x_0 is called the orbit of the dynamical system under T. The "time average" of the integrand $B(x)$ with respect to the orbit is defined as $\langle B(x_i)\rangle_N := (1/N) \sum_{i=1}^{N} B(x_i)$. A natural question to ask is whether the time average will converge or not as $N \to \infty$. More importantly, will it converge to the ensemble average I? The Birkhoff theorem [7] says that the time average of an integrable function $B(x)$ will converge to an integrable function $\bar{B}(x)$ almost surely, and $\mathbf{E}[B(X)] = \mathbf{E}[\bar{B}(X)]$, where the expectation is taken with respect to the measure μ. In general, $\bar{B}(x)$ is a function of the initial seed x_0. If a measure-preserving dynamical system has the property that every integrable function $B(x)$ has a constant time average, then it is called an *ergodic dynamical system*. By the Birkhoff theorem, this constant must agree with the ensemble average I. That is, $\langle B(x_i)\rangle_N \to \mathbf{E}[B(X)]$ pointwise as $N \to \infty$. In this section, we will focus on a special type of ergodic system, which has "chaotic" behavior in the sense of Lasota and Mackey [13] that (1) it has a dense orbit in the space, and (2) the orbits are unstable, meaning that orbits arising from different x_0, even if arbitrarily close to each other, grow apart exponentially.

For example, the *doubling map*

$$T_d(x) = \begin{cases} 2x & \text{if } 0 \le x < 0.5, \\ 2x - 1 & \text{if } 0.5 \le x < 1, \end{cases}$$

defined on $\Omega = [0, 1)$ is known to be chaotic. The invariant pdf $\rho(x)$ is the uniform distribution on $0 \le x \le 1$, and 0 elsewhere. The doubling map is related to many other chaotic dynamical systems, such as the Chebyshev dynamical system. The Chebyshev dynamical system of order p is defined on the domain $\Omega = [-1, 1]$ with the mapping $T_p(y) = \cos(p \arccos(y))$, where p is a positive integer. The mapping $T_p(y)$ is in fact the pth order Chebyshev polynomial of the first kind. The Chebyshev dynamical system of order 2 is related to the doubling map via the relation $y_i = \cos(2\pi x_i)$, where $x_i = T_d(x_{i-1})$.

9.15 Chaotic Monte Carlo Simulation

A chaotic MC simulation is a MC simulation with a PR sequence replaced by a chaotic sequence [2]. More specifically, let T be a chaotic mapping, and $\rho(x)$ its invariant pdf. We first draw a seed x_0 from the invariant pdf $\rho(x)$, and use the chaotic mapping T to generate the sequence $\{x_0, x_1, \dots\}$ by $x_{i+1} = T(x_i)$, $i = 0, 1, \dots$ where N is the sample size. The "time-average"$\langle B(x_i)\rangle_N = (1/N) \sum_{i=1}^{N} B(x_i) \to I$ will converge to the integral I as N approaches infinity [7].

9.15.1 Statistical and Dynamical Correlation

The greatest distinction between conventional and chaotic MC simulation is that the chaotic sequence has correlation between samples. For conventional MC simulation, good PR number generators produce near iid samples. Correlation between samples is generally considered to be a bad thing as it may decrease the convergence rate of the simulation. However, if we select the chaotic mapping carefully, the correlation between samples may actually improve the convergence rate for certain integrands. In the following, we show how the correlation can affect the variance of the approximation error.

For measure-preserving dynamical systems, any measurable function $B(x)$ on Ω forms a stationary random process, $\{B(x_k)\}_{k \in N}$, where $B(x_k) = B(T^k(x_0))$. For simplicity, denote $B(x_k)$ by B_k and $\langle B(x_i)\rangle_N$ by $\langle B\rangle_N$. Define the autocorrelation function $R(k) = \mathbf{E}(B_{k+i} - I)(B_i - I)$, where the expectation is taken with respect to the invariant pdf $\rho(x)$, and $i = 1, 2, \ldots$ is arbitrary. The variance of the approximation error $\langle B \rangle_N - I$ is given by

$$\sigma_N^2 := \mathbf{E}\left[\left(\langle B \rangle_N - I\right)^2\right] = \frac{1}{N}\text{var}[B] + \frac{2}{N^2}\sum_{k=1}^{N}(N-k)R(k).$$

The first term on the right-hand side is called the *statistical correlation*, which depends on the integrand $B(x)$ and the pdf $\rho(x)$. The second term is called the *dynamical correlation*, which depends on the integrand as well as on the chaotic sequence [2]. Clearly, for iid random samples $\{x_0, x_1, \ldots\}$ we have $R(k) = 0$, and hence the above expression reduces to the conventional case, where the convergence rate is $1/N$. If there are positive correlations between samples, the variance of the approximation error σ_N^2 will increase. On the other hand, negative correlations between samples might decrease σ_N^2. It is therefore natural to ask what is the best achievable convergence rate of chaotic MC simulation. This leads to the notion of the super-efficiency of the chaotic MC simulation considered in the next subsection.

9.15.2 Super-Efficient Chaotic MC Simulation

We can rewrite the variance of the approximation error as

$$\sigma_N^2 = \underbrace{\frac{1}{N}\left(\text{var}[B] + 2\sum_{k=1}^{N} R(k)\right)}_{\eta} - \frac{2}{N^2}\sum_{k=1}^{N} kR(k).$$

This shows that the convergence rate of σ_N^2 has two contributors, one decaying as $1/N$ and the other as $1/N^2$. Eventually, the convergence rate will be dominated by $1/N$, which suggests that the chaotic MC simulation has the same asymptotic performance as the standard MC simulation.

However, if the dynamical system introduces the right amount of negative correlation such that $\eta = 0$, the convergence rate becomes $1/N^2$, which is a huge improvement over the conventional MC simulation. To obtain $\eta = 0$, and hence the convergence rate $1/N^2$, one should properly combine the sequence correlation with the integrand [20]. We say the chaotic MC simulation is *super-efficient* (SE) if the variance of the approximation error decays as $1/N^2$, for $N \to \infty$. Umeno [20] also showed that the condition $\eta = 0$ for super-efficiency is necessary as well as sufficient.

Example 9.1 Consider the integrand $A(x) = -8x^4 + 8x^2 + x - 1/\pi\sqrt{1-x^2}$ defined on p. 1447 of [20], which satisfies the SE condition under the Chebyshev dynamical system of order $p = 2$ and $p = 4$. Fig. 9.13 shows the results of applying the chaotic MC simulation to find the integral of $A(x)$ using Chebyshev chaotic mappings, and compares their convergence rates with the conventional MC simulation using uniform PR samples. The numerical results verify that the chaotic MC simulation is super-efficient when $p = 2$ and $p = 4$. On the other hand, both conventional and chaotic MC simulation with $p = 5$ have convergence rate $1/N$.

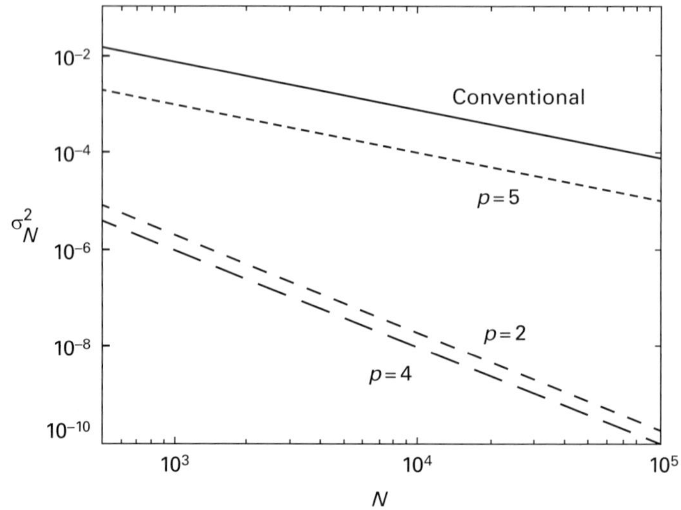

Figure 9.13 Variance σ_N^2 vs sample size N for conventional and Chebyshev mappings of order $p = 2, 4, 5$

9.15.3 Condition for Super-Efficiency

The super-efficiency condition of $\eta = 0$ does not explicitly suggest any way to achieve it. Umeno [20] first gave a characterization of super-efficiency in terms of the coefficients of the generalized Fourier series of the modified integrand for Chebyshev and piecewise linear dynamical systems, but the results in [20] did not make clear whether those conclusions are also applicable to other dynamical systems. Yao [9] established the connection between super-efficiency and the Lebesgue spectrum of ergodic theory [7], which puts the super-efficient condition derived in Umeno's work in a general framework, and generalizes Umeno's result to a wide range of dynamical systems, namely those with a Lebesgue spectrum. This observation helps us explain the super-efficiency systematically and hopefully leads to practical algorithms to achieve $\eta = 0$. Now, we briefly introduce the concept of the Lebesgue spectrum and the important characterization of super-efficiency in terms of the Lebesgue spectrum.

Definition. Let Λ be an index set and $N_0 = \{0, 1, 2, \ldots\}$. A dynamical system with mapping T is said to have countable one-sided Lebesgue spectrum if there exists an orthogonal basis containing the constant function 1, and the collection of functions $\{f_{\lambda,j}(x)|\lambda \in \Lambda, j \in N_0\}$ satisfy $f_{\lambda,j}(T(x)) = f_{\lambda,j+1}(x)$ for all λ and j, where the index λ labels the classes and j labels the functions within each class. The Koopman operator induced by the transformation T is defined as $U_T f(x) := f(T(x))$. It is an isometry and the orthogonal basis function can be rewritten as $U_T f_{\lambda,j} = f_{\lambda,j+1}$, which means U_T has invariant subspaces $W_\lambda = \text{span}(f_{\lambda,0}, f_{\lambda,1}, \ldots)$ generated by $f'_{\lambda,0}$s. Therefore, the "least element" of the invariant subspace W_λ is the generating vector $f_{\lambda,0}$. Note that the dynamical system has a one-sided Lebesgue spectrum if and only if it is exact (see e.g. [7] and [17]). If a dynamical system has a Lebesgue spectrum, then it is strongly mixing [13] and hence is chaotic in the sense of Lasota and Mackey [13]. All the dynamical systems we consider have Lebesgue spectrum. Since $\{1\} \cup \{f_{\lambda,j}(x) | \lambda \in \Lambda, j \in {}_0\}$ forms a complete orthogonal basis on the square-integrable functions $L^2(\Omega)$, the generalized Fourier series expansion of an integrand $B(x)$ can be written as $B(x) = b_{0,0} + \sum_{\lambda\in\Lambda} \sum_{j\in_0} b_{\lambda,j} f_{\lambda,j}(x)$, where $b_{0,0}$ is the coefficient corresponding to the constant function 1, which is just the integral I of $B(x)$.

THEOREM 9.2 *Consider a dynamical system which has a Lebesgue spectrum $\{f_{\lambda,j}(x) | \lambda \in \Lambda, j \in {}_0\}$ indexed by the sets Λ. The associated chaotic MC simulation is super-efficient if and only if $d_\lambda := \sum_{j=0}^{\infty} b_{\lambda,j} = 0$ for all $\lambda \in \Lambda$, where $B(x) = b_{0,0} + \sum_{\lambda\in\Lambda} \sum_{j\in_0} b_{\lambda,j} f_{\lambda,j}(x)$ is the generalized Fourier series of $B(x) = A(X)/\rho(x)$.*

Proof: The autocorrelation function of the chaotic sequence can be written as

$$\begin{aligned} R(n) &= \mathbf{E}\left[\sum_{\lambda\in\Lambda} \sum_{j=0}^{\infty} b_{\lambda,j} f_{\lambda,j}(T^n x) \sum_{\nu\in\Lambda} \sum_{i=0}^{\infty} b_{\lambda,i} f_{\lambda,i}(x)\right] \\ &= \sum_{\nu,\lambda\in\Lambda} \sum_{i,j=0}^{\infty} b_{\lambda,j} b_{\lambda,i} \mathbf{E}\left[f_{\lambda,j+n}(x) f_{\lambda,i}(x)\right] \\ &= \sum_{\lambda\in\Lambda} \sum_{j=0}^{\infty} b_{\lambda,j} b_{\lambda,j+n}. \end{aligned}$$

Then η can be expressed as

$$\eta = \sum_{\lambda\in\Lambda} \left(\sum_{j=0}^{\infty} b_{\lambda,j}^2 + 2 \sum_{k=1}^{N} \sum_{j=0}^{\infty} b_{\lambda,j} b_{\lambda,j+k} \right) = \sum_{\lambda\in\Lambda} \left(\sum_{j=0}^{\infty} b_{\lambda,j}^2 + 2 \sum_{j=0}^{\infty} \sum_{i=j+1}^{j+N} b_{\lambda,j} b_{\lambda,i} \right).$$

As N goes to infinity,

$$\eta = \sum_{\lambda\in\Lambda} \left(\sum_{j=0}^{\infty} b_{\lambda,j}^2 + 2 \sum_{i>j}^{\infty} b_{\lambda,j} b_{\lambda,i} \right) = \sum_{\lambda\in\Lambda} \left(\sum_{j=0}^{\infty} b_{\lambda,j} \right)^2.$$

Therefore, $\eta = 0$ if and only if $\sum_{j=0}^{\infty} b_{\lambda,j} = 0$ for each $\lambda \in \Lambda$.

Thus, the explicit condition for super-efficiency is that the sum of coefficients in each class λ be zero. We say that an integrand $A(x)$ is *super-efficient* (under the dynamical system with mapping T and invariant pdf ρ) if $d_\lambda = 0$ for all $\lambda \in \Lambda$ holds for all $B(x) = A(x)/\rho(x)$.

Example 9.2 Consider a variant of the integrand $A(x)$ in Example 9.1 given by

$$A(x) = \frac{-8x^4 + 8x^2 + (1+)x - 1}{\pi\sqrt{1-x^2}} = B_\epsilon(x)\rho(x).$$

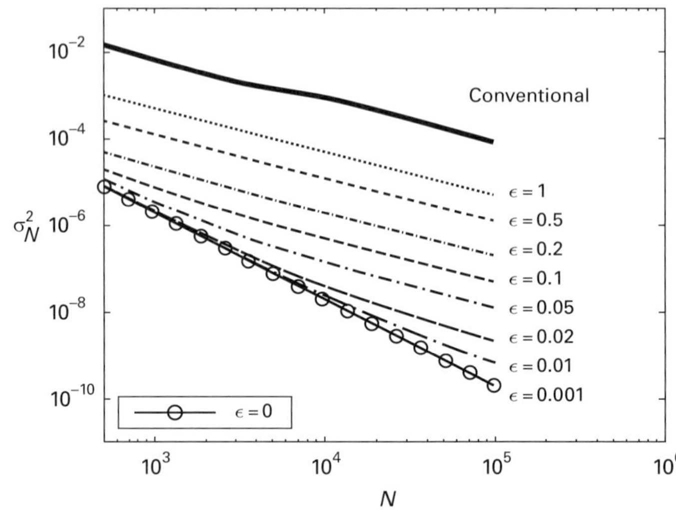

Figure 9.14 The variance of the approximation error σ_N^2 versus the number of samples N

Under the Chebyshev dynamical system, $B_\epsilon(x)$ can be expanded as $B_\epsilon(x) = (1 + \epsilon)T_1(x) - T_4(x)$. If $p = 2$, the coefficients of the generalized Fourier series are $B_\epsilon(x) = (1 + \epsilon)T_1(x) - T_4(x)$, and zero otherwise. The sum of coefficients are $d_1 = \epsilon$ and $d_\epsilon = 0$ for $\epsilon \neq 0$. Therefore, $A(x)$ is super-efficient if and only if $\epsilon = 0$. When $\epsilon \neq 0$, we have a "mismatched" SE MC simulation, which appears to be super-efficient for small N but gradually loses super-efficiency as N increases, as shown in Fig. 9.14. The slope of the conventional MC simulation curve is -1, indicating its $1/N$ behavior. Otherwise, the slope of the super-efficient MC simulation is -2 because σ_N^2 decays like $1/N^2$. Between these two extremes are the mismatched SE MC simulations with different sizes of ϵ. For $\epsilon = 0.001$, the curve is almost identical to the super-efficient curve. As ϵ becomes larger, the slope of the mismatched SE MC simulations gradually increases as N becomes larger. It is now clear why the integrand in Example 9.15 under the chaotic mappings T_2 and T_4 (but not T_5) is super-efficient. When $p = 2$ and $p = 4$, the sum of coefficients in the class $\lambda = 1$ is $d_1 = 1 - 1 = 0$, and all other λ are zero. Hence, the chaotic MC simulations under both chaotic mappings are super-efficient. On the other hand, when the chaotic MC simulation has the same convergence rate as the conventional MC simulation, because it does not satisfy the super-efficiency condition.

9.15.4 Multi-Dimensional Dynamical Systems

Note that the characterization of super-efficiency $d_\lambda = 0$ holds regardless of the dimension of the domain Λ as long as the system has a Lebesgue spectrum. Nevertheless, high-dimensional dynamical systems may arise naturally through the product of multiple one-dimensional dynamical systems. In this section, we show that the

Lebesgue spectrum of the product dynamical system has a special structure, and we derive the corresponding necessary and sufficient condition for super-efficiency. For simplicity, we consider two one-dimensional dynamical systems $(\Omega_1, \mathcal{A}_1, \mu_1, T_1)$ and $(\Omega_2, \mathcal{A}_2, \mu_2, T_2)$. The product dynamical system $(\Omega, \mathcal{A}, \mu, T)$ is defined as the product probability space $(\Omega_1 \times \Omega_2, \mathcal{A}_1 \otimes \mathcal{A}_2, \mu_1 \otimes \mu_2)$ with the mapping $T(x, y) = T_1(x)T_2(y)$. It is not difficult to show that the product space is also a measure-preserving dynamical system [17]. Suppose both the dynamical systems $(\Omega_1, \mathcal{A}_1, \mu_1, T_1)$ and $(\Omega_2, \mathcal{A}_2, \mu_2, T_2)$ have a Lebesgue spectrum with a basis function $\mathcal{B}_1 = \{1\} \cup \{f^{(1)}_{\lambda_1, j_1}(x) | \lambda_1 \in \Lambda_1, j_1 \in N_0\}$ and $\mathcal{B}_2 = \{1\} \cup \{f^{(2)}_{\lambda_2, j_2}(y) | \lambda_2 \in \Lambda_2, j_2 \in N_0\}$, respectively. The complete orthogonal basis on $(\Omega_1 \times \Omega_2, \mathcal{A}_1 \otimes \mathcal{A}_2, \mu_1 \otimes \mu_2)$ is $\mathcal{B} = \mathcal{B}_1 \times \mathcal{B}_2$, which can be written explicitly as

$$\{1\} \cup \{f^{(1)}_{\lambda_1, j_1}(x)\} \cup \{f^{(2)}_{\lambda_2, j_2}(y)\} \cup \{f^{(1)}_{\lambda_1, j_1}(x) f^{(2)}_{\lambda_2, j_2}(y)\} \quad \text{for } \lambda_1 \in \Lambda_1, \lambda_2 \in \Lambda_2, j_1, j_2 \in N_0.$$

Because of the function taking the value of 1 the above expression becomes very messy. It gets even more cumbersome for higher-dimensional spaces. For notational convenience, we define the redundant functions $f_{0,j}(x) := 1 \quad \text{for all } j - 0, 1, 2, \ldots$ In this way, the constant function 1 can be indexed by $(0, j)$ for any non-negative j. To make sense of this definition, we require that $b_{0,j} = 0$ for all $j > 0$ and $b_{0,0} = I$ is the integral of $B(X)$.

Clearly, the action of U_T on the basis function yields

$$U_T f^{(1)}_{\lambda_1, j_1}(x) f^{(2)}_{\lambda_2, j_2}(y) = f^{(1)}_{\lambda_1, j_1}(T_1(x)) f^{(2)}_{\lambda_2, j_2}(T_2(y)) = f^{(1)}_{\lambda_1, j_1+1}(x) f^{(2)}_{\lambda_2, j_2+1}(y)$$

for all $\lambda_1 \in \Lambda_1$, $\lambda_2 \in \Lambda_2$, and $j_1, j_2 \in N_0$. That is, the index of the basis function changes from $(\lambda_1, \lambda_2, j_1, j_2)$ to $(\lambda_1, \lambda_2, j_1 + 1, j_2 + 1)$ after applying U_T to the basis function. The least element in each invariant subspace associated with U_T has the form $f^{(1)}_{\lambda_1, j_1}(x)\, f^{(2)}_{\lambda_2, j_2}(y)$, $\quad j_1, j_2 \in N_0, j_1 = 0, j_2 = 0$. That is, at least one of the index j_1 and j_2 of the least element must be zero so that no other function can precede it under U_T.

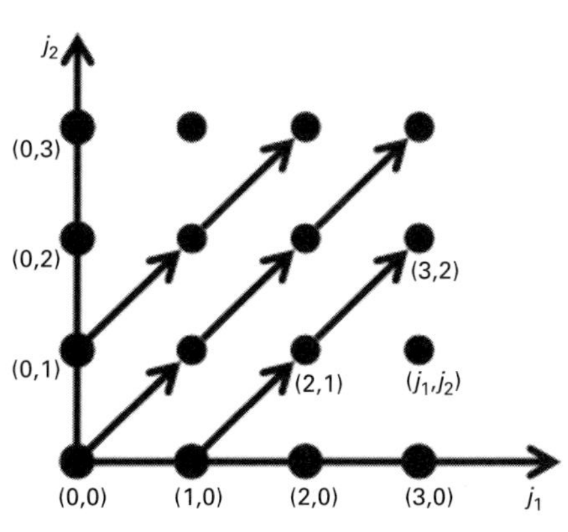

Figure 9.15 An illustration of the Lebesgue spectrum for 2D systems

Fig. 9.15 illustrates a Lebesgue spectrum for 2D systems. Given λ_1 and λ_2, each circle represents a basis function indexed by $(\lambda_1, \lambda_2, j_1, j_2)$. Applying T to the basis function increases both of j_1 and j_2 by 1. The least elements are on the boundary. They generate the invariant subspace. Therefore, we define the index set $\Lambda = \Lambda_1 \times \Lambda_2$ and $J = \{(j_1, j_2) \in N_0 : j_1, j_2 \in N_0, j_1 = 0, j_2 = 0\}$ to index the least elements in each of the invariant subspaces associated with U_T. The generalized Fourier expansion of an integrable function $B(x, y)$ is given by $B(x, y) = \sum_{\lambda\in\Lambda} \sum_{\mathbf{j}\in J} \sum_{k=0}^{\infty} b_{\lambda,\mathbf{j}+k\mathbf{1}} f_{\lambda,\mathbf{j}+k\mathbf{1}}(x, y)$, where $\mathbf{1}$ denotes the vector with all unity components. Similarly, for d-dimensional space, we define $\Lambda = \prod_{i=1}^{d} \Lambda_i$ and $J = \{(j_1, j_2, \ldots, j_d) \in N_0^d : j_1 = j_2 = \cdots j_d = 0\}$. The necessary and sufficient condition for super-efficiency is given by $d_{\lambda},\mathbf{j} = \sum_{k=0}^{\infty} b_{\lambda},\mathbf{j}+ k\mathbf{1} = 0$ for each $\lambda \in \Lambda$ and $\mathbf{j} \in J$.

9.16 Conclusion

Section 9.1 provided some of the historical backgrounds of chaos in non-linear dynamical systems. Section 9.2 introduced the evolution toward bifurcations for the simple logistic and the Lorenz non-linear dynamical systems. Section 9.3 considered the state-space aspects of a chaotic orbit. Section 9.4 treated the chaotic behaviors from time-series data. Section 9.4.2 introduced the fractal dimensions of chaotic attractors. Section 9.4.3 introduced the pseudo-random sequences generated by non-linear chaotic generators. Section 9.5 introduced the use of chaotic sequences for CDMA communication purposes. Section 9.10 considered the performances of the chaotic CDMA systems. Section 9.13 introduced the use of a certain class of chaotic sequences for "super-efficient" Monte Carlo applications.

9.17 References

A non-technical "best-seller" for the lay-person on chaos appeared in [1]. E.N. Lorenz noticed "chaotic" behavior in his research on using non-linear differential equations to model meteorological phenomena [2]. Simple chaotic behavior was observed in the logistic non-linear dynamical equation by R.M. May [3]. Detailed aspects of bifurcations and chaos in various non-linear dynamical systems can be found in [23]. Fractal aspects of chaotic attractors were introduced by [4]. Invariant measures of chaotic maps and pseudo-random sequences generated by non-linear dynamical systems can be found in [23], [24], and [25]. Applications of non-linear chaotic dynamical systems for CDMA were considered by [7], [6], [5], and [14]. Minimum interference power properties of certain chaotic sequences in a CDMA setting were treated by [8] and [26]. Detailed analysis of performance of these "optimum" chaotic sequences for CDMA systems using ergodic theory were given in [15], [16], [17], [9], and [19]. Super-efficient chaos-based Monte Carlo simulation was introduced by [20] and detailed discussions were presented in [22]. Although not included in this chapter, important relationships between chaos and turbulence theory are known [27] as well as explicit testing of deterministic chaos from time-series [28].

[1] J. Gleick, *Chaos: The Amazing Science of the Unpredictable*, Random House, 1988.

[2] E.N. Lorenz, "Deterministic Nonperiodic Flow," *Jour. of Atmospheric Science*, 1963, pp. 130–141.

[3] R.M. May, "Simple Mathematical Models with Very Complicated Dynamics," *Nature*, 1976, vol. 261, pp. 45–67.

[4] B.B. Mandelbrot, *The Fractal Geometry of Nature*, Freeman, 1982.

[5] J. Schweizer and M. Hasler, "Multiple Access Communications Using Chaotic Signals," *Proc. ISCAS*, 1996, pp. 108–111.

[6] T. Kodha and A. Tsuneda, "Statistics of Chaotic Binary Sequences," *IEEE Trans. Information Theory*, Jan. 1997, vol. 43, pp. 104–112.

[7] A. Abel, A. Bauer, K. Kelber, and W. Schwarz, "Chaotic Codes for CDMA Applications," *Proc. ECCTD*, 1997, pp. 306–311.

[8] G. Mazzini, R. Rovatti, and G. Setti, "Interference Minimization by Auto-Correlation Shaping in Asynchronous DS-CDMA Systems: Chaos-Based Spreading Is Nearly Optimal," *Electronics Letters*, June 1999, vol. 35, pp. 1054–1055.

[9] C.C. Chen, K. Yao, K. Umeno, and E. Biglieri, "Design of Spread Spectrum Sequences Using Chaotic Dynamical Systems and Ergodic Theory," *IEEE Trans. on Circuits and Systems, I*, Sept. 2001, vol. 48, pp. 1110–1113.

[10] M.B. Pursley, "Performance Evaluation for Phased-Coded Spread-Spectrum Multiple-Access Communication – Part I: System Analysis," *IEEE Trans. Communications*, 1977, vol. 25, pp. 795–799.

[11] K. Yao, "Error Probability of Asynchronous Spread Spectrum Multiple Access Communication Systems," *IEEE Trans. Communications*, August 1977, vol. 25, pp. 803–809.

[12] M.B. Pursley and D.V. Sarwate, "Performance Evaluation for Phased-Coded Spread-Spectrum Multiple-Access Communication – Part II: Code Sequence Aanalysis," *IEEE Trans. Communications*, 1977, vol. 25, pp. 800–803.

[13] A. Lasota, and M.C. Mackey, *Chaos, Fractals, and Noise: Stochastic Aspects of Dynamics*, Springer-Verlag, 1994.

[14] K. Umeno and K.I. Kitayama, "Improvement of SNR with Chaotic Spreading Sequences for CDMA," *IEEE Information Theory Workshop*, South Africa, June, 1999.

[15] R.L. Adler, and T.J. Rivlin, "Ergodic and Mixing Properties of Chebyshev Polynomials," *Proc. American Math. Soc.*, 1964, vol. 15, pp. 794–796.

[16] V.I. Arnold and A. Avez, *Ergodic Problems of Classical Mechanics*, W.A. Benjamin, 1968.

[17] D.S. Broomhead, J.P. Huke, and M.R. Muldoon, "Codes for Spread Spectrum Applications Generated Using Chaotic Dynamical Systems," *Dynamics and Stability of Systems*, 1999, vol. 14, no. 1, pp. 95–105.

[18] D.V. Sarwate and M.B. Pursley, "Crosscorrelation Properties of Pseudorandom and Related Sequences," *Proceedings of the IEEE*, 1980, vol. 68, no. 5, pp. 593–619.

[19] C.C. Chen, K. Yao, and E. Biglieri, "Optimal Spread Spectrum Sequences – Constructed from Gold Codes, *Proceedings of the IEEE Globecom*, Nov. 2000, pp. 867–871.

[20] K. Umeno, "Chaotic Monte Carlo Computation: A Dynamical Effect of Random-Number Generations," *Japan Journal of Applied Physics*, 2000, vol. 39, part 1, no. 3A, pp. 1442–1456.

[21] S. Ulam, "On the Monte Carlo Methods," *Proc. 2nd Symposium on Large Scale Digital Calculating Machinery*, 1952, pp. 207–212.

[22] K. Yao and K. Umeno, "Superefficient Monte Carlo Simulation," in *Simulation Technologies in Networking and Communications*, ed. A.K. Parthan *et al.*, CRC, 2015, Chapter 3, pp. 69–92.
[23] R.C. Hilborn, *Chaos and Nonlinear Dynamics*, Oxford, 1994.
[24] F. Takens, *Dynamical Systems and Turbulence*, ed. D.A. Rand and L.S. Young, *Lecture Notes in Mathematics*, vol. 898, Springer-Verlag, 1981.
[25] P. Grassberger and I. Procaccia, "Characterization of Strange Attractors," *Physical Review Letters*, 1983, vol. 50, pp. 346–349.
[26] C.C. Chen, E. Biglieri, and K. Yao, "Design of Spread Spectrum Sequences Using Ergodic Theory," *International Symposium on Information Theory*, Sorrento, Italy, June 2000, p. 379.
[27] U. Frisch, *Turbulence – The Legacy of A.N. Kolmogorov*, Cambridge University Press, 1995.
[28] J.B. Gao and Z.M. Zheng, "Direct Dynamical Test for Deterministic Chaos and Optimal Embedding of a Chaotic Time Series," *Phys. Rev. E*, 1994, vol. 49, pp. 3807–3814.

9.18 Exercises

1. When you have time, read the general science book, *Chaos*: *The Amazing Science of the Unpredictable*, by J. Gleick, Random House, 1988.
2. Consider an estimator

$$\widehat{\Theta}\,(X_1, \ldots, X_n)$$

of the r.v. $\{X_1, X_2, \ldots, X_n\}$ with a joint pdf $p_n(x_1, \ldots, x_n)$ which depends on a deterministic scalar parameter θ. Based on these samples, we want to find the minimum variance of this estimator to the parameter θ.
3. Consider some computational investigation of the logistic map of (9.2). Show one obtains: period 2 solutions for $A_1 = 3 < A \leq A_2 = 3.449$ (i.e., show that in the steady state the sequence takes only two values); period 4 solutions for $A_2 < A \leq A_3 = 3.544$ (i.e., show that in the steady state the sequence takes only four values); period 8 solutions for $A_3 < A \leq A_4 = 3.564$ (i.e., show that in the steady state the sequence takes only eight values).
4. Consider the two-dimensional Heron map. It is the simplest 2D non-linear dynamical system with possible chaotic behavior.

$$x_{n+1} = 1 - ax_n^2 + y_n; \;\; y_{n+1} = bx_n, \;\; n = 0, 1, \ldots$$

   a. Take $a = 0.9$, $b = 0.3$ with $x_0 = 0.8$, $y_0 = 0.8$. Show for these parameters, the map has period-2 behavior.
   b. Take $a = 1.4$, $b = 0.3$ with $x_0 = 0.8$, $y_0 = 0.8$. Show for these parameters, the map is chaotic. Show the phase plot of y vs x over the region of $-0.4 \leq y \leq 0.4; \; -1.5 \leq x \leq 1.5$.

5. Consider the function $f(x) = 27rx^2(1-x)/16$, $0 \leq x \leq 1$, $0 \leq r \leq 4$. Find the two intersecting points of this function with the line of $y = x$. Take $r = 2.9$ and start at $x_0 = 0.3$; find its fixed point. Now, take $r = 2.9$ and start at $x_0 = 0.25$; find its fixed point.
6. Show that the invariant measure of the logistic map is given by $p(x) = 1/[\pi\sqrt{x(1-x)}]$, $0 \leq x \leq 1$.

10 Computational Linear Algebra

Basic concepts of linear algebra constitute some of the most fundamental and useful mathematical tools needed in the analysis and design of algorithms and processing in modern communication and radar systems. While linear algebra has been considered and used in all areas of pure and applied science and engineering, we are particularly interested in the modern computational and algorithmic aspects of linear algebra in this chapter. All the topics considered in estimation, filtering, approximation, spectral analysis, adaptive systems, Kalman filtering, and systolic array algorithmic and architectural formulations of these problems, all depend crucially on the basic computational linear algebra concepts considered here. In Section 10.1, we consider the direct method for the solution of a system of linear equations, while the Gaussian elimination procedure is given in Section 10.2. In many practical estimation and approximation problems, when the number of equations is larger than the number of unknowns, then the classical normal equation approximation approach can be used and is given in Section 10.3. However, the second-order effect (i.e., $A^T A$) of the data can lead to numerical instability for situations with large original condition numbers. Thus, various more modern techniques of decomposition (or factorization) have been considered to tackle this problem. In Section 10.4, triangular and Cholesky decompositions are presented. Section 10.5, considers the QR factorization method, which is known to be one of the numerically stable techniques. Specific QR factorization techniques based on Gram–Schmidt orthogonalization, modified Gram–Schmidt orthogonalization, Givens transformation, and Householder transformation are given in Sections 10.6–10.9, respectively. The use and advantages of the QR technique for the solution of a linear system of equations are described in Section 10.10. In Section 10.11, basic motivations and properties of singular value decomposition (SVD), which is often considered to be one of the most fundamental decompositions in linear algebra, are given. The use and advantages of the SVD technique for the solution of a linear system of equations are considered in Section 10.12. In practice, in order to fully utilize the advantages of the SVD approach for the solution of a linear system of equations, as well as for various estimation, filtering, identification, and approximation problems, we need to have operational methods to determine the matrix effective rank by computing the number of significant singular values of the matrix in the presence of disturbances. Various approaches used to address this issue are considered in Section 10.13.

10.1 Solution of a System of Linear Equations

In *direct methods* of solving

$$Ax = b,$$

where A is a known $n \times n$ matrix, x is an unknown $n \times 1$ vector, and b is a known $n \times 1$ vector, we consider explicit procedures with which after a finite number of iterations, the exact solution (disregarding rounding operation) is obtained. This is to be contrasted to *iterative methods*, which yield sequences of approximating solutions that converge to the exact solution as the number of iterations goes to infinity.

Consider a triangular system of equations

$$Ux = b,$$

where U is upper triangular (i.e., $u_{ij} = 0, i > j$) and has the form

$$\begin{aligned} u_{11}x_1 + \cdots + u_{1(n-1)}x_{n-1} + u_{1n}x_n &= b_1, \\ &\vdots \\ u_{(n-1)(n-1)}x_{n-1} + u_{(n-1)n}x_n &= b_{n-1}, \\ u_{nn}x_n &= b_n. \end{aligned} \tag{10.1}$$

Assume $u_{ii} \neq 0$ for all i. Then we can solve

$$\begin{aligned} x_n &= b_n/u_{nn}, \\ x_{n-1} &= \{b_{n-1} - u_{(n-1)n}x_n\}/u_{(n-1)(n-1)}, \\ \vdots &= \vdots \\ x_1 &= \{b_1 - u_{1n}x_n - \cdots - u_{12}x_2\}/u_{11}. \end{aligned} \tag{10.2}$$

The above algorithm for solving an upper triangular system of equations is called a *back substitution*, since we start by solving for the last variable x_n. The known x_n is used to solve for x_{n-1}. Then x_n and x_{n-1} are used to solve for x_{n-2}. Finally, $x_i, i = n, \ldots, 2$ are used to solve for x_1. A general expression for (10.2) is given by

$$x_i = \left\{ b_i - \sum_{j=i+1}^{n} u_{ij}x_j \right\} / u_{ii}, i = n, n-1, \ldots, 1. \tag{10.3}$$

For a lower triangular matrix L, with elements l_{ij}, where $l_{ij} = 0, i < j$, a triangular systems of equations is given by

$$Lx = b.$$

The *forward substitution* approach starts by solving for the first variable x_1 and then using it to solve for x_2. Finally, $x_i, i = 1, \ldots, n-1$ are used to solve for x_n. A general expression is then given by

$$x_i = \left\{ b_i - \sum_{j=1}^{i-1} l_{ij}x_j \right\} / l_{ii}, i = 1, \ldots, n. \tag{10.4}$$

In either (10.3) or (10.4), there are n divisions, and

$$\sum_{i=1}^{n}(i-1) = \frac{1}{2}n(n-1) \simeq \frac{1}{2}n^2 \tag{10.5}$$

additions and multiplications.

10.2 Gaussian Elimination Procedure

Consider the equations

$$\begin{aligned} 2x_1 + 4x_2 - 2x_3 &= 6, \\ x_1 - x_2 + 5x_3 &= 0, \\ 4x_1 + x_2 - 2x_3 &= 2. \end{aligned} \tag{10.6}$$

Multiply the first equation by $\frac{1}{2}$ and subtract from the second equation and multiply the first equation by 2 and subtract from the third equation which yields

$$\begin{aligned} 2x_1 + 4x_2 - 2x_3 &= 6, \\ -3x_2 + 6x_3 &= -3, \\ -7x_2 + 2x_3 &= -10. \end{aligned} \tag{10.7}$$

The coefficient 2 of x_1 in the first equation of (10.6) is called a *pivot* in the first stage of the elimination. Similarly, multiply the second equation by $7/3$ and subtract from the third equation to yield

$$\begin{aligned} 2x_1 + 4x_2 - 2x_3 &= 6, \\ -3x_2 + 6x_3 &= -3, \\ -12x_3 &= -3. \end{aligned} \tag{10.8}$$

The coefficient -3 of x_2 in the second equation of (10.7) is the pivot in the second stage of the elimination.

Equation (10.8) is an upper triangular system of equations of the form in Equation (10.1). Thus, the backward substitution of (10.1)–(10.3) can be used. Let the matrices of (10.6)–(10.8) be denoted by

$$A = A_1 = \begin{pmatrix} 2 & 4 & -2 \\ 1 & -1 & 5 \\ 4 & 1 & -2 \end{pmatrix}, \; A_2 = \begin{pmatrix} 2 & 4 & -2 \\ 0 & -3 & 6 \\ 0 & -7 & 2 \end{pmatrix}, \; A_3 = \begin{pmatrix} 2 & 4 & -2 \\ 0 & -3 & 6 \\ 0 & 0 & -12 \end{pmatrix}. \tag{10.9}$$

From elementary matrix theory,

$$A_2 = M_1 A_1, \quad M_1 = \begin{pmatrix} 1 & 0 & 0 \\ -\frac{1}{2} & 1 & 0 \\ -2 & 0 & 1 \end{pmatrix}, \tag{10.10}$$

$$A_3 = M_2 A_2, \quad M_2 = \begin{pmatrix} 1 & 0 & 0 \\ 0 & 1 & 0 \\ 0 & -\frac{7}{3} & 1 \end{pmatrix},$$

$$A_3 = M_2 M_1 A. \tag{10.11}$$

In order to solve $Ax = b$, consider

$$A_3 x = M_2 M_1 A x = M_2 M_1 b = c. \tag{10.12}$$

Since M_1 and M_2 are lower triangular and non-singular, M_1^{-1} and M_2^{-1} exist and are also lower triangular. Thus, from (10.11),

$$\begin{aligned} A &= M_1^{-1} M_2^{-1} A_3 \\ &= LU, \end{aligned} \tag{10.13}$$

where

$$L = M_1^{-1} M_2^{-1}, \quad U = A_3, \tag{10.14}$$

with L being a lower triangular matrix and U being an upper triangular matrix. For this example,

$$M_1^{-1} = \begin{pmatrix} 1 & 0 & 0 \\ \frac{1}{2} & 1 & 0 \\ 2 & 0 & 1 \end{pmatrix}, \quad M_2^{-1} = \begin{pmatrix} 1 & 0 & 0 \\ 0 & 1 & 0 \\ 0 & \frac{7}{3} & 1 \end{pmatrix}. \tag{10.15}$$

These M_i matrices are called elementary lower triangular matrices.

An *elementary lower triangular* matrix M_k of order n and index k is an identity matrix of order n with some non-zero elements in the kth column below the diagonal. It has the form

$$M_k = I_n - m_k e_k^T, \tag{10.16}$$

$$e_i^T m_k = 0, i = 1, \ldots, k, \tag{10.17}$$

$$m_k = (0, \ldots, 0, c_{k+1}, \ldots, c_n)^T, \tag{10.18}$$

$$e_i = (0, \ldots, 0, 1, 0, \ldots, 0)^T, \tag{10.19}$$

where all elements in the right-hand side of (10.19) are zero except for a one at the ith position. The inverse of M_k in (10.16) has the form of

$$M_k^{-1} = I_n + m_k e_k^T, \tag{10.20}$$

since

$$\begin{aligned} {M_k}^{-1} M_k &= (I_n - m_k e_k^T)(I_n + m_k e_k^T) \\ &= I_n - m_k e_k^T + m_k e_k^T - m_k (e_k^T m_k) e_k^T \\ &= I_n. \end{aligned}$$

The purpose of M_k is that it can be used to introduce zero in the matrix of a system of equations in the Gaussian elimination procedure.

When A is non-singular, the matrix A of $Ax = b$, can be reduced to an upper triangular form by

$$A_n = M_{n-1}M_{n-2}\cdots M_1 A. \tag{10.21}$$

At the ith stage, if the pivot (a_{ii}) is not zero, we can proceed with the elimination. If $a_{ii} = 0$, two possibilities exist. First, if there are some $a_{ji} \neq 0$, for some $j > i$, then interchange the jth row with the ith row of the matrix as well as interchange the jth and ith components of the vector b. Then proceed with the elimination. Second, if $a_{ji} = 0$, for all $j \geq i$, then the matrix is singular and the original $Ax = b$ has no unique solution.

In practice, the "partial pivoting" scheme considers at stage i, all a_{ji}, with $j \geq i$ in the ith column . It uses the largest $|a_{ji}|$ for pivoting by interchanging the ith and jth row of the matrix as well as the ith and jth components of b. Then proceed with the elimination. In the Gaussian elimination procedure, if at any step the leading coefficient of a row (i.e., pivot) is zero, that row can be interchanged with another row with a non-zero pivot. The "complete pivoting" scheme at stage i uses a pivot given by the largest element in magnitude among all the remaining elements a_{jk}, with $j \geq i$ and $k \geq i$, by interchanging rows and columns.

10.3 Normal Equation Approach to Linear System of Equations

Consider a system of m equations with n unknowns in $x_1, \ldots, x_n$

$$Ax = b, \tag{10.22}$$

where

$$A = \begin{pmatrix} a_{11} & \ldots & a_{1n} \\ \vdots & & \vdots \\ a_{m1} & \ldots & a_{mn} \end{pmatrix}, x = \begin{pmatrix} x_1 \\ \vdots \\ x_n \end{pmatrix}, b = \begin{pmatrix} b_1 \\ \vdots \\ b_m \end{pmatrix}. \tag{10.23}$$

Assume A and b are known and we want to solve for x. If $m = n$ and A is non-singular, then A^{-1} exists and in theory we have

$$x = A^{-1}b. \tag{10.24}$$

In practice, various more practical methods including various versions of Gaussian elimination methods can be used to solve (10.22).

In many practical problems, we encounter the $m \geq n$ case where there are more equations obtained from many measurements than unknowns.

Example 10.1 Consider $m = 2$ and $n = 1$ with

$$\begin{aligned} x_1 &= b_1, \\ 2x_1 &= b_2. \end{aligned} \tag{10.25}$$

or

$$\begin{pmatrix} 1 \\ 2 \end{pmatrix} [x_1] = \begin{pmatrix} b_1 \\ b_2 \end{pmatrix}. \tag{10.26}$$

Clearly, if b is proportional to $A = [1, 2]^T$, then x_1 has a unique solution. Otherwise x_1 has no exact solution.

If there is no unique x that satisfies (10.22), then we want an x that minimizes the norm of the *residual* r defined by

$$r = Ax - b = \begin{pmatrix} r_1 \\ \vdots \\ r_m \end{pmatrix}, \tag{10.27}$$

and

$$\| r \| = \| Ax - b \| = \left(\sum_{i=1}^{m} r_i^2 \right)^{\frac{1}{2}}. \tag{10.28}$$

Let the optimum x that minimized the norm in (10.28) be denoted by $\hat{x}$. Then

$$\| A\hat{x} - b \| = \min_x \| Ax - b \| . \tag{10.29}$$

From (10.28) and (10.29), for a given $b \epsilon R^m$, we want to find a point Ax in the range space in R^m closest to b. Geometrically, we want to project b onto the space $L(a_1, \ldots, a_n)$ where $A = [a_1, \ldots, a_n]$. That is, we want $(A\hat{x} - b) \perp Az$ for all $z \epsilon R^n$. Thus,

$$(Az)^T (A\hat{x} - b) = z^T (A^T A\hat{x} - A^T b) = 0, \quad z \epsilon R^n. \tag{10.30}$$

Since (10.30) is valid for all $z \epsilon R^n$, we have the *normal equation*

$$A^T A\hat{x} = A^T b. \tag{10.31}$$

If the columns of A are linearly independent (i.e., rank(A) $= n \leq m$), then $A^T A$ is invertible and

$$\hat{x} = (A^T A)^{-1} A^T b. \tag{10.32}$$

The projection of b onto the range of A is

$$b_p = A\hat{x} = A(A^T A)^{-1} A^T b. \tag{10.33}$$

Example 10.2 Consider

$$A = \begin{pmatrix} 1 & 2 \\ 1 & 5 \\ 0 & 0 \end{pmatrix}, \quad b = \begin{pmatrix} 4 \\ 3 \\ 9 \end{pmatrix}.$$

Then

$$A^T A = \begin{pmatrix} 2 & 7 \\ 7 & 29 \end{pmatrix}, \quad (A^T A)^{-1} = \frac{1}{9} \begin{pmatrix} 29 & -7 \\ -7 & 2 \end{pmatrix},$$

$$b_p = \begin{pmatrix} 1 & 2 \\ 1 & 5 \\ 0 & 0 \end{pmatrix} \frac{1}{9} \begin{pmatrix} 29 & -7 \\ -7 & 2 \end{pmatrix} \begin{pmatrix} 1 & 1 & 0 \\ 2 & 5 & 0 \end{pmatrix} \begin{pmatrix} 4 \\ 3 \\ 9 \end{pmatrix} = \begin{pmatrix} 4 \\ 3 \\ 0 \end{pmatrix},$$

$$\| \hat{r} \| = \| A\hat{x} - b \| = 9.$$

Example 10.3 Consider an $m \times n$ matrix A in (10.34) with $m \geq n$. Suppose ϵ is significant but $1 + \epsilon^2$ is not significant. Then A in (10.34) is a matrix of rank n, while $A^T A$ is an $n \times n$ matrix of rank 1.

$$A = \begin{pmatrix} 1 & 1 & 1 & \dots & 1 & 1 \\ \epsilon & & & & & \\ & \epsilon & & & \text{0s} & \\ & & \epsilon & & & \\ & \text{0s} & & & \epsilon & \\ & & & & & \epsilon \\ & & \text{0s} & & & \end{pmatrix}, \tag{10.34}$$

$$A^T A = \begin{pmatrix} 1+\epsilon^2 & & \text{1s} & \\ & 1+\epsilon^2 & & \\ & & 1+\epsilon^2 & \\ & \text{1s} & & 1+\epsilon^2 \end{pmatrix} \approx \begin{pmatrix} 1 & 1 & \dots & 1 \\ 1 & 1 & \dots & 1 \\ \vdots & & & \vdots \\ 1 & 1 & \dots & 1 \end{pmatrix}. \tag{10.35}$$

Example 10.4 Consider Example 10.3 with $m = 4$ and $n = 3$.

$$A = \begin{pmatrix} 1 & 1 & 1 \\ \epsilon & 0 & 0 \\ 0 & \epsilon & 0 \\ 0 & 0 & \epsilon \end{pmatrix}, \quad b = \begin{pmatrix} 1 \\ 0 \\ 0 \\ 0 \end{pmatrix}. \tag{10.36}$$

Then

$$A^T A = \begin{pmatrix} 1+\epsilon^2 & 1 & 1 \\ 1 & 1+\epsilon^2 & 1 \\ 1 & 1 & 1+\epsilon^2 \end{pmatrix}, \quad A^T b = \begin{pmatrix} 1 \\ 1 \\ 1 \end{pmatrix}. \tag{10.37}$$

From (10.31),

$$(A^T A)\hat{x} = A^T b. \tag{10.38}$$

For this simple case, by inspection we obtain the true optimum solution as

$$\hat{x} = [1, 1, 1]^T / (3 + \epsilon^2). \tag{10.39}$$

Now, suppose we have an $\epsilon = 10^{-5}$ on a single precision (e.g., 8 digits) floating point processor. Then $\epsilon^2 = 10^{-10}$ is significant but

$$1 + \epsilon^2 = 1.0000000001 \approx 1 \tag{10.40}$$

and the 3×3 matrix $A^T A$ has all 1s. Thus, $A^T A$ is singular and no solution from the normal equation approach (i.e., (10.32)) is possible.

The somewhat exaggerated results of Examples 10.3 and 10.4 show that the effects due to finite precision computations can be considerably worse for $A^T A$ as compared to A. This observation is a prime motivating factor in finding direct methods (i.e., working with the original A) instead of the normal equation approach (i.e., using the inverse of $A^T A$) to solve the $Ax = b$ problem of (10.22).

10.4 Triangular Decomposition

We have seen in the Gaussian elimination procedure that a non-singular square matrix A can be reduced to

$$A = LU, \tag{10.41}$$

where

$$A_n = M_{n-1}, \ldots, M_1 A, \tag{10.42}$$

with upper triangular A_n and lower triangular M_i. Then

$$L = M_1^{-1} M_2^{-1} \cdots M_{n-1}^{-1}, \tag{10.43}$$

and

$$U = A_n. \tag{10.44}$$

Since an upper and a lower triangular matrix is much simpler than the general matrix A, it may be advantageous to work with L and U instead of A. For example, if we need $A^{-1} = U^{-1}L^{-1}$, then finding U^{-1} and L^{-1} and multiplying them are easier than finding the original A^{-1}. Of course, we note that A^{-1} is not needed in the actual solution of x in $Ax = b$.

In general, the LU decomposition of A given by

$$A = LU,$$

has no unique L and U. Indeed, consider a non-singular diagonal matrix D, then $L^* = LD$ is still lower triangular and $U^* = D^{-1}U$ is still upper triangular. Thus, $A = L^*U^*$ is still an LU decomposition of A. Furthermore, $A = LDU$ is an LDU decomposition provided L is unit lower triangular (i.e., L has unity diagonal), D is diagonal, and U is unit upper triangular.

Three different LU decompositions are possible by considering the diagonal D in the LDU decomposition differently. In the Crout decomposition, D is associated with L to give

$$A = L^*U = (LD)U, \tag{10.45}$$

where U has unity diagonal. In the Doolittle decomposition, D is associated with U to give

$$A = LU^* = L(DU), \tag{10.46}$$

where L has unity diagonal. When A is symmetric and has a unique LDU decomposition, it is called the *generalized Cholesky decomposition* and has the form

$$A = LDL^T, \tag{10.47}$$

where L^T is the transpose of L. If the diagonal elements of D are positive, then it can be expressed as

$$A = (LD^*)(D^*L^T) = L^*L^{*T}. \tag{10.48}$$

The diagonal elements of L^* are of course the same as those of L^{*T}. The decomposition of (10.48) is called the *Cholesky decomposition* of A.

In general, a symmetric matrix A need not have a Cholesky decomposition. Suppose $A = LL^T$ and L is lower triangular. Let the $(1,1)$ element of A be a_{11} and the $(1,1)$ element of L be l_{11}. Then we have $a_{11} = l_{11}^2 \geq 0$. Thus, if a symmetric A has a negative a_{11} element, it cannot have a Cholesky decomposition.

However, every symmetric positive-definite matrix A has a Cholesky decomposition. The matrix A is positive-definite if $(x, Ax) = x^T Ax > 0$ and non-negative-definite if $(x, Ax) = x^T Ax \geq 0$ for any $x \neq 0$. For any $n \times n$ matrix B, let

$$A = B^T B.$$

Then A is symmetric and non-negative-definite as shown by

$$(x, Ax) = (x, B^T Bx) = (Bx, Bx) = (B^T Bx, x) = (A^T x, x) \geq 0.$$

If $\text{rank}(B) = n$, then A is positive-definite. This follows from $x \neq 0$ implies $y = Bx \neq 0$ and

$$x^T Ax = x^T B^T Bx = (Bx, Bx) > 0.$$

THEOREM 10.1 (Cholesky Decomposition Theorem) *If A is symmetric and positive-definite, then there is a unique lower triangular matrix L with positive diagonal elements such that $A = LL^T$.*

Cholesky Decomposition Algorithm

Consider the decomposition of a symmetric positive-definite matrix

$$A = [a_{ij}]_{i,j=1,\ldots,n} = LL^T,$$
$$L = [l_{ij}]_{i,j=1,\ldots,n}.$$

1. Initialize with $l_{11} = \sqrt{a_{11}}$.
   For $i = 2, 3, \ldots, n$, do
2. $l_{ij} = (a_{ij} - \sum_{k=1}^{j-1} l_{ik} l_{jk})/l_{jj}, \quad j = 1, 2, \ldots, i-1.$
3. $l_{ii} = \left\{ a_{ii} - \sum_{k=1}^{i-1} l_{ik}^2 \right\}^{\frac{1}{2}}.$
4. $l_{i-1,j} = 0, \quad j = i, i+1, \ldots, n.$

In the above expressions, the sum is taken to be zero if the upper limit in the sum is less than the lower limit.

The inverse of a lower triangular matrix L is also lower triangular and is given simply by

$$L^{-1} = [l^*_{ij}]i, j = 1, \ldots, n,$$
$$L^{-1}L = I,$$
$$l^*_{ii} l_{ii} = 1, \quad i = 1, \ldots, n,$$
$$\sum_{k=1}^{i} l^*_{ik} l_{kj} = 0, \quad i = 2, 3, \ldots, n; \quad j = i - 1, \ldots, 1.$$

Inversion Algorithm for L.

1. Initialize with $l^*_{11} = 1/l_{11}$.
   For $i = 2, 3, \ldots, n$,
2. $l^*_{ii} = 1/l_{ii}$.
3. $l^*_{ij} = (-\sum_{k=j+1}^{i} l^*_{ik} l_{kj})/l_{jj}, \quad j = i - 1, i - 2, \ldots, 1.$
4. $l^*_{(i-1)j} = 0, j = i, i + 1, \ldots, n.$

In detection, estimation, and signal processing, Cholesky decomposition can be used to perform whitening operation on colored observations. Let A be the positive-definite covariance matrix of a real-valued zero-mean wide-sense stationary random vector $x = (x_1, \ldots, x_n)^T$. Let $A = LL^T$, where L^{-1} is the whitening filter of x. Specifically, denote

$$y = (y_1, \ldots, y_n)^T = L^{-1}x = \begin{pmatrix} l^*_{11} & 0 & 0 & \ldots & 0 \\ l^*_{21} & l^*_{22} & 0 & \ldots & 0 \\ \vdots & & & & \vdots \\ l^*_{n1} & & \ldots & & l^*_{nn} \end{pmatrix} \begin{pmatrix} x_1 \\ \vdots \\ x_n \end{pmatrix}.$$

Then

$$y_i = \sum_{j=1}^{i} l^*_{ij} x_j, \quad i = 1, \ldots, n,$$

is a causal sequence of r.v. with a covariance given by

$$\begin{aligned} E\{yy^T\} &= E\{L^{-1}xx^T(L^{-1})^T\} = L^{-1}E\{xx^T\}(L^{-1})^T \\ &= L^{-1}A(L^{-1})^T = L^{-1}LL^T(L^{-1})^T = I. \end{aligned}$$

Thus, y is a "white" sequence as desired.

10.5 QR Factorization

Previously, we observed that the Gaussian elimination procedure provided one method of decomposing a non-singular square matrix A into an upper triangular matrix U by applying a sequence of lower triangular matrices $\cdots L_2 L_1$ to A. In the QR factorization method, a similar reduction is performed using a sequence of orthogonal (or unitary for the complex-valued matrix A) matrices. (We note, QR factorization or decomposition is related to, but not to be confused with, the QR algorithm to be used in eigenvalue and singular value evaluations).

Consider n real-valued column vectors of dimension $m \geq n$,

$$q_1 = (q_{11}, q_{21}, \ldots, q_{m1})^T, \ldots, q_n = (q_{1n}, \ldots, q_{mn})^T. \tag{10.49}$$

This system of vectors is mutually orthogonal if

$$q_i^T q_j = (q_i, q_j) = \sum_{k=1}^{m} q_{ki} q_{kj} = 0, \quad i \neq j. \tag{10.50}$$

Furthermore, this system of vectors is orthonormal if in addition to (10.50) we have

$$q_i^T q_i = (q_i, q_i) = \| q_i \|^2 = \sum_{k=1}^{m} q_{ki}^2 = 1, \quad i = 1, \ldots, n. \tag{10.51}$$

Let $Q = [q_{ij}]_{i=1,\ldots,m;\, j=1,\ldots,n}$ be an $m \times n$ real-valued matrix also denoted by

$$Q = [q_1, \ldots, q_n], \tag{10.52}$$

where each column q_i is of the form given in (10.49). Q is said to have orthonormal columns if the system of column vectors $(q_1, \ldots, q_n)$ are orthonormal. From (10.50) and (10.51), we see

$$Q^T Q = \begin{pmatrix} q_1^T \\ \vdots \\ q_n^T \end{pmatrix} (q_1, \ldots, q_n) = I_n, \tag{10.53}$$

where I_n is an $n \times n$ identity matrix. In general, an $m \times n$ matrix with orthonormal columns needs not have orthogonal rows.

THEOREM 10.2 *Let A be a real-valued $m \times n$ matrix with $m \geq n$ and n linearly independent columns (i.e.,* rank$(A) = n$*). Then A has a unique QR decomposition (QRD) of the form*

$$A = QR, \tag{10.54}$$

where Q is a real-valued $m \times n$ matrix with orthonormal columns and R is a real-valued $n \times n$ upper triangular matrix with positive diagonal elements.

Proof: From $0 < \| Ax \|^2 = (Ax, Ax) = (A^T Ax, x) = x^T (A^T A) x$, $A^T A$ is a symmetric positive-definite matrix. From Theorem 10.1 on Cholesky decomposition, $A^T A = LL^T$, when L is a lower triangular matrix with positive diagonal elements. Let $L^T = R$ be an upper triangular matrix with positive diagonal elements. Denote $Q = AR^{-1}$ or $A = QR$ and

$$Q^T Q = R^{-T} A^T A R^{-1} = R^{-T} L L^T R^{-1} = R^{-T} R^T R R^{-1} = I_n. \tag{10.55}$$

We note, if $m = n$ in all our previous discussions, then a real-valued matrix Q with orthonormal columns also has orthonormal rows. Specifically, for $m = n$, Q is non-singular and Q^{-1} exists. Then from (10.55),

$$Q^T = Q^T QQ^{-1} = Q^{-1}. \tag{10.56}$$

By denoting

$$Q = \begin{pmatrix} Q_1 \\ \vdots \\ Q_n \end{pmatrix}, Q^T = [Q_1^T, \ldots, Q_n^T], \tag{10.57}$$

and using (10.56), we have

$$QQ^{-1} = QQ^T = \begin{pmatrix} Q_1 \\ \vdots \\ Q_n \end{pmatrix} [Q_1^T, \ldots, Q_n^T] = I_n. \tag{10.58}$$

This shows Q has orthonormal rows. An $n \times n$ real-valued matrix Q satisfying

$$Q^T Q = QQ^T = I_n \tag{10.59}$$

is an *orthogonal* matrix. Clearly, from (10.56) and (10.59) for an orthogonal matrix, $Q^T = Q^{-1}$. Furthermore, for an orthogonal matrix Q and any $n \times 1$ vector x,

$$\| x \|^2 = (x, x) = (Q^T Qx, x) = x^T Q^T Qx = (Qx, Qx) = \| Qx \|^2 . \tag{10.60}$$

Thus, the square of the norm of any vector (i.e., the energy of x) is invariant under an orthogonal transformation.

10.6 Gram–Schmidt Orthogonalization Procedure

Consider a set of n real-valued linearly independent column vectors of dimension m denoted by $(a_1, a_2, \ldots, a_k), k = 1, \ldots, n$ with $m \geq n$. The Gram–Schmidt orthonormalization procedure provides a simple method of obtaining an orthonormal set of vectors $(q_1, q_2, \ldots, q_k)$ which has the same linear span as the $(a_1, \ldots, a_k)$, for $k = 1, \ldots, n$. Let $L(q_1, \ldots, q_k) = L(a_1, \ldots, a_k)$ denote the equivalence of the two linear spans.

Gram–Schmidt Orthonormalization Algorithm

1. Set

$$q_1 = a_1 / \| a_1 \| . \tag{10.61}$$

2. For $k = 2, \ldots, n$, do

$$b_k = a_k - \sum_{j=1}^{k-1} (q_j^T a_k) q_j, \tag{10.62}$$

$$q_k = b_k / \| b_k \| . \tag{10.63}$$

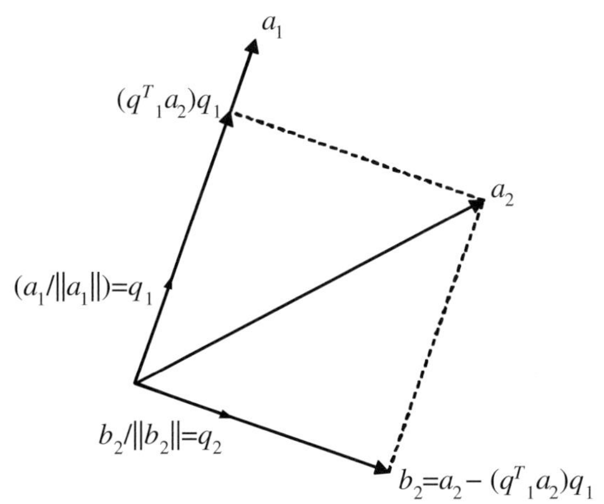

Figure 10.1 Gram–Schmidt orthonormalization procedure for $n = 2$

Clearly, (10.61) is a simple normalization to obtain unity norm for q_1. In (10.62), b_k contains the part of a_k not in $L(q_1, \ldots, q_{k-1}) = L(a_1, \ldots, a_{k-1})$. Furthermore, since the original $(a_1, \ldots, a_k)$ vectors are linearly independent, $a_k \notin L(a_1, \ldots, a_{k-1})$ and $b_k \neq 0$. Finally, (10.63) is again a simple normalization to obtain unity norm for q_k. A simple demonstration of the steps of (10.61)–(10.63) for the $n = 2$ case is given in Fig. 10.1.

The above Gram–Schmidt orthogonalization procedure can be used to perform the QR decomposition of Theorem 10.2. Consider

$$A = QR, \tag{10.64}$$

or

$$[a_1, a_2, \ldots, a_n] = [q_1, q_2, \ldots, q_n] \begin{pmatrix} r_{11} & r_{12} & \ldots & r_{1n} \\ 0 & r_{22} & & r_{2n} \\ \vdots & \vdots & & \vdots \\ 0 & 0 & & r_{nn} \end{pmatrix}. \tag{10.65}$$

Direct evaluation in (10.65) shows

$$a_k = \sum_{j=1}^{k} r_{jk} q_j, k = 1, \ldots, n. \tag{10.66}$$

By using (10.63) in (10.62), we have

$$\| b_k \| q_k = a_k - \sum_{j=1}^{k-1} (q_j^T a_k) q_j, \tag{10.67}$$

or

$$a_k = \sum_{j=1}^{k-1} (q_j^T a_k) q_j + \| b_k \| q_k, k = 2, \ldots, n. \tag{10.68}$$

From (10.61), we have

$$a_1 = \| a_1 \| q_1. \tag{10.69}$$

By comparing (10.66) with (10.68) and (10.69), we see

$$r_{11} = \| a_1 \| , r_{kk} = \| b_k \| , k = 2, \ldots, n, \tag{10.70}$$

$$r_{jk} = q_j^T a_k, j = 1, \ldots, k-1, k = 2, \ldots, n. \tag{10.71}$$

For reasons that will become clearer in the next section, we may be only interested in obtaining a set of orthogonal vectors $(b_1, \ldots, b_k)$ such that $L(b_1, \ldots, b_k) = L(a_1, \ldots, a_k), k = 1, \ldots, n.$

Gram–Schmidt Orthogonalization Algorithm

1. For $k = 1, \ldots, n$, do

$$b_k = a_k - \sum_{j=1}^{k-1} (b_j^T a_k) b_j / \| b_j \|^2 . \tag{10.72}$$

For $k = 1$, the upper limit of the sum is less than the lower limit and no summation is performed.

10.7 Modified Gram–Schmidt Orthogonalization Procedure

Under infinite precision evaluations, the Gram–Schmidt (G-S) orthogonalization (or orthonormalization) procedure is suitable for performing the QR decomposition. However, under the more realistic constraint of finite precision evaluations, the modified Gram–Schmidt (M-G-S) orthogonalization procedure is superior. Furthermore, this modified procedure is no more costly in computational complexity.

Modified Gram–Schmidt Orthogonalization Algorithm

For $k = 1, \ldots, n$,

1. Set

$$b_k^{(1)} = a_k. \tag{10.73}$$

2. For $j = 1, \ldots, k-1$,

$$b_k^{(j+1)} = b_k^{(j)} - (b_j^T b_k^{(j)}) b_j / \| b_j \|^2 . \tag{10.74}$$

3. Set

$$b_k = b_k^{(k)}. \tag{10.75}$$

In order to illustrate the differences of the G-S and M-G-S orthogonalization procedures under finite precision computations, consider the following simple but instructive example. Let $A = [a_1, a_2, a_3]$ with

$$a_1 = \begin{pmatrix} 1 \\ \epsilon \\ 0 \\ 0 \end{pmatrix}, a_2 = \begin{pmatrix} 1 \\ 0 \\ \epsilon \\ 0 \end{pmatrix}, a_3 = \begin{pmatrix} 1 \\ 0 \\ 0 \\ \epsilon \end{pmatrix}. \tag{10.76}$$

We shall model our computations as finite precision floating point operations in which ϵ and ϵ^2 are valid and usable numbers, but $(1 + \epsilon^2)$ is representable only as 1. For example, under single precision floating point computations with 8 significant digits, let $\epsilon = 10^{-5}$. Then ϵ, $\epsilon^2 = 10^{-10}$, and $1 + \epsilon = 1.00001$ are all valid numbers but $1 + \epsilon^2 = 1.0000000001$ becomes 1.

First, apply the G-S orthogonalization algorithm to (10.76).

$$b_1 = a_1, \parallel b_1 \parallel^2 = 1 + \epsilon^2 = 1, \tag{10.77}$$

$$b_2 = a_2 - (a_2^T b_1)b_1 / \parallel b_1 \parallel^2 = a_2 - a_1 = (0, -\epsilon, \epsilon, 0)^T, \parallel b_2 \parallel^2 = 2\epsilon^2, \tag{10.78}$$

$$b_3 = a_3 - (a_3^T b_2)b_2 / \parallel b_2 \parallel^2 - (a_3^T b_1)b_1 / \parallel b_1 \parallel^2 = a_3 - a_1 = (0, -\epsilon, 0, \epsilon)^T, \tag{10.79}$$

$$\parallel b_3 \parallel^2 = 2\epsilon^2.$$

The crucial test is how orthogonal are the (b_1, b_2, b_3)? The magnitudes of the normalized inner products of these vectors under the G-S orthogonal algorithm are

$$|b_1^T b_2| / \parallel b_1 \parallel \parallel b_2 \parallel = \epsilon/\sqrt{2}, \tag{10.80}$$

$$|b_1^T b_3| / \parallel b_1 \parallel \parallel b_3 \parallel = \epsilon/\sqrt{2}, \tag{10.81}$$

$$|b_2^T b_3| / \parallel b_2 \parallel \parallel b_3 \parallel = 1/2. \tag{10.82}$$

Thus, starting with three (a_1, a_2, a_3) linearly independent (but almost linearly dependent) vectors of (10.76), (10.80)–(10.82) showed b_1 and b_2 as well as b_1 and b_3 are quite orthogonal but b_2 and b_3 are not orthogonal at all.

Now, apply the M-G-S orthogonalization algorithm on (10.76). For $k = 1$,

$$b_1 = b_1^{(1)} = a_1, \parallel b_1 \parallel^2 = 1 + \epsilon^2 = 1. \tag{10.83}$$

For $k = 2$,

$$b_2^{(1)} = a_2. \tag{10.84}$$

$$b_2 = b_2^{(2)} = b_2^{(1)} - (b_1^T b_2^{(1)})b_1 / \parallel b_1 \parallel^2 = a_2 - a_1 = (0, -\epsilon, \epsilon, 0)^T, \parallel b_2 \parallel^2 = 2\epsilon^2. \tag{10.85}$$

For $k = 3$,

$$b_3^{(1)} = a_3, \tag{10.86}$$

$$b_3^{(2)} = b_3^{(1)} - (b_1^T b_3^{(1)})b_1 / \parallel b_1 \parallel^2 = (0, -\epsilon, 0, \epsilon)^T, \tag{10.87}$$

$$b_3 = b_3^{(3)} = b_3^{(2)} - (b_2^T b_3^{(2)})b_2 / \parallel b_2 \parallel^2 = (0, -\epsilon/2, -\epsilon/2, \epsilon)^T, \parallel b_3 \parallel^2 = 3\epsilon^2/2. \tag{10.88}$$

The magnitudes of the normalized inner products of these vectors under the M-G-S orthogonal algorithm are given by

$$|b_1^T b_2| / \parallel b_1 \parallel \parallel b_2 \parallel = \epsilon/\sqrt{2}, \tag{10.89}$$

$$|b_1^T b_3| / \parallel b_1 \parallel \parallel b_3 \parallel = \epsilon/\sqrt{6}, \tag{10.90}$$

$$|b_2^T b_3| / \parallel b_2 \parallel \parallel b_3 \parallel = (\epsilon^2/2 - \epsilon^2/2)/(2\epsilon^2)^{1/2}(3\epsilon^2/2)^{1/2} = 0. \tag{10.91}$$

Thus, we can consider b_1 and b_2 as well as b_1 and b_3 to be nearly orthogonal while b_2 and b_3 are also orthogonal.

The original M-G-S orthogonalization algorithm given in (10.73)–(10.75) can be modified such that the resulting vectors are not only orthogonal but are also normalized.

M-G-S Orthonormalization Algorithm

For $k = 1, \ldots, n$,

1. Set

$$a_k^{(1)} = a_k. \tag{10.92}$$

2. For $i = 1, \ldots, n$,

$$q_i = a_i^{(i)} / \parallel a_i^{(i)} \parallel, \tag{10.93}$$

$$a_j^{(i+1)} = a_j^{(i)} - (q_i^T a_j^{(i)}) q_i, \; j = i+1, \ldots, n. \tag{10.94}$$

We shall need shortly the fact that under infinite precision computation

$$\begin{aligned} q_i^T a_j^{(i)} &= q_i^T a_j^{(i-1)} - (q_{i-1}^T a_j^{(i-1)}) q_i^T q_{i-1} = q_i^T a_j^{(i-1)} - 0 \\ &= q_i^T a_j^{(i-2)} - 0 = \cdots = q_i^T a_j^{(1)} = q_i^T a_j = r_{ij}. \end{aligned} \tag{10.95}$$

Indeed, by requiring the algorithm in (10.92)–(10.94) to have in-placed computation, we have the following simpler algorithm.

M-G-S Orthonormalization Algorithm (In-Place Version)

For $k = 1, \ldots, n$,

1. Set

$$a_k = a_k / \parallel a_k \parallel. \tag{10.96}$$

2. Evaluate

$$a_j = a_j - (a_j^T a_k) a_k, \; j = k+1, \ldots, n. \tag{10.97}$$

Now, consider the order of updating the R matrix under G-S and M-G-S orthogonalization procedures. In the G-S approach, at each iteration, for a given a_k in (10.62), we remove all dependent components of $(q_1, \ldots, q_{k-1})$ from a_k. The coefficients of q_j are given by $r_{jk}, j = 1, \ldots, k-1$, as shown in (10.71). This means that under G-S iteration, we update on the kth column of R. Of course, any finite precision errors in the evaluation of r_{jk} are not known or compensated for in the evaluation of r_{pk}, for $j < p \leq k-1$, in (10.62). On the other hand, in the M-G-S approach, we set $q_1 = a_1/ \parallel a_1 \parallel$ and immediately remove that q_1 component from $a_j^{(1)}, j = 2, \ldots, n$, in (10.94). In general, for a given orthonormal vector q_i, as given in (10.93), the removal of that component from $a_j^{(i)}$, for $j = i + 1, \ldots, n$, in (10.94) has the coefficient value of $q_i^T a_j^{(i)}$ (equal to r_{ij} as given in (10.95) under infinite precision). We note that under M-G-S iteration, we update in the ith row of R. Of course, under finite precision operations, all the terms with a value of zero in (10.95) are no longer valid and information due to the ill-effects of earlier $q_p^T a_j^{(n)}$ compuations are all used at each iteration. This is one explanation for the advantages obtained in (10.89)–(10.91) for M-G-S procedure as compared to that obtained in (10.68)–(10.70) for G-S procedure.

corollary 10.1 *Let A be a real-valued $m \times n$ matrix with $m \geq n$ and n linearly independent columns. Then A has a $B\tilde{R}$ decomposition of the form $A = B\tilde{R}$, where $B = [b_1, b_2, \ldots, b_n]$ is an $m \times n$ matrix of orthogonal (i.e., not necessarily orthonormal) column vectors and $\tilde{R}$ is an $n \times n$ upper triangular matrix. Furthermore, $R = D\tilde{R}$, where R is an $n \times n$ upper triangular matrix of the QR decomposition, $A = QR$, and $D = \mathrm{diag}[||b_1||, \ldots, ||b_n||]$.*

Proof: From the G-S or M-G-S method, decompose the matrix A by

$$A = [b_1, b_2, \ldots, b_n] \begin{pmatrix} \tilde{r}_{11} & \tilde{r}_{12} & \cdots & \tilde{r}_{1n} \\ 0 & \tilde{r}_{22} & & \tilde{r}_{2n} \\ \vdots & \vdots & & \vdots \\ 0 & 0 & & \tilde{r}_{nn} \end{pmatrix} = B\tilde{R}, \tag{10.98}$$

where b_i is orthogonal and $|\tilde{r}_{ii}| > 0$. Then

$$[\frac{b_1}{||b_1||}, \ldots, \frac{b_n}{||b_n||}] = BD^{-1} = Q, \tag{10.99}$$

where $Q^T Q = I_n$ and $B^T B = D^2$. Thus,

$$\begin{aligned} A = B\tilde{R} &= [\frac{b_1}{||b_1||}, \ldots, \frac{b_n}{||b_n||}] \begin{pmatrix} \tilde{r}_{11}||b_1|| & \tilde{r}_{12}||b_1|| & \cdots & \tilde{r}_{1n}||b_1|| \\ 0 & \tilde{r}_{22}||b_2|| & & \tilde{r}_{2n}||b_2|| \\ \vdots & \vdots & & \vdots \\ 0 & 0 & & \tilde{r}_{nn}||b_n|| \end{pmatrix} \\ &= BD^{-1}D\tilde{R} = QR, \end{aligned} \tag{10.100}$$

where $r_{ij} = \tilde{r}_{ij}||b_i||$. Thus, $R = D\tilde{R}$.

10.8 Givens Orthogonal Transformation

A *Givens* orthogonal matrix transformation $G = [g_{ij}]$ is any $m \times m$ matrix consisting of an identity matrix I_m except for the values of the matrix at the (i,i), (j,j), (i,j), and (j,i) locations which are given by

$$g_{ii} = g_{jj} = c = \cos\theta, \tag{10.101}$$

$$g_{ij} = -s = -\sin\theta = -g_{ji}, \tag{10.102}$$

for any $1 \le i \ne j \le m$ and any $0 \le \theta < 2\pi$. A typical G is given in (10.103) as

$$G = G(i,j) = \begin{array}{c|cccccccccc|}
 & & & & i & & & & j & & \\
\hline
 & 1 & & & & & & & & & \\
 & & 1 & & & & & & & & \\
 & & & 1 & & & & & & & \\
i & & & & c & & & & -s & & \\
 & & & & & 1 & & & & & \\
 & & & & & & \ddots & & & & \\
 & & & & & & & 1 & & & \\
j & & & & s & & & & c & & \\
 & & & & & & & & & 1 & \\
 & & & & & & & & & & 1 \\
\hline
\end{array}\,. \tag{10.103}$$

First, we show G is an orthogonal matrix (i.e., $G^{-1} = G^T$). Clearly, the transpose of G is

$$G^T = \begin{array}{c|cccccccccc|}
 & & & & i & & & & j & & \\
\hline
 & 1 & & & & & & & & & \\
 & & 1 & & & & & & & & \\
 & & & 1 & & & & & & & \\
i & & & & c & & & & s & & \\
 & & & & & 1 & & & & & \\
 & & & & & & \ddots & & & & \\
 & & & & & & & 1 & & & \\
j & & & & -s & & & & c & & \\
 & & & & & & & & & 1 & \\
 & & & & & & & & & & 1 \\
\hline
\end{array}\,. \tag{10.104}$$

Then GG^T is a diagonal matrix given by

$$GG^T = \mathrm{diag}[1, \ldots, 1, (c^2 + (-s)^2), 1, \ldots, 1, (s^2 + c^2), 1, \ldots, 1] = I_m, \tag{10.105}$$

and $G^{-1} = G^T$. Consider an $m \times 1$ column vector $a = a^{(0)} = [a_1^{(0)}, \ldots, a_m^{(0)}]^T$. Then

$$a^{(1)} = Ga^{(0)} = \begin{array}{c|c|}
\hline
 & a_1^{(0)} \\
 & a_2^{(0)} \\
 & \vdots \\
i & ca_i^{(0)} - sa_j^{(0)} \\
 & \vdots \\
j & sa_i^{(0)} + ca_j^{(0)} \\
 & \vdots \\
 & a_m^{(0)} \\
\hline
\end{array}\,. \tag{10.106}$$

By picking θ appropriately, we want to make

$$a_j^{(1)} = sa_i^{(0)} + ca_j^{(0)} = 0. \tag{10.107}$$

Specifically, if $a_i^{(0)} \neq 0$, then

$$\begin{aligned} -ca_j^{(0)} &= sa_i^{(0)}, \\ \tan\theta &= \frac{s}{c} = -\frac{a_j^{(0)}}{a_i^{(0)}}, \end{aligned} \tag{10.108}$$

or

$$\theta = \tan^{-1}[-a_j^{(0)}/a_i^{(0)}]. \tag{10.109}$$

From (10.106)–(10.109), we see that if $a_i^{(0)} \neq 0$, we can make $a_j^{(1)} = 0$. Thus, by taking $i = 1$ and $j = 2$, we make $a_2^{(1)} = 0$. Next, we take $i = 1$ and $j = 3$ and make $a_3^{(2)} = 0$. Proceeding this way, we can use a series of, at most, $(m-1)$ Givens transformations to reduce the vector $a^{(0)}$ to a vector with all zero elements except for the first component.

$$G^{(m-1)} \cdots G^{(2)}G^{(1)}a^{(0)} = a^{(m-1)} = (a_1^{(m-1)}, 0, \ldots, 0)^T. \tag{10.110}$$

Example 10.5 Consider $a = a^{(0)} = (1,2,3)^T$. Then $||a^{(0)}||^2 = 1 + 4 + 9 = 14$. At iteration 1,

$$G^{(1)} = \begin{bmatrix} c_1 & -s_1 & 0 \\ s_1 & c_1 & 0 \\ 0 & 0 & 1 \end{bmatrix},$$

$$\begin{aligned} a_2^{(1)} &= s_1 + 2c_1 = 0, \\ -2 &= \tan\theta_1, \theta_1 = -1.10715, \\ \cos\theta_1 &= 0.44721, \sin\theta_1 = -0.89443, \\ a_1^{(1)} &= c_1 - 2s_1 = 2.23607, \\ a^{(1)} &= (2.23607, 0, 3)^T, ||a^{(1)}||^2 = 5 + 9 = 14. \end{aligned}$$

At iteration 2,

$$G^{(2)} = \begin{bmatrix} c_2 & 0 & -s_2 \\ 0 & 1 & 0 \\ s_2 & 0 & c_2 \end{bmatrix},$$

$$\begin{aligned} a_3^{(2)} &= 2.23607s_2 + 3c_2 = 0, \\ \tan\theta_2 &= -3/2.23607, \theta_2 = -0.93027, \\ \cos\theta_2 &= 0.59761, \sin\theta_2 = -0.80178, \\ a_1^{(2)} &= 2.23607c_2 - 3s_2 = 3.74166, \\ a^{(2)} &= (3.74166, 0, 0)^T, ||a^{(2)}||^2 = 14. \end{aligned}$$

$$G = G^{(2)}G^{(1)} = \begin{bmatrix} c_1c_2 & -c_2s_1 & -s_2 \\ s_1 & c_1 & 0 \\ c_1s_2 & -s_1s_2 & c_2 \end{bmatrix}, \tag{10.111}$$

$$a^{(2)} = Ga^{(0)} = (3.74166, 0, 0)^T.$$

In summary, for this example, G in (10.111) is an orthogonal matrix expressed as the product of two Givens transformation matrices that forces the resulting vector $a^{(2)}$ to have zero values in the second and third components.

Consider an $m \times n$ matrix A, where $m \geq n$, denoted by

$$A = [a_1, a_2, \ldots, a_n]. \tag{10.112}$$

We need at most $(m - 1)$ Givens transformations to make the elements of $a_{i1} = 0$, $2 \leq i \leq m$; $(m - 2)$ transformations to make $a_{i2} = 0$, $3 \leq i \leq m$, $\ldots$, and $(m - n)$ transformations to make $a_{in} = 0$, $n + 1 \leq i \leq m$. Thus, a total of

$$(m-n)+(m-n+1)+\cdots+(m-1) = n(m-n)+\sum_{i=1}^{n-1} i = n(m-n)+\frac{n(n-1)}{2} = M. \tag{10.113}$$

This shows a sequence of M Givens transformations can reduce A to an upper triangular matrix R. That is, let

$$G = G^{(M)} \cdots G^{(2)}G^{(1)}, \tag{10.114}$$

then

$$GA = R \tag{10.115}$$

or

$$A = G^{-1}R. \tag{10.116}$$

Since $Q = G^{-1} = G^T = G^{(1)^T}G^{(2)^T} \cdots G^{(M)^T}$ is an orthogonal transform, then (10.116) shows an explicit QR decomposition where Q is obtained from a cascade of Givens transformations. In practice, due to the large number of terms needed in (10.113), Givens transformations are often used for QR decomposition only when the original matrix A is sparse.

Givens transformation has simple obvious geometric rotation interpretations. Since the 1s on the diagonal of G have no effect, we need only to consider the effects of

$$G_0 = \begin{bmatrix} \cos\theta & -\sin\theta \\ \sin\theta & \cos\theta \end{bmatrix}, \tag{10.117}$$

$$G_0^{-1} = \begin{bmatrix} \cos\theta & \sin\theta \\ -\sin\theta & \cos\theta \end{bmatrix}, \tag{10.118}$$

for any $0 \le \theta < 2\pi$. Consider G_0 operating on the components of an arbitrary 2×1 vector A with components x and y denoted by

$$A = \begin{bmatrix} x \\ y \end{bmatrix}. \tag{10.119}$$

Denote θ_0 by

$$\tan\theta_0 = \frac{y}{x}, \tag{10.120}$$

then

$$y = \sqrt{x^2 + y^2}\sin\theta_0, \tag{10.121}$$

$$x = \sqrt{x^2 + y^2}\cos\theta_0. \tag{10.122}$$

Then the Givens rotated vector A_1 is defined by

$$A_1 = \begin{bmatrix} x_1 \\ y_1 \end{bmatrix} = G_0 \begin{bmatrix} x \\ y \end{bmatrix} = \begin{bmatrix} x\cos\theta - y\sin\theta \\ x\sin\theta + y\cos\theta \end{bmatrix}. \tag{10.123}$$

Thus,

$$\begin{aligned} \tan\theta_1 &= \frac{y_1}{x_1} = \frac{\sqrt{x^2+y^2}\,[\cos\theta_0 \sin\theta + \sin\theta_0 \cos\theta]}{\sqrt{x^2+y^2}\,[\cos\theta_0 \cos\theta - \sin\theta_0 \sin\theta]} \\ &= \frac{\sin(\theta_0 + \theta)}{\cos(\theta_0 + \theta)} = \tan(\theta_0 + \theta), \end{aligned} \tag{10.124}$$

and the new rotated angle θ_1 is given by

$$\theta_1 = \theta_0 + \theta. \tag{10.125}$$

This shows the multiplication of A on the left by G_0 results in a rotation of θ radians on the A vector as shown in Fig. 10.2a. If we replace G_0 by G_0^{-1} in (10.118), then the rotation on A is by $-\theta$ radians.

Similarly, consider an arbitrary 1×2 vector

$$B = [x, y], \tag{10.126}$$

where (10.120)–(10.122) still hold. Then the Givens rotated vector B_1 is defined by

$$B_1 = [x_1, y_1] = [x, y]G_0 = [x\cos\theta + y\sin\theta, \; -x\sin\theta + y\cos\theta]. \tag{10.127}$$

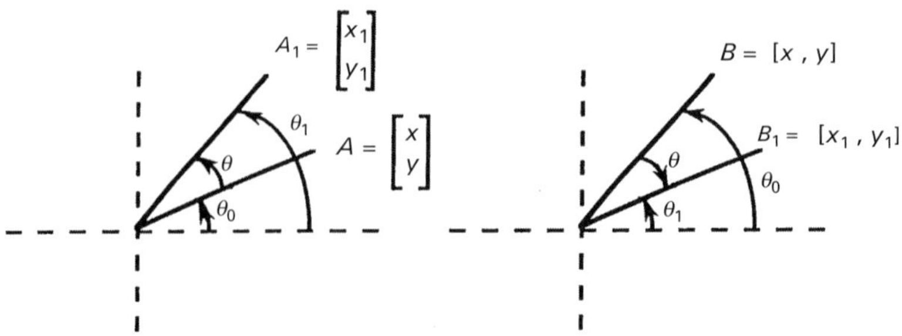

Figure 10.2 (a) Givens rotation of a column vector (b) Givens rotation of a row vector

Then

$$\tan\theta_1 = \frac{-x\sin\theta + y\cos\theta}{x\cos\theta + y\sin\theta} = \frac{\sin(\theta_0 - \theta)}{\cos(\theta_0 - \theta)} = \tan(\theta_0 - \theta), \tag{10.128}$$

and

$$\theta_1 = \theta_0 - \theta. \tag{10.129}$$

This shows the multiplication of B on the right by G_0 results in a rotation of $-\theta$ radians on the B vector as shown in Fig. 10.2b. If we replace G_0 by G_0^{-1} in (10.118), then the rotation of B is by θ radians.

Example 10.6 Recursive evaluation of sinusoids. Suppose we want to evaluate $\cos m\theta_0$ and $\sin m\theta_0$ recursively from known $\cos(m-1)\theta_0$, $\sin(m-1)\theta_0$, $\cos\theta_0$, and $\sin\theta_0$. From

$$\begin{aligned}\sin m\theta_0 &= \sin((m-1)\theta_0 + \theta_0) \\ &= \sin(m-1)\theta_0 \cos\theta_0 + \cos(m-1)\theta_0 \sin\theta_0,\end{aligned} \tag{10.130}$$

$$\begin{aligned}\cos m\theta_0 &= \cos((m-1)\theta_0 + \theta_0) \\ &= \cos(m-1)\theta_0 \cos\theta_0 - \sin(m-1)\theta_0 \sin\theta_0.\end{aligned} \tag{10.131}$$

These two equations can be expressed as

$$[\sin m\theta_0, \cos m\theta_0] = [\sin(m-1)\theta_0, \cos(m-1)\theta_0] \begin{bmatrix} \cos\theta_0 & -\sin\theta_0 \\ \sin\theta_0 & \cos\theta_0 \end{bmatrix}. \tag{10.132}$$

Since the matrix on the r.h.s. of (10.132) is the Givens orthogonal rotation matrix G_0 in (10.117), the iterative algorithm of (10.132) is numerically stable.

Finally, a Givens transformation can transform two arbitrary (non-collinear) real-valued column vectors a and b to two orthogonal vectors α and β such that $L(a,b) = L(\alpha,\beta)$. Then

$$[\alpha,\beta] = [a,b] \begin{bmatrix} \cos\theta & -\sin\theta \\ \sin\theta & \cos\theta \end{bmatrix}, \tag{10.133}$$

$$\alpha = a\cos\theta + b\sin\theta, \tag{10.134}$$

$$\beta = -a\sin\theta + b\cos\theta. \tag{10.135}$$

For α to be orthogonal to β, we need

$$0 = \alpha^T \beta = -(a^T a - b^T b)\sin\theta\cos\theta + a^T b(\cos^2\theta - \sin^2\theta). \tag{10.136}$$

There are various ways to obtain θ satisfying (10.136). One approach is to take

$$p = a^T b, \tag{10.137}$$

$$q = a^T a - b^T b, \tag{10.138}$$

$$v = (4p^2 + q^2)^{1/2}, \tag{10.139}$$

$$\cos\theta = \left[\frac{v+q}{2v}\right]^{1/2}, \quad q \geq 0,$$

$$\sin\theta = \frac{p}{v\cos\theta}, \quad q \geq 0, \tag{10.140}$$

$$\sin\theta = \operatorname{sgn}(p)\left[\frac{v-q}{2v}\right]^{1/2}, \quad q < 0,$$

$$\cos\theta = \frac{p}{v\sin\theta}, \quad q < 0. \tag{10.141}$$

10.9 Householder Transformation

Householder transformation is an efficient orthogonal transformation method of reducing an $m \times n$ matrix A to an upper triangular matrix. In contrast to the Givens transformation approach (where each transformation only results in one additional zero along a column), now in one orthogonal transformation, all the elements of a given column vector are reduced to zero except for the first element.

In R^n, every vector $x \in R^n$ can be expressed uniquely as

$$x = x_m + x_m^{\perp}, x_m \in M, x_m^{\perp} \in M^{\perp}, \tag{10.142}$$

for any subspace M of R^n. P is a *projection* operator onto a subspace M if $Px = P(x_m + x_m^{\perp}) = x_m$ for all $x \in R^n$. Necessary and sufficient conditions for an operator on R^n to be a projection operator are

$$P^2 = P, \tag{10.143}$$

$$P = P^T. \tag{10.144}$$

Let u be any unit norm column vector in R^n. Then

$$U = uu^T \tag{10.145}$$

is a projection operator onto $M = L(u)$ and satisfies (10.143) and (10.144). Indeed,

$$U^{\perp} = I - U \tag{10.146}$$

is the projection operator onto the complement $M^{\perp}$.

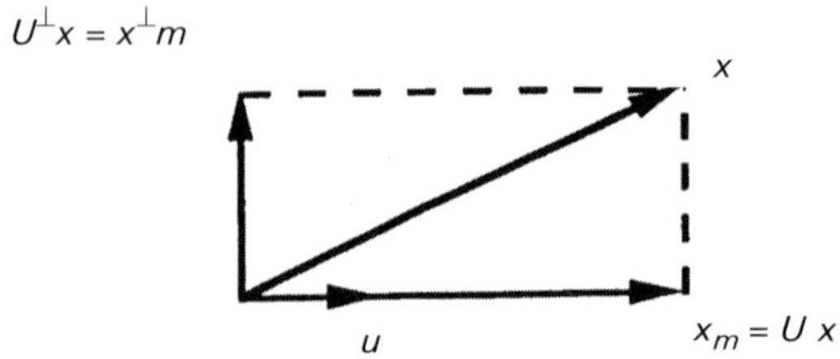

Figure 10.3 Projection operator U operating on x

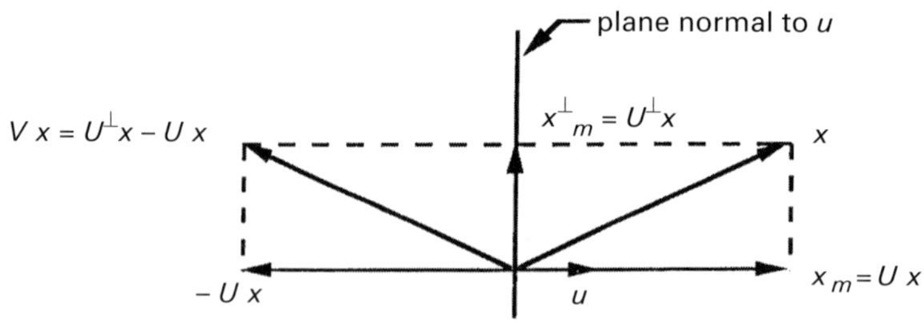

Figure 10.4 Reflection operator V on vector x

Example 10.7 Let u be a unit norm vector in R^n. Define $M = L(u)$. Then for any $x = x_m + x_m^{\perp}$ in R^n, $x_m = cu$ for some real-valued c as shown in Fig. 10.3.

Then U defined by (10.145) is indeed a projection onto M since

$$Ux = uu^T(x_m + x_m^{\perp}) = u[u^T x_m + u^T x_m^{\perp}] = uc = x_m,$$

and $U^{\perp}$ defined by (10.146) is a projection operator onto $M^{\perp}$ since

$$U^{\perp}x = (I - uu^T)(x_m + x_m^{\perp}) = (x_m + x_m^{\perp}) - x_m = x_m^{\perp}.$$

Now, consider the operator

$$V = I - 2uu^T = (I - uu^T) - uu^T = U^{\perp} - U. \tag{10.147}$$

The operator V in (10.147) is called a *reflection* operator and is symmetric and orthogonal. From (10.147), $V^T = V$ and

$$\begin{aligned} V^T V &= (I - 2uu^T)(I - 2uu^T) \\ &= I - 4uu^T + 4uu^T uu^T = I - 4uu^T + 4uu^T = I. \end{aligned}$$

Example 10.8 Consider the operator V operating on the vector x of Example 10.7. As shown in Figure 10.4, $Vx = U^{\perp}x - Ux$ represents a reflection of the vector about the plane normal to u.

A Householder operator H is a special class of reflection operator V such that Hx has zeros in all elements of the resulting column vector except the first. That is, Hx is proportional to $e_1 = (1,0,\ldots,0)^T$. Geometrically, if $x \neq constant \times e_1$, there is a plane such that upon reflection about that plane, $Hx \in L(e_1)$.

THEOREM 10.3 *For any $n \times 1$ vector $a = (a_1,\ldots,a_n)^T \neq \sigma e_1$ where σ is a scalar, there exists a unique $n \times n$ Householder operator H defined by*

$$H = I - 2uu^T, \tag{10.148}$$

with

$$u = \frac{v}{||v||}, \tag{10.149}$$

$$v = a + \sigma e_1, \tag{10.150}$$

$$\sigma = (\text{sgn } a_1)||a||, \tag{10.151}$$

such that

$$Ha = -\sigma e_1. \tag{10.152}$$

Proof: Since $||u|| = 1$, then H of (10.148) is a reflection operator. In particular,

$$\begin{aligned}\frac{||v||^2}{2} &= \frac{(a+\sigma e_1)^T(a+\sigma e_1)}{2}\\ &= \frac{(a^Ta + 2\sigma a^T e_1 + \sigma^2)}{2}\\ &= \frac{(2a^Ta + 2\sigma a^T e_1)}{2}\\ &= a^Ta + \sigma a_1.\end{aligned}$$

Then

$$\begin{aligned}Ha &= [I - 2\frac{vv^T}{||v||^2}]a\\ &= [I - 2\frac{1}{||v||^2}(a+\sigma e_1)(a+\sigma e_1)^T]a\\ &= a - \frac{(a+\sigma e_1)(a^Ta + \sigma a_1)}{(a^Ta+\sigma a_1)}\\ &= a - (a+\sigma e_1) = -\sigma e_1.\end{aligned}$$

Now, the Householder transformation can be used to perform QR decomposition on an $m \times n$ matrix A.

THEOREM 10.4 *Let $A = A^{(0)}$ be an $m \times n$ matrix with $m \geq n$ of rank n. Consider*

$$A^{(k)} = P_k A^{(k-1)}, \qquad k = 1,\ldots,n, \tag{10.153}$$

where

$$P_k = \begin{bmatrix} I_{k-1} & 0 \\ 0 & H_{m-k+1} \end{bmatrix}, \qquad k = 1,\ldots,n, \tag{10.154}$$

is an $m \times m$ orthogonal matrix with H_{n-k+1} being a Householder matrix of dimension $(m-k+1) \times (m-k+1)$ of the form of (10.148). Then

$$A^{(n)} = \begin{bmatrix} R \\ 0 \end{bmatrix}, \tag{10.155}$$

where R is an $n \times n$ upper triangular matrix and 0 is an $(m-n) \times n$ matrix of all zeros.

Proof: At step k, assume the $m \times n$ matrix $A^{(k)}$ is upper triangular with the upper-left submatrix α_k of dimension $(k-1) \times (k-1)$ also being upper triangular as shown in

$$A^{(k)} = \begin{bmatrix} \alpha_k & \beta_k \\ 0 & \mu_k \end{bmatrix}. \tag{10.156}$$

From (10.153) and (10.154),

$$A^{(k+1)} = \begin{bmatrix} \alpha_k & \beta_k \\ 0 & H_{m-k+1}\mu_k \end{bmatrix}. \tag{10.157}$$

By taking the first column of μ_k to be of the form of vector a of Theorem 10.3, then (10.152) shows the elements in the first column of $H_{m-k+1}\mu_k$ below the first component are all zero. By continuing in this manner, the result of (10.155) is obtained. Furthermore,

$$\begin{aligned} P_k^T P_k &= \begin{bmatrix} I_{k-1} I_{k-1} & 0 \\ 0 & (I_{m-k+1} - 2uu^T)(I_{m-k+1} - 2uu^T) \end{bmatrix} \\ &= \begin{bmatrix} I_{k-1} & 0 \\ 0 & I_{m-(k-1)} \end{bmatrix} \\ &= I_m, \qquad k = 1, \cdots, n. \end{aligned}$$

Thus, $P_k, k = 1, \ldots, n$, is a sequence of orthogonal matrices and $H = P_n \cdots P_1$.

10.10 QR Decomposition Approach to Linear System of Equations

Consider a real-valued $m \times n$ matrix A with $m \geq n$ and all the columns are linearly independent (i.e., rank $(A) = n$). Then from the QR decomposition of A in Section 10.5, we can find an $m \times m$ orthogonal matrix $Q = [Q_1, Q_2]$, where Q_1 is $m \times n$ and Q_2 is $m \times m - n$, such that

$$Q^T A = \begin{bmatrix} Q_1^T \\ Q_2^T \end{bmatrix} A = R = \begin{array}{c} \\ n \\ m-n \end{array} \begin{array}{c} n \\ \begin{bmatrix} R_1 \\ R_2 \end{bmatrix} \end{array} \tag{10.158}$$

is an $m \times n$ matrix that can be expressed as an $n \times n$ upper triangular matrix R_1 (with non-zero diagonal elements) and an all zero $(m-n) \times n$ matrix R_2. Then the least-squares problem becomes

$$||Ax - b||^2 = ||Q^T(Ax - b)||^2 = ||Q^T Ax - Q^T b||^2 = ||Rx - f||^2, \tag{10.159}$$

where R is given by (10.158) and the $m \times 1$ matrix f is given by

$$f = \begin{bmatrix} Q_1^T \\ Q_2^T \end{bmatrix} b = \begin{matrix} & 1 \\ n & \\ m-n & \end{matrix} \begin{bmatrix} c \\ d \end{bmatrix}. \tag{10.160}$$

In (10.159), we used the basic property that the norm of any vector is invariant with respect to any orthogonal transformation. From (10.158), (10.159) becomes

$$||Ax - b||^2 = ||[(R_1x)^T, (R_2x)^T]^T - [c^T, d^T]^T||^2 = ||R_1x - c||^2 + ||d||^2. \tag{10.161}$$

Since R_1 is a non-singular upper triangular square matrix, the back substitution method (i.e., the second part of the Gaussian elimination method) can be used to solve for the exact solution of

$$R_1\hat{x} = c, \tag{10.162}$$

where

$$c = Q_1^T b, \; d = Q_2^T b. \tag{10.163}$$

Let $\hat{x}$ denote the solution of (10.162). Thus, the least-squares problem reduces to

$$\begin{aligned} \min ||Ax - b||^2 &= ||A\hat{x} - b||^2 = ||R\hat{x} - f||^2 \\ &= ||R_1\hat{x} - c||^2 + ||d||^2 = ||d||^2. \end{aligned} \tag{10.164}$$

Equation (10.164) shows that the optimum least-squares solution is indeed given by the solution $\hat{x}$ of (10.162) and the minimum residual norm squared is given simply by $||d||^2$.

As far as the least-squares problem is concerned (i.e., (10.159)–(10.164)), any of the QR decomposition techniques including *G-S*, *M-G-S*, Givens transformation, and Householder transformation as considered in Sections 10.6–10.9 are equally valid.

Furthermore, from the $B\tilde{R}$ decomposition (Corollary 10.1) representation of an $m \times n$ matrix A with $m \geq n$ and rank $(A) = n$, then the least-squares estimation solution $\hat{x}$ of $Ax \approx b$, for an $m \times n$ vector b, is given by the exact solution of $\tilde{R}\hat{x} = y$, where y is the solution of $B^T b = D^2 y$ with $D^2 = B^T B$. This result follows from the sequence of arguments

$$\begin{aligned} &A^T(A\hat{x} - b) = 0 \Leftrightarrow \tilde{R}^T B^T(B\tilde{R}\hat{x} - b) = 0 \Leftrightarrow \\ &\tilde{R}^T(D^2\tilde{R}\hat{x} - B^T b) = 0 \Leftrightarrow D^2\tilde{R}\hat{x} = B^T b \Leftrightarrow \\ &\tilde{R}\hat{x} = D^{-2}B^T b = y. \end{aligned} \tag{10.165}$$

Example 10.9 Consider the 4×3 matrix A and 3×1 vector b of (10.36) in Example 10.4 of Section 10.3. By using the M-G-S method to perform the $B\tilde{R}$ decomposition (see (10.98)–(10.100)), we have

$$B = \begin{bmatrix} 1 & 0 & 0 \\ \epsilon & -\epsilon & -\epsilon/2 \\ 0 & \epsilon & -\epsilon/2 \\ 0 & 0 & \epsilon \end{bmatrix}, \tilde{R} = \begin{bmatrix} 1 & 1 & 1 \\ 0 & 1 & \frac{1}{2} \\ 0 & 0 & 1 \end{bmatrix}. \tag{10.166}$$

Furthermore,

$$D^2 = B^T B = \begin{bmatrix} 1 & -\epsilon^2 & -\epsilon^2/2 \\ -\epsilon^2 & 2\epsilon^2 & 0 \\ -\epsilon^2/2 & 0 & 3\epsilon^2/2 \end{bmatrix}. \tag{10.167}$$

By using

$$B^T b = D^2 y, \tag{10.168}$$

we have

$$\begin{aligned} [1,0,0]^T &= [y_1 - \epsilon^2 y_2 - \epsilon^2 y_3/2, \epsilon^2(-y_1 + 2y_2), \epsilon^2(-y_1 + 3y_3)/2]^T \\ &= [y_1, \epsilon^2(-y_1 + 2y_2), \epsilon^2(-y_1 + 3y_3)/2]^T, \end{aligned} \tag{10.169}$$

which yields

$$y = [1, 1/2, 1/3]^T. \tag{10.170}$$

Thus, the solution of $\hat{x}$ from

$$\tilde{R}\hat{x} = y \tag{10.171}$$

yields

$$\hat{x} = [1/3, 1/3, 1/3]^T. \tag{10.172}$$

We note this solution agrees closely with the true solution of

$$\hat{x} = [1, 1, 1]^T / \ (3 + \epsilon^2) \tag{10.173}$$

given earlier in (10.39).

10.11 Singular Value Decomposition

Singular value decomposition (SVD) is one of the most basic decomposition results in linear algebra. While there are many ways to present the SVD concept, we shall consider it from the point of view of orthonormalization in the domain and range of a given matrix A. Initially, consider a full rank $m \times n$ matrix A with $m \geq n$. Let $\{v_1, \ldots, v_n\}$ be n orthonormal column vectors in R^n. Then every $x \in R^n$ can be expressed as

$$x = \sum_{i=1}^{n} x_i v_i, \tag{10.174}$$

with

$$x_i = v_i^T x, \qquad i = 1, \ldots, n. \tag{10.175}$$

If $\{v_1, \ldots, v_n\}$ are not orthonormal, then the coefficients $\{x_i\}$ are not obtained easily as in (10.175).

Consider the set $\{w_1, \ldots, w_n\}$ of m-dimensional column vectors where

$$w_i = Av_i \in R^m, \qquad i = 1, \ldots, n. \tag{10.176}$$

Then the transformation of any column vector $x \in R^n$ becomes

$$y = Ax = A\left(\sum_{i=1}^{n} x_i v_i\right) = \sum_{i=1}^{n} x_i A v_i = \sum_{i=1}^{n} x_i w_i, \tag{10.177}$$

where y is an m-dimensional column vector in R^m. While every $x \in R^n$ has an equivalent representation $\{x_1, \ldots, x_n\}$ w.r.t. $\{v_1, \ldots, v_n\}$ given by the simple relationships of (10.175), every y in the range of A also has the representation $\{x_1, \ldots, x_n\}$ w.r.t. $\{w_1, w_2, \ldots, w_n\}$. However, in general $\{w_1, \ldots, w_n\}$ need not be orthonormal, and then there is no simple relationship between y and $\{x_i\}$.

Suppose that not only is $\{v_i, \ldots, v_n\}$ an orthonormal set in R^n, but $\{w_1, \ldots, w_n\}$ is also an orthogonal set in R^m. That is

$$w_i^T w_j = \begin{cases} s_i^2 > 0, & i = j \\ 0, & i \neq j \end{cases}. \tag{10.178}$$

By normalization, we can obtain an orthonormal set $\{u_1, \ldots, u_n\}$ in R^m by setting

$$u_i = \frac{1}{s_i} w_i, \qquad i = 1, \ldots, n, \tag{10.179}$$

where

$$u_i^T u_j = \begin{cases} 1, & i = j \\ 0, & i \neq j \end{cases}, \tag{10.180}$$

with $s_i > 0$. Then (10.177) can be expressed as

$$y = \sum_{i=1}^{n} x_i w_i = \sum_{i=1}^{n} x_i s_i u_i = \sum_{i=1}^{n} y_i u_i, \tag{10.181}$$

with

$$y_i = x_i s_i = u_i^T y, \qquad i = 1, \ldots, n. \tag{10.182}$$

Since $\{u_1, \ldots, u_n\}$ is an orthonormal set, the $\{y_i\}$ in (10.182) is simply related to y as $\{x_i\}$ is to x in (10.175). From (10.181) and (10.182), y becomes

$$\begin{aligned} y &= \sum_{i=1}^{n} x_i s_i u_i = \sum_{i=1}^{n} u_i s_i v_i^T x \\ &= [u_1 s_1, u_2 s_2, \ldots, u_n s_n] \begin{bmatrix} v_1^T \\ v_2^T \\ \vdots \\ v_n^T \end{bmatrix} x \\ &= [u_1, u_2, \ldots, u_n] \begin{bmatrix} s_1 & & & 0 \\ & s_2 & & \\ & & \ddots & \\ 0 & & & s_n \end{bmatrix} \begin{bmatrix} v_1^T \\ v_2^T \\ \vdots \\ v_n^T \end{bmatrix} x = Ax. \end{aligned} \tag{10.183}$$

Since (10.183) is valid for every $x \in R^n$, then

$$A = USV^T, \tag{10.184}$$

where

$$U = [u_1, \ldots, u_n], \tag{10.185}$$

$$V = [v_1, \ldots, v_n], \tag{10.186}$$

and

$$S = \begin{bmatrix} s_1 & & & 0 \\ & s_2 & & \\ & & \ddots & \\ 0 & & & s_n \end{bmatrix}. \tag{10.187}$$

We note U is an $m \times n$ matrix with orthonormal columns vectors, V is an $n \times n$ matrix with orthonormal column vectors, and S is an $n \times n$ diagonal matrix with positive diagonal elements. Equation (10.184) is the expression of singular value decomposition (SVD) of a full rank $m \times n$ matrix A with $m \geq n$ under the assumption of the joint orthonormality of $\{v_1, \ldots, v_n\}$ and orthogonality of $\{Av_i, \ldots, Av_n\}$.

If in addition, the above matrix A is also a square, symmetric, and positive-definite $n \times n$ real-valued matrix, then (10.184) becomes

$$A = USV^T = A^T = VSU^T. \tag{10.188}$$

Then

$$[u_1, \ldots, u_n] = U = V = [v_1, \ldots, v_n], \tag{10.189}$$

and

$$\frac{1}{s_i} Av_i = \frac{1}{s_i} w_i = u_i = v_i, \qquad i = 1, \ldots, n. \tag{10.190}$$

Thus, the orthonormal basis for the domain of A is identical to the orthonormal basis of the range of A. Furthermore, (10.190) yields the eigenvalue equation of the matrix A given by

$$Av_i = s_i v_i, \qquad i = 1, \ldots, n. \tag{10.191}$$

Thus far, from (10.174)–(10.191) we have only considered some heuristic arguments on orthonormal vectors related to SVD. Now, we state an SVD theorem applicable to any real-valued matrix and provide an explicit algorithm for the solution of this problem. An equivalent SVD for the complex-valued A matrix also exists.

THEOREM 10.5 (SVD) *For any arbitrary real-valued $m \times n$ matrix A of rank $r \leq$ $\min(m, n)$, there exists an orthogonal $n \times n$ matrix V, an orthogonal $m \times m$ matrix U, an $m \times n$ "diagonal matrix" Σ of the form*

$$\Sigma = \begin{bmatrix} S_r & 0 \\ 0 & 0 \end{bmatrix}, \tag{10.192}$$

where the diagonal matrix $S_r = \mathrm{diag}(s_1, \ldots, s_r)$ *with the* s_i *being real-valued and called singular values (SV) and satisfying*

$$s_1 \geq s_2 \geq \cdots \geq s_r > s_{r+1} = \cdots = s_{\min(m,n)} = 0, \tag{10.193}$$

such that

$$A = U\Sigma V^T. \tag{10.194}$$

Proof: There are several possible methods of proving the SVD of A in (10.194). We shall consider this decomposition based on orthogonalization via the Givens plane rotation approach.

Case I. $m \geq n$

The basis of (10.194) is that we want to find an $n \times n$ orthogonal matrix V such that

$$[w_1, \ldots, w_n] = W = AV \tag{10.195}$$

with the m-dimensional column vectors w_i being orthogonal. That is

$$w_i^T w_j = \begin{cases} s_i^2 \geq 0, & i = j \\ 0, & i \neq j \end{cases}. \tag{10.196}$$

Then denote

$$u_i = \frac{1}{s_i} w_i, \qquad i = 1, \ldots, n, \tag{10.197}$$

with

$$u_i^T u_j = \begin{cases} 1, & i = j \\ 0, & i \neq j \end{cases}. \tag{10.198}$$

If the rank of A (i.e., r) is less than n, then only the first r number of s_i are positive and only r of the u_i are defined by (10.197), the remaining $(n - r)u_i$ can be chosen arbitrarily to satisfy (10.198). While the orthogonal matrix condition of V indicates

$$V^T V = VV^T = I_n, \tag{10.199}$$

the condition of (10.198) yields

$$U^T U = I_n, \tag{10.200}$$

which is equivalent to having the columns of U being orthonormal. In general, the rows of U need not be orthonormal. That is, in general

$$UU^T \neq I_m. \tag{10.201}$$

But (10.197) indicates

$$W = US = [u_1, \ldots, u_n] \begin{bmatrix} s_1 & & & 0 \\ & s_2 & & \\ & & \ddots & \\ 0 & & & s_n \end{bmatrix}, \tag{10.202}$$

where $s_{r+1} = \cdots = s_n = 0$. Then (10.195) and (10.202) show that $US = AV$, and by using (10.199) we obtain

$$A = USV^T. \tag{10.203}$$

Clearly, the crucial step in the above discussion is to find an orthogonal matrix V capable of producing (10.195) with orthogonal w_i. Indeed, we shall find a cascade of Givens transformations $G_k, k = 1, \ldots, N$, such that

$$A_{k+1} = A_k G_k, \qquad k = 1, \ldots, N, \tag{10.204}$$

with $A_1 = A$ and $W = A_{N+1}$ having column vectors that are essentially mutually orthogonal. Then the cascade of orthogonal Givens transformations yield the desired orthogonal

$$V = G_1 G_2, \ldots, G_N. \tag{10.205}$$

As in Section 10.8 on Givens orthogonal transformation, each G_k is of the form

$$G = G(i,j) = \begin{array}{cc} & \begin{array}{cccccccccc} & & & i & & & & j & & \end{array} \\ \begin{array}{c} \\ \\ \\ i \\ \\ \\ \\ j \\ \\ \\ \end{array} & \left[\begin{array}{cccccccccc} 1 & & & & & & & & & \\ & 1 & & & & & & & & \\ & & 1 & & & & & & & \\ & & & c & & & & -s & & \\ & & & & 1 & & & & & \\ & & & & & \ddots & & & & \\ & & & & & & 1 & & & \\ & & & s & & & & c & & \\ & & & & & & & & 1 & \\ & & & & & & & & & 1 \end{array} \right] \end{array} \tag{10.206}$$

where $s = \sin\theta$ and $c = \cos\theta$. Denote the $m \times n$ matrices A_{k+1} and A_k as n column vectors by

$$A_{k+1} = [a_1, \ldots, a_{i-1}, \alpha, a_{i+1}, \ldots, a_{j-1}, \beta, a_{j+1}, \ldots, a_n], \tag{10.207}$$

$$A_k = [a_1, \ldots, a_{i-1}, a, a_{i+1}, \ldots, a_{j-1}, b, a_{j+1}, \ldots, a_n]. \tag{10.208}$$

From (10.204)–(10.208), we see all the columns of A_{k+1} and A_k are identical except for the ith and jth columns. For these two columns, we have

$$\alpha = a\cos\theta + b\sin\theta, \tag{10.209}$$

$$\beta = -a\sin\theta + b\cos\theta, \tag{10.210}$$

as in (10.134) and (10.135). By picking the rotation angles θ such that α and β are orthogonal, then

$$0 = \alpha^T \beta = -(a^T a - b^T b)\sin\theta\cos\theta + a^T b(\cos^2\theta - \sin^2\theta). \tag{10.211}$$

Since (10.211) has the same form as that of (10.136), then the results given in (10.137)–(10.141) for the solution of $\cos\theta$ and $\sin\theta$ can be used in (10.209) and (10.210).

Since (10.209)–(10.210) is an orthogonal transformation, then the sum of the square of the norm of the original vectors $\{a, b\}$ equals that of the vectors $\{\alpha, \beta\}$ after the transformation

$$||\alpha||^2 + ||\beta||^2 = ||a||^2 + ||b||^2. \tag{10.212}$$

Furthermore, by direct evaluation we have

$$(\alpha, a_k)^2 + (\beta, a_k)^2 = (a, a_k)^2 + (b, a_k)^2, \qquad k \neq i, k \neq j. \tag{10.213}$$

Now, consider in place computation of the Givens transformation in (10.204). As before, let the generic terms of A be denoted by $[a_1, \ldots, a_n]$. At the kth iteration, consider a measure of the amount of non-orthogonalization in A_k before the transformation as given by

$$z^{(k)} = \sum_{i=1}^{n-1} \sum_{j=i+1}^{n} (a_i^T a_j)^2. \tag{10.214}$$

If $z^{(k)} = 0$, then all the columns of A_k are orthogonal. Since only the ith and jth columns of A_k are involved in the kth iteration, by using the properties of (10.212) and (10.213), we have

$$z^{(k+1)} - (\beta^T \alpha)^2 = z^{(k)} - (b^T a)^2. \tag{10.215}$$

But by design, $(\beta^T \alpha) = 0$ in (10.211). Thus,

$$z^{(k+1)} = z^{(k)} - (b^T a)^2, \qquad k = 1, \ldots, N. \tag{10.216}$$

Since $(b^T a)^2$ is always non-negative, then $z^{(k)}$ is a monotonically decreasing positive-valued sequence. Details on its convergence to zero are omitted here.

We note, that in an iteration, if the ith and jth column vectors are made orthogonal, in a subsequent iteration in making the ith and mth column vectors orthogonal, the original ith and jth column vectors need no longer to be orthogonal. However, in light of (10.216), after each iteration, even though some of the vectors need not be completely orthogonal, the total measure of non-orthogonality given by $z^{(k)}$ will be decreasing and repeated iterations will lead to the desired solution of orthogonal column vectors.

We have shown the SVD of a real-valued $m \times n$ matrix A where $m \geq n$ has the form given in (10.203). If we pad an all zero $(m - n) \times n$ matrix below the S matrix given in (10.202), then it is in the form of Σ of (10.192). Furthermore, with the padding of the $(m - n) \times n$ matrix below the S matrix, we can augment the $m \times n$ matrix U in (10.202) with $m - n$ additional orthonormal vectors to the right so that $U = [u_1, \ldots, u_n, u_{n+1}, \ldots, u_m]$ is an $m \times m$ orthogonal matrix. Thus, the SVD of (10.203) is equivalent to the general form of SVD in (10.194).

Case II. $m < n$

When $m < n$, we can define $B = A^T$. Then B is an $n \times m$ matrix with $n > m$. Thus, the SVD result of Case I applies. Thus, $A^T = B = U_B \Sigma_B V_B^T$. Then $A = V_B \Sigma_B^T U_B^T$ has the general form of SVD in (10.193).

In a sequential processing computer, a sweep can be defined by

$$G_0 = \prod_{i=1}^{n-1} \prod_{j=i+1}^{n} G(i,j), \tag{10.217}$$

where $G(i, j)$ are the Givens transformations given in (10.206). In each sweep, we can start with $i = 1$ and $j = 2, 3, \ldots, n$, then $i = 2$ and $j = 3, \ldots, n$, and finally end with $i = n - 1$ and $j = n$. Thus, there are a total number of $n(n-1)/2$ iterations per sweep. In a parallel processing computer, we can perform the Givens rotations

$$G(i,j), \ldots, G(s,t), \ldots, G(x,y), \ldots, \qquad i \neq j \neq s \neq t \neq x \neq y, \tag{10.218}$$

simultaneously. Thus, the throughput rate of the overall convergence of this algorithm should be increased significantly.

Now, consider some simple relationships among the eigenvalues and eigenvectors of $A^T A$ and AA^T to the singular values and singular vectors of an arbitrary $m \times n$ real-valued matrix A with $m \geq n$. First, we note $A^T A$ is a symmetric and non-negative definite $n \times n$ matrix. Indeed, for any $x \in R^n$,

$$(x, A^T Ax) = (Ax, Ax) = (A^T Ax, x) = ||Ax||^2 \geq 0. \tag{10.219}$$

Similarly, AA^T is a symmetric and non-negative definite $m \times m$ matrix. However, from (10.203)

$$A^T A = (USV^T)^T (USV^T) = VSU^T USV^T = VSI_n SV^T = VS^2 V^T. \tag{10.220}$$

By multiplying $A^T A$ from the right by V, we obtain

$$(A^T A)V = VS^2 V^T V = VS^2 I_n = VS^2. \tag{10.221}$$

But (10.221) is equivalent to

$$(A^T A)v_i = s_i^2 v_i, \qquad i = 1, 2, \ldots, n. \tag{10.222}$$

This shows that $\{s_1^2, \ldots, s_n^2\}$ are the eigenvalues and $\{v_1, \ldots, v_n\}$ are the eigenvectors of $A^T A$. Similar to the above arguments,

$$(AA^T)u_i = s_i^2 u_i, \qquad i = 1, 2, \ldots, n. \tag{10.223}$$

Thus, $\{s_1^2, \ldots, s_n^2\}$ are the eigenvalues and $\{u_1, \ldots, u_n\}$ are the eigenvectors of the $n \times n$ AA^T matrix. We note, since AA^T is an $m \times m$ symmetric and non-negative definite matrix, the remaining $m - n$ eigenvalues are all zero. Similarly, when A is an $m \times n$ matrix with $n > m$, all the above results still hold with $i = 1, \ldots, m$ in (10.222) and (10.223) with the remaining $n - m$ eigenvalues of AA^T being all zero. While the results of (10.222) and (10.223) are interesting in relating the eigenvalues of $A^T A$ and AA^T to the singular values of A, from the numerical point of view, it is not suggested to use (10.222) or (10.223) to obtain the singular values by taking the square roots of the eigenvalues of $\{s_1^2, \ldots, s_n^2\}$.

10.12 SVD Approach to Linear System of Equations

As considered in Section 10.11, if the real-valued $m \times n$ matrix A with $m \geq n$ has full rank (i.e., rank $(A) = n$ or equivalently all the columns of A are linearly independent), then the QR decomposition approach to the least-squares solution of $Ax \approx b$ is simple and is considered to be numerically satisfactory. In practice, often the rank of A is unknown and may not be full rank. While the QR decomposition approach can still be modified with some complications, the SVD approach is generally recognized to be the most general theoretically and numerically satisfactory approach for the linear least-squares problem.

THEOREM 10.6 *Consider any real-valued $m \times n$ matrix A of rank $r \leq \min\{m,n\}$. Let b be any $m \times 1$ real-valued vector. Let*

$$\hat{x} = A^{+}b, \tag{10.224}$$

where the $n \times m$ matrix A^{+} is defined by

$$A^{+} = V\Sigma^{+}U^{T}, \tag{10.225}$$

with the $n \times m$ matrix

$$\Sigma^{+} = \begin{array}{c} \left[\begin{array}{cc} S_r^{-1} & 0 \\ 0 & 0 \end{array} \right] \begin{array}{c} r \\ n-r \end{array} \\ \begin{array}{cc} r & m-r \end{array} \end{array}, \tag{10.226}$$

$$S_r = \operatorname{diag}[s_1, s_2, \ldots, s_r], \tag{10.227}$$

and the $m \times m$ matrix U and the $n \times n$ matrix V are used in the SVD of A as given by

$$A = U\Sigma V^{T}, \tag{10.228}$$

with the $m \times n$ matrix

$$\Sigma = \begin{array}{c} \left[\begin{array}{cc} S_r & 0 \\ 0 & 0 \end{array} \right] \begin{array}{c} r \\ m-r \end{array} \\ \begin{array}{cc} r & n-r \end{array} \end{array}, \tag{10.229}$$

satisfying

$$s_1 \geq s_2 \geq \cdots \geq s_r > 0. \tag{10.230}$$

Then A^{+} is called the Moore–Penrose pseudo-inverse of A, and $\hat{x}$ is the optimum solution of the linear least-squares problem in the sense of

$$||A\hat{x} - b|| \leq ||Ax - b||, \; x \in R^{n},$$

and $||\hat{x}|| \leq ||x||$ for all x satisfying

$$||Ax - b|| = ||A\hat{x} - b||. \tag{10.231}$$

We want to show consistency of the $\hat{x}$ in (10.224) to that obtained in the normal equation approach with $m \geq n$ and rank $(r) = n$. From (10.224)–(10.226),

$$\hat{x} = A^+ b = V\Sigma^+ U^T b = V[S_n^{-1} \ \ 0]U^T b = VS_n^{-1}[u_1 \cdots u_n]^T b. \tag{10.232}$$

From (10.31) and (10.32), we have the normal equation solution

$$\hat{x}^{NE} = (A^T A)^{-1} A^T b. \tag{10.233}$$

From (10.228)–(10.230) on the SVD of A, we have

$$A^T = V\Sigma^T U^T, \tag{10.234}$$

$$A^T A = V\Sigma^T U^T U \Sigma V^T = V\Sigma^T \Sigma V^T = VS_n^2 V^T, \tag{10.235}$$

$$(A^T A)^{-1} = VS_n^{-2} V^T, \tag{10.236}$$

$$(A^T A)^{-1} A^T = VS_n^{-2} V^T V \Sigma^T U^T = VS_n^{-2}[S_n \ \ 0]U^T = VS_n^{-1}[u_1 \cdots u_n]^T. \tag{10.237}$$

Substituting (10.237) into (10.233) shows $\hat{x}^{NE}$ of (10.233) equals $\hat{x}$ of (10.232). Thus, under infinite precisions, there is no difference in the two approaches. However, as can be seen in the following Example 10.10, under finite precisions, the SVD approach yields a more stable solution than that under the normal equation approach.

Example 10.10 Consider

$$A = \begin{bmatrix} 1 & 11 & 100.9999999 \\ 1 & 12 & 102 \\ 1 & 13 & 103 \\ 1 & 14 & 104 \end{bmatrix}, \quad b = \begin{bmatrix} 0.676800 \\ 0.229619 \\ -0.217562 \\ -0.664743 \end{bmatrix}. \tag{10.238}$$

Direct evaluation of the SV of A shows $s_1 = 206.5435$, $s_2 = 1.948945$, and $s_3 = 6.085054 \times 10^{-10}$. Then A has an effective rank of 2 and its Σ^+ is denoted by

$$\Sigma_2^+ = \begin{bmatrix} S_2^{-1} & 0 & 0 \\ 0 & 0 & 0 \end{bmatrix}, \tag{10.239}$$

where

$$S_2^{-1} = \text{diag}(1/s_1, 1/s_2). \tag{10.240}$$

From (10.239) and (10.240), the least-squares estimate assuming an effective rank of 2 is

$$\hat{x}_2 = V\Sigma_2^+ U^T b = [6.348788 \times 10^{-3}, -0.5092860, 0.06210493]^T. \tag{10.241}$$

By assuming rank 3 for A, we have

$$S_3^{-1} = \text{diag}(1/s_1, 1/s_2, 1/s_3), \tag{10.242}$$

and $\Sigma_3^+ = [S_3^{-1} \ \ 0]$. Then

$$\hat{x}_3 = V\Sigma_3^+ U^T b = [6.34873 \times 10^{-3}, -0.5092860, 0.0621049]^T. \tag{10.243}$$

Now, consider a slight perturbation on b of (10.238). Let

$$\tilde{b} = [0.68200000, 0.2220000, -0.2190000, -0.6610000]^T. \tag{10.244}$$

Then for the same A with the assumption of rank 2, we have

$$\tilde{x_2} = V\Sigma_2^+ U^T \tilde{b} = [6.34621 \times 10^{-3}, -0.509079, 0.0620795]^T, \tag{10.245}$$

and with the assumption of rank 3, we have

$$\tilde{x_3} = V\Sigma_3^+ U^T \tilde{b} = [1.65 \times 10^7, 1.83333 \times 10^5, -1.83333 \times 10^5]^T. \tag{10.246}$$

Clearly, by comparing (10.245) with (10.241), the rank 2 assumption yields a much more stable solution than the rank 3 assumption, as can be seen in (10.246) and (10.243).

10.13 Effective Rank Determination by SVD

The determination of the rank of a given matrix is a basic problem of interest as is the solution of a linear system of equations in Section 10.12 as well as various approximations, least-squares, and spectral estimations considered in Chapter 11. As considered in Section 10.12, SVD provides a fundamental approach toward the determination of the rank of a matrix by counting the number of significant SV. However, when the observed matrix consists of the true matrix of interest corrupted by some disturbances (such as round-off errors or other additive noises), determination of the effective rank of the observed matrix by considering the significant small and the insignificant large SV becomes more difficult.

A simple approach to this problem is to consider an additive perturbation model given by

$$B = A + E, \tag{10.247}$$

where the original $m \times n$ matrix A is assumed to satisfy $m \geq n$ and have rank $r \leq n \leq m$, while E is an $m \times n$ full rank perturbation matrix with small values. Let the SV of B be given by

$$\beta_1 \geq \beta_2 \geq \cdots \geq \beta_r \geq \beta_{r+1} \geq \cdots \geq \beta_n > 0,$$

and $\{\beta_{r+1}, \ldots, \beta_n\}$ will usually be small but not zero. Since β_r may also be small, the practical problem is to determine this value.

Example 10.11 Consider a 3×3 matrix A given by

$$A = \begin{bmatrix} 3 & 2 & 5 \\ 1 & 5 & 6 \\ 2 & 1 & 3 \end{bmatrix}. \tag{10.248}$$

Since the third column is the sum of the first two columns, A is of rank 2. Indeed, the SV of A are given by $[10.342, 2.653, 2.363 \times 10^{-17}]$, and thus it is simple to conclude that A is of rank 2. Now, consider an observed matrix B modeled by (10.247) as given by

$$B = \begin{bmatrix} 2.998 & 1.999 & 5.001 \\ 1.001 & 4.982 & 5.998 \\ 2.001 & 1.001 & 2.990 \end{bmatrix}.$$

Direct evaluation of the determinant of B yields a value of -0.176 and by applying the Gaussian elimination procedure to B, we obtain

$$\tilde{B} = \begin{bmatrix} 2.998 & 1.999 & 5.001 \\ 0 & 4.3146 & 4.3282 \\ 0 & 0 & -0.0136 \end{bmatrix},$$

which clearly indicates B is a matrix of rank 3. Yet, the SV of B are given by $\{10.331, 2.644, 0.0064\}$, which suggests that indeed B may be the result of a matrix of rank 2 disturbed by some small-valued perturbation matrix.

The determination of the effective rank t of the observed matrix B using its SV can be based on various criteria.

Criterion 1.

$$\beta_1 \geq \beta_2 \geq \cdots \geq \beta_t > \delta_1 \geq \beta_{t+1} \geq \cdots \geq \beta_n.$$

Criterion 2.

$$(\beta_t/\beta_1) > \delta_2 > (\beta_{t+1}/\beta_1).$$

Criterion 3.

$$\beta_t \gg \beta_{t+1}.$$

Criterion 4.

$$\beta_{t+1}^2 + \beta_{t+2}^2 + \cdots + \beta_n^2 < \delta_4.$$

Criterion 5.

$$\sqrt{[\beta_1^2 + \beta_2^2 + \cdots + \beta_t^2]/[\beta_1^2 + \beta_2^2 + \cdots + \beta_n^2]} > \delta_5.$$

Intuitively, all these five (and possibly others) appear reasonable and indeed may work in various cases. However, the threshold values of δ_2, δ_4, and δ_5 do not appear to be based on any explicit analytical expressions but are selected on an ad hoc basis. In this section, we shall use some results from the theories of perturbations of singular values of matrices and from statistical significance tests to provide an analytical approach on the construction of the threshold δ_1 needed in the approach of Criterion 1.

Consider an $m \times n$ real-valued matrix A with elements $\{a_{ij}\}$ and SV satisfying

$$\alpha_1 \geq \alpha_2 \geq \cdots \geq \alpha_k > 0,$$

where $k \leq n \leq m$. Then the Frobenius norm of A is given by

$$||A||_F = \sqrt{\sum_{i=1}^{m}\sum_{j=1}^{n}|a_{ij}|^2} = \sqrt{\sum_{i=1}^{k}\alpha_i^2}, \tag{10.249}$$

the 2-norm of A is given by

$$||A||_2 = \max_x \frac{||Ax||_2}{||x||_2} = \alpha_1, \tag{10.250}$$

and the 2-norm of $x = (x_1, \ldots, x_n)^T$ is given by

$$||x||_2 = \sqrt{(x_1^2 + \cdots + x_n^2)}. \tag{10.251}$$

THEOREM 10.7 *For any real-valued $m \times n$ matrix A with column vectors $\{a_1, \ldots, a_n\}$, and norms defined in (10.249)–(10.251), the following inequalities are valid*

$$\max|a_{ij}| \leq \max||a_j||_2 \leq ||A||_2 \leq ||A||_F \leq \sqrt{n}\max||a_j||_2 \leq \sqrt{mn}\max|a_{ij}|. \tag{10.252}$$

THEOREM 10.8 *Let A, B, and E be $m \times n$ real-valued matrices with $B = A + E$. Denote their respective SV by α_i, β_i, and ϵ_i, $i = 1, \ldots, k$, $k \leq \min(m,n)$, each set labeled in non-increasing order. Then*

$$|\beta_i - \alpha_i| \leq \epsilon_1 = ||E||_2, \; i = 1, \ldots, k. \tag{10.253}$$

From Theorem 10.8, if A is an $m \times n$ matrix of rank r, where $r \leq \min(m,n)$, E is an $m \times n$ perturbation matrix with a 2-norm $||E||_2 = \epsilon_1$, and $B = A + E$, then

$$|\beta_{r+1} - \alpha_{r+1}| = \beta_{r+1} \leq \epsilon_1 = ||E||_2. \tag{10.254}$$

Hence, if $\beta_r > \epsilon_1$, then

$$\beta_1 \geq \beta_2 \geq \cdots \geq \beta_r > \epsilon_1 \geq \beta_{r+1} \geq \cdots \geq \beta_n. \tag{10.255}$$

As a result of (10.255), we can give the following definition for the effective rank of the matrix B.

For any $m \times n$ matrix $B = A + E$, the *effective rank* of B is defined to be r, when

$$\beta_r > \epsilon_1 \geq \beta_{r+1}, \tag{10.256}$$

where $1 \leq r \leq \min(m,n)$ and $\epsilon_1 = ||E||_2$ is the 2-norm of E.

Clearly, the above definition of an effective rank given in (10.256) is consistent with that of Criterion 1 in which $\delta_1 = \epsilon_1$. In light of this definition, a simple sufficient condition for the determination of the effective rank of B in terms of the SV of A and of E is considered in Theorem 10.9.

THEOREM 10.9 *Let A, B, and E be $m \times n$ matrices as defined above, with the 2-norm of E denoted by ϵ_1. If $\alpha_r > 2\epsilon_1$, then $\beta_r > \epsilon_1 \geq \beta_{r+1}$ and B has effective rank r.*

While Theorem 10.9 guarantees the determination of the effective rank of B if $\alpha_r > 2\epsilon_1$, in practice we may not know the SV $\{\alpha_i\}$ of A, and the elements of E are not necessarily fully known. Thus, in order to circumvent this problem, we shall derive upper and lower bounds on ϵ_1 to be used in (10.256) for the effective rank determination under various cases of interest.

First, consider the finite precision case where the elements $\{e_{ij}\}$ of E satisfy

$$-\Delta/2 \le e_{ij} \le \Delta/2, \qquad 1 \le i \le m, 1 \le j \le n. \tag{10.257}$$

Here Δ may be the step size of an A/D converter or may be the round-off errors in some computations. In general, Δ can be taken to be some small number such that the equality is attained at least for some e_{ij}. Let e_j denote the jth column of E. Then from Theorem 10.7 and $\max |e_{ij}| = \Delta/2$ and

$$\max ||e_j||_2 = \max \sqrt{[\sum e_{ij}^2]} \le \sqrt{m}\Delta/2,$$

we obtain

$$\Delta/2 \le ||E||_2 = \epsilon_1 \le \sqrt{mn}\Delta/2. \tag{10.258}$$

Example 10.12 Consider the 3×3 matrix B in Example 10.11. By comparison to the matrix A in (10.247), we have $\Delta/2 = 0.018$. Then (10.258) yields

$$0.018 \le \epsilon_1 \le 0.054.$$

Since the SV of B of are $\{10.331, 2.644, 0.0064\}$, then

$$\beta_2 = 2.644 > 0.054 \ge \epsilon_1 \ge 0.018 > \beta_3 = 0.0064,$$

this shows the effective rank of B is predicted correctly as $r = 2$.

In many practical situations, the perturbation matrix E arises from noises in the measurement modeled by

$$b_{ij} = a_{ij} + e_{ij}, \qquad 1 \le i \le m, 1 \le j \le n. \tag{10.259}$$

While there are many possible models for the r.v. of $\{e_{ij}\}$, the simplest and possibly most useful is to assume $\{e_{ij}\}$ to be i.i.d. Gaussian r.v. with zero-mean and variance σ^2. Let $K = \{e_{ij} : -k \le e_{ij} \le k\}$ and $K' = \{e_{ij} : e_{ij} < -k \text{ or } e_{ij} > k\}$. From the theory of statistical test, if α denotes the level of significance of the test, then

$$\text{Prob}(K') = \text{Prob}(|e_{ij}| > k) \le \alpha. \tag{10.260}$$

For i.i.d. Gaussian $\{e_{ij}\}$, then

$$\text{Prob}(K') = 2\text{Prob}(e_{ij} > k) = 2(1 - \Phi(k/\sigma)), \tag{10.261}$$

where $\Phi(\cdot)$ is the zero-mean and unit variance Gaussian distribution function. From (10.260) and (10.261), then $(1 - \Phi(k/\sigma)) \leq \alpha/2$, or

$$k \geq \sigma \Phi^{-1}(1 - (\alpha/2)). \tag{10.262}$$

From the Gaussian distribution table, for $0.01 \leq \alpha \leq 0.05$, (10.262) yields approximately $2\sigma \leq k \leq 2.6\sigma$. The operational significance of these results for any e_{ij} is that with 0.99 probability of confidence, we have

$$\max |e_{ij}| = 2.6\sigma, \tag{10.263}$$

and with 0.95 probability confidence, we have

$$\max |e_{ij}| = 2\sigma. \tag{10.264}$$

The maximization in (10.263) or (10.264) is with respect to the values of the realization of the r.v. $|e_{ij}|$ for any i and j. Thus, by using a sufficiently large k, we can have high probability confidence that $|e_{ij}| \leq k$. By using the first and fifth inequalities of (10.252) in Theorem 10.7 and (10.262), we obtain

First Bounds

$$k \leq \epsilon_1 \leq \sqrt{mn}k. \tag{10.265}$$

For large value of mn and σ, the lower and upper bounds in (10.265) may not be tight and indeed may become useless. However, by using the second and fourth inequalities in (10.252), we can obtain tighter bounds. For any j, denote

$$S^2 = ||e_j||_2^2 = \sum_{i=1}^{m} |e_{ij}|^2. \tag{10.266}$$

Then S^2/σ^2 has a chi-square distribution with m degrees of freedom. Therefore, for a level of significance α, a constant c (corresponding to k of (10.260)) can be found such that

$$\text{Prob}(S^2/\sigma^2 > c) \leq \alpha. \tag{10.267}$$

Various values of c as a function of the level of significance α and the degree of freedom m can be found. From Theorem 10.7 and (10.266) and (10.267), we have

Second Bounds

$$\sqrt{c}\sigma \leq \epsilon_1 \leq \sqrt{nc}\sigma. \tag{10.268}$$

Finally, by using the second and third inequalities of (10.252), we obtain an even tighter bound on the r.h.s. of ϵ_1. Let

$$S_F^2 = \sum_{i=1}^{m} \sum_{j=1}^{n} |e_{ij}|^2. \tag{10.269}$$

Then as before, S_F^2/σ^2 has a chi-square distribution of mn degree of freedom. Then as in (10.267), for a given α, a constant c_F can be found such that

$$\text{Prob}(S_F^2/\sigma^2 > c_F) \leq \alpha. \tag{10.270}$$

By using (10.269) and (10.270), and Theorem 10.7, we obtain

Third Bounds

$$\sqrt{c}\sigma \leq \epsilon_1 \leq \sqrt{c_F}\sigma. \tag{10.271}$$

While the bounds in (10.271) are tighter than those of (10.265) and (10.268), we can simplify the r.h.s. of (10.271) for large values of mn. Specifically, when $mn > 30$, a simple (but good) approximation of the sample variance of e_{ij} yields

$$S_F^2/mn = \frac{1}{mn}\left[\sum_{i=1}^{m}\sum_{j=1}^{n}|e_{ij}|^2\right] \approx \sigma^2. \tag{10.272}$$

By using (10.271), (10.272) becomes

Third Bounds (Modified)

$$\sqrt{c}\sigma \leq \epsilon_1 \leq \sqrt{mn}\sigma. \tag{10.273}$$

In order to determine the tightness and usefulness of all these bounds, consider the following example.

Example 10.13 Let A be a 7×7 matrix of rank 4 given by

$$A = \begin{bmatrix} 3 & 2 & 1 & 7 & 4 & 5 & 3 \\ 1 & 4 & 2 & 6 & 5 & 10 & 3 \\ 8 & 1 & 5 & 13 & 5 & 7 & 0 \\ 4 & 2 & 7 & 15 & 11 & 11 & 4 \\ 1 & 2 & 1 & 3 & 2 & 5 & 1 \\ 2 & 1 & 3 & 5 & 3 & 5 & 0 \\ 3 & 10 & 1 & 5 & 2 & 21 & 1 \end{bmatrix}.$$

The SV of A are given by $\{39.8, 15.99, 6.23, 3.00, 1.09 \times 10^{-15}, 5.63 \times 10^{-17}, 6.89 \times 10^{-17}\}$. For a level of confidence of $\alpha = 0.05$ and from (10.264), (10.265), (10.268), and (10.273), the upper and lower bounds of ϵ_1 (denoted by ϵ_U and ϵ_L), are given by

1 First Bounds

$$\epsilon_{L1} = 2\sigma \leq \epsilon_1 \leq 14\sigma = \epsilon_{U1}. \tag{10.274}$$

2 Second Bounds

$$\epsilon_{L2} = 3.75\sigma \leq \epsilon_1 \leq 9.92\sigma = \epsilon_{U2}. \tag{10.275}$$

3 Third Bounds (Modified)

$$\epsilon_{L3} = 3.75\sigma \leq \epsilon_1 \leq 7\sigma = \epsilon_{U3}. \tag{10.276}$$

Fori noise variance σ^2 of 0.1, 0.01, and 0.001, Table 10.1 shows six threshold bounds given by (10.274)–(10.276). For later comparisons, the 3σ values are also tabulated.

Now consider the observed matrix $B = A + E$, where E is a noise matrix with i.i.d. Gaussian r.v. of zero-mean and variance σ^2. We evaluate the SV of B for 200 simulation

Table 10.1 Threshold bounds for Example 10.13

σ^2	0.1		0.01		0.001	
3σ	0.948		0.3		0.0948	
Bounds	ϵ_L	ϵ_U	ϵ_L	ϵ_U	ϵ_L	ϵ_U
First	0.632	4.427	0.200	1.400	0.063	0.442
Second	1.185	3.137	0.375	0.992	0.118	0.313
Third (Mod.)	1.185	2.213	0.375	0.700	0.118	0.221

Table 10.2 Rank estimation of matrix B for Example 10.13

Criterion	Noise Variance		
	$\sigma^2 = 0.1$	$\sigma^2 = 0.01$	$\sigma^2 = 0.001$
1: First Bound $\epsilon_1 = \epsilon_{U1}$	: $r = 3$:200 times	: $r = 4$:200 times	: $r = 4$:200 times
1: Second Bound $\epsilon_1 = \epsilon_{U2}$	$r = 3$:101 times $r = 4$:99 times	: $r = 4$:200 times	: $r = 4$:200 times
1: Third Bound (Mod.) $\epsilon_1 = \epsilon_{U3}$	: $r = 4$:200 times	: $r = 4$:200 times	: $r = 4$:200 times
1: 3σ Bound $\epsilon_1 = 3\sigma$	$r = 4$:157 times $r = 5$:43 times	$r = 4$:155 times $r = 5$:45 times	$r = 4$:154 times $r = 5$:46 times
3	$r = 2$:2 times $r = 4$:98 times $r = 5$:11 times $r = 6$:89 times	: $r = 4$:171 times : $r = 6$:29 times	: $r = 4$:186 times : $r = 6$:14 times
5	$r = 3$:200 times	$r = 3$:200 times	$r = 3$:200 times

runs of the noise. Table 10.2 shows the estimated rank of B based on Criterion 1 (for the first, second, third (modified), and 3σ threshold bounds given in Table 10.1), as well as those based on Criterion 3 with the largest ratio of SVs and Criterion 5 with $\delta_5 = 0.99$. Results based on Criteria 2 and 4 are not included in the comparisons of Table 10.2, since the explicitly needed values of δ_2 and δ_4 are unknown and unclear.

As can be seen in Table 10.2, this example shows that under Criterion 1, a meaningful threshold ϵ_1 such that

$$\beta_4 > \epsilon_1 \geq \beta_5$$

can be found and thus provide a practical method for the determination of the effective rank of the observation matrix B as being 4. While taking ϵ_1 equal to ϵ_{U1} and ϵ_{U2} are adequate for small values of the noise, only the threshold value of ϵ_{U3} is able to predict the correct effective rank under all considered noise conditions. As noted before, the actual applications of Criteria 2 and 4 are not clear since the needed ϵ_2 and ϵ_4 are not available. At least for this example, Criterion 3 (i.e., rank selected as t if $\beta_t/\beta_{t+1} = \max(\beta_j/\beta_{j+1})$) and Criterion 5 (i.e., $\epsilon_5 = 0.99$) are not adequate to predict the correct rank of the observed matrix.

10.14 Conclusion

In Section 10.1, we considered the direct methods for the solution of a system of linear equations. In Section 10.2, the Gaussian elimination procedure was introduced in support of the direct methods of the previous section. The normal equation approach to the solution of an over-determined linear system of equations was treated in Section 10.3. Section 10.4 introduced triangular decomposition of the A matrix with related LU decompositions and the Cholesky decomposition. Section 10.5 considered the useful QR factorization, which included the Gram–Schmidt orthogonalization procedure in Section 10.6, the modified G-S orthogonalization procedure in Section 10.7, the Givens orthogonal transformation of Section 10.8, and finally the Householder transformation in Section 10.9. The application of the QR decomposition to the solution of a linear system of equations was considered in Section 10.10. Section 10.11 introduced the singular value decomposition, while its application to the solution of a linear system of equations was given in Section 10.12. The practical issue on the effective rank of a matrix using SVD was considered in Section 10.13.

10.15 References

There are many elementary and advanced books on the basic concepts of linear algebra covered in this chapter. At the introductory textbook level, we have [1], more advanced books include [2] and [3]. All the materials in Section 10.1–Section 10.12 can be found in [2] and [3]. The rank determination of Section 10.13 appeared in [4].

[1] G. Strang, *Linear Algebra and Its Applications*, Academic Press, 1980.

[2] G.H. Golub and C.F. Van Loan, *Matrix Computations*, 3rd edn., J. Hopkins University Press, 2016.

[3] D.S. Watkins, *Fundamentals of Matrix Computations*, J. Wiley, 2002.

[4] K. Konstantinides and K. Yao, "Statistical Analysis of Effective Singular Values in Matrix Rank Determination," *IEEE Trans. on Acoustics, Speech, and Signal Processing*, May 1988, pp.757–763.

10.16 Exercises

1. There are several ways to verify the positive-definiteness of any real-valued square matrix A.
   a. A necessary and sufficient condition for the positive-definiteness of A is that all the principal minors, that is, n determinants of the $1 \times 1, \ldots, n \times n$, upper-left corner of the $A_S = (A + A^T)/2$ matrix are positive.

b. A necessary and sufficient condition for the positive-definiteness of A is that $x'Ax$ is expressible as the sum of the squares of $\{x_i\}$ with non-negative coefficients and the square of sums or differences of $\{x_i,\ x_j\}$ with non-negative coefficients.

Use both methods to verify the positive-definiteness of

$$A = \begin{bmatrix} 2 & 2 \\ 1 & 3 \end{bmatrix}.$$

2. There are several ways to verify the positive-definiteness of any real-valued square matrix A.

a. A necessary and sufficient condition for the positive-definiteness of A is that all the principal minors, that is, n determinants of the $1 \times 1, \ldots, n \times n$, upper-left corner of the $A_S = (A + A^T)/2$ matrix are positive.

b. A necessary and sufficient condition for the positive-definiteness of A is that $x'Ax$ is expressible as the sum of the squares of $\{x_i\}$ with non-negative coefficients and the square of sums or differences of $\{x_i,\ x_j\}$ with non-negative coefficients.

Use both methods to verify the positive-definiteness of

$$A = \begin{bmatrix} 2 & -1 & 0 \\ -1 & 2 & -1 \\ 0 & -1 & 2 \end{bmatrix}.$$

3. Consider

$$A = \begin{bmatrix} 1 & 1 & 2 \\ 2 & 3 & 5 \\ 3 & 5 & 7.9999 \end{bmatrix}.$$

a. Find the U, V, and Σ of the SVD of A.

b. Define $B = AA^T$ and $C = A^T A$. Find the eigenvalues and eigenvectors of B and C. How are they related to the SVD of A?

c. Find the normal equation pseudo inverse A^+_{normal} of A. Find the rank-2 Moore-Penrose pseudo-inverse A^+_{MP} of A.

4. Consider a 3×3 matrix

$$A = \begin{bmatrix} 2 & 4 & 5 \\ 2 & 6 & 8 \\ 3 & 4 & 9 \end{bmatrix}.$$

Find the QR decomposition of A using Gram–Schmidt, modified Gram–Schmidt, Givens, and Householder transformations. Show the intermediate steps in your computations. No derivation is needed.

5. Write a computer code to implement the Jacobi–Hestennes SVD algorithm (as discussed in Section 10.11). Find the S, U, and V of A_1.

$$A_1 = \begin{bmatrix} 16 & 6 & 1 \\ 15 & 7 & 13 \\ 20 & 8 & 5 \\ 5 & 8 & 5 \end{bmatrix}.$$

Show the orthogonality property at the end of each sweep. Stop when $z^{(k)} \leq 10^{-10}$.

6. Write a computer code to implement the Jacobi–Hestennes SVD algorithm (as discussed in Section 10.11). Find the S, U, and V of A_2.

$$A_2 = \begin{bmatrix} 1 & 1 & 1 & 1 \\ 2 & 4 & 6 & 8 \\ 1 & 3 & 5 & 7 \end{bmatrix}.$$

Show the orthogonality property at the end of each sweep. Stop when $z^{(k)} \leq 10^{-10}$.

7. Define the condition number, $\text{cond}(A) = \|A\| \|A^{-1}\|$, for some matrix norm $\| \ \|$. Each matrix norm will yield a different condition number.

   a. Define the l_2 norm of the matrix A by $||A||_2 = \max_x\{||Ax||_2/||x||_2\}$. Show $\text{cond}_2(A) = \|A\|_2 \ \|A^{-1}\|_2 = \sigma_{\max}/\sigma_{\min}$. Note the l_2 norm of a vector $||x||_2 = (\sum_{i=1}^{n} |x_i|^2)^{0.5}$. Hint: From the SVD of A, what unit norm x will maximize $||Ax||_2$?

   b. Using the matrix A in Exercise 2, find cond(A) using: $\| \ \|_1$, $\| \ \|_2$, and $\| \ \|_\infty$. Hint: Use the norm(A,1), norm(A,2), and norm(A,inf) functions defined in Matlab.

   c. Another possible definition of the condition number of A is $\text{cond}_{2'}(A) = (\max_x\{\|Ax\|_2/\|x\|_2\})(\max_x\{\|A^{-1}x\|_2/\|x\|_2\})$. Take 10 randomly chosen x and evaluate $\|Ax\|_2/\|x\|_2$ and $\|A^{-1}x\|_2/\|x\|_2$ and pick the largest value in $\{\|Ax\|_2/\|x\|_2\}$ and the largest value in $\{\|A^{-1}x\|_2/\|x\|_2\}$ to use in $\text{cond}_{2'}(A)$. How does your randomly chosen x generated $\text{cond}_{2'}(A)$ compare with the theoretical $\text{cond}_2(A) = \|A\|_2 \ \|A^{-1}\|_2$? Note: Each person's $\text{cond}_{2'}(A)$ will be different since the ten x vectors are chosen randomly.

11 Applications of LS and SVD Techniques

In this chapter, we consider the use of least-squares (LS) and singular value decomposition (SVD) techniques in various applications. Section 11.1 considers the flight load measurement problem in which loads are applied on the wing of an aircraft and strain gauge measurements are obtained to identify the parameters characterizing the structure. These parameters are obtained as the least-squares solution of a system of linear equations. Depending on modeling the load matrix as a linear function of the measurement matrix or the measurement matrix as a linear function of the load matrix, different systems of linear equations are obtained. The classical normal equation approach imposes restrictions on the allowable number of gauges versus the number of loads in order to obtain valid least-squares solutions. We show that the SVD approach always yields a valid least-squares solution in all cases without imposing any restrictions. The relation to the minimum norm normal equation approach is also discussed. In Section 11.2, we first consider the relationship between least-squares (LS) and total least-squares (TLS) solutions. Correspondence analysis (CA) is also used in various applied data and clustering analyses. It is known that SVD forms a basic operation in both TLS and CA analysis. Upon an appropriate pre-processing operation of centering the data matrix, we show that by minimizing the energy of the perturbation imposed on the data matrix, the TLS and CA solutions are equivalent. Some simple examples are given to illustrate these issues. In Section 11.3, the rank reduction property of the SVD is applied to the forward-backward linear prediction estimation technique used in the maximum entropy spectral estimation problem. In Section 11.4, the rank reduction property of the SVD is used to perform complexity reduction in a class of finite impulse response (FIR) filters. In Section 11.5, the SVD technique is used to perform detection of subspace variations for performing segmentation of voiced and unvoiced speech sounds. This approach is useful for detecting glottal closure in speech generation. Section 11.6 considers the relationship among the input update rate, the rate of convergence of the Jacobi–SVD algorithm, and the quality of the SVD processed outputs. A real-time non-stationarity indicator of the observed data in terms of their singular values is presented. Applications to a non-stationary DOA problem are considered.

11.1 Flight Load Measurement Solution Based on SVD Technique

Consider a practical aircraft load measurement system identification problem. The physical problem deals with the wing surface of an aircraft which is constantly experiencing different loadings during the flight. The ability to estimate these in-flight loadings are essential to the understanding and design processes of the wing structure. Strain gauges are mounted on different parts of the wing which are sensitive to the loads. In order to relate the gauges' outputs to the loadings on the wing surface, a pre-flight calibration procedure is performed. The calibration stage is simply a procedure to obtain the gauges' outputs when a set of known wing loads are applied to the wing structure. In Fig. 11.1, we apply m known shears S_i at location (x_i, y_i), $i = 1, \ldots, m$ on the aircraft structure and obtain the measurements from n strain gauges. From these known m shear and location values, the associated bending moments B_i and torques T_i (defined by equations (11.4) and (11.6)) are also determined and denoted as the load matrix L. In Fig. 11.2, the system identification calibration stage, performed on the ground, uses the $m \times 3$ load matrix L as the input to the aircraft wing structure considered as "blackbox" system with the output given by the $m \times n$ strain gauge measurement matrix M. From these input and output data, we obtain a set of parameters that characterizes the structure. In Fig. 11.3, during the flight measurement stage, from the $1 \times n$ in-flight gauge measurements $\tilde{M}$, we can then estimate the desired 1×3 in-flight loads $\tilde{L}$.

There are two fundamental and intuitively equally justifiable linear approaches arbitrarily denoted as Approach 1 and Approach 2, that are applicable to the load measurement problem. In Approach 1, we model the load value matrix L as dependent linearly on the influence coefficient value matrix M measured by the gauges. In Approach 2, we model M as dependent linearly on L. In general, these matrices are rectangular, thus it is not immediately clear that these two approaches are equivalent. Historically, all the work was based on Approach 1. Now, we shall show that these two approaches are indeed equivalent in all cases, which can be proved by the use of the SVD method. The classical normal equation approach for the least-squares solution (also called the linear

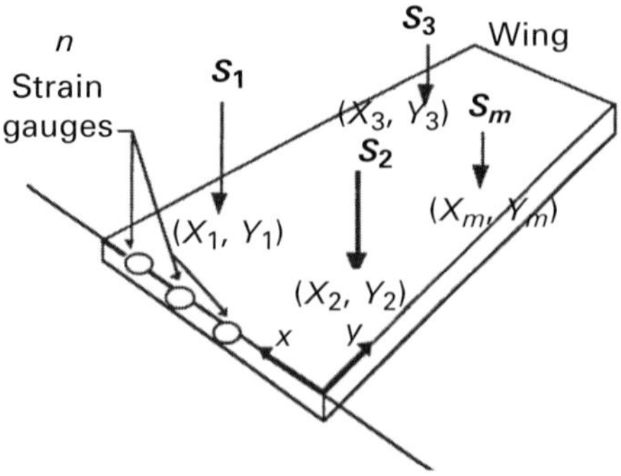

Figure 11.1 Load measurement using strain gauges in calibration stage

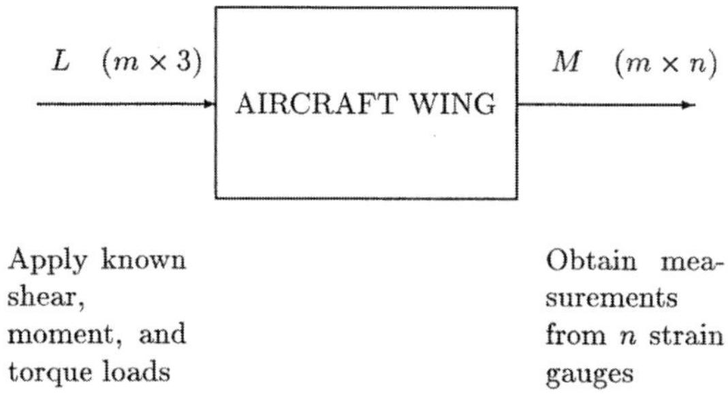

Figure 11.2 Calibration (system ID) stage (on ground)

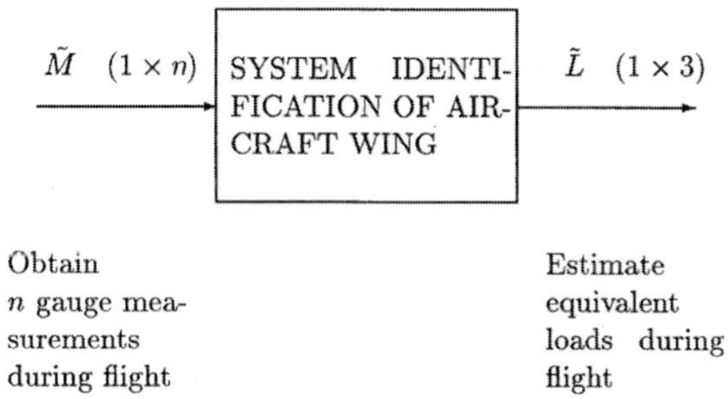

Figure 11.3 Flight measurement stage (in the air)

regression technique) was not capable of handling both of these approaches under all conditions. In particular, in Approach 1 this technique was limited to the case in which the number of gauges n was less than or equal to the number of loads m. On the other hand, it was required that n were greater than or equal to m if Approach 2 was chosen.

At the most basic level of understanding, of course, it is theoretically important to know the equivalence of these two seemingly different approaches that yield the desired results. In practice, it is often the case that the number of gauges n is greater than or equal to the number of applied loads m. However, there are certain conditions in which we want to consider more gauges than the number of loads in the calibration stage. The classical normal equation approach (i.e., Approach 1) is not possible since $M^T M$ needed in the processing is singular.

When the data from the gauges are quite linearly independent, then there is no significant numerical difference between the use of the SVD technique or the normal equation technique. However, for highly dependent data, there can be significant advantages for the SVD technique. Detailed numerical computations based on practical observed gauge measurements and load values are necessary to verify their differences. The crucial point is that in all cases the SVD approach is always computationally more costly as well as numerically more stable. For typical dimensions encountered in the load measurement problems, the additional computational cost of the SVD approach is not of significant concern, when we perform only a few LS computations. However, when we perform the LS computations repeatedly, then the higher SVD computational cost may be unacceptable. In the following, we assume that all the matrices are full rank, unless otherwise noted.

11.1.1 Approach 1 – Linear Dependence of Load Values on Gauge Values

Consider the $m \times 3$ load matrix

$$L = [L_1, L_2, L_3], \tag{11.1}$$

where

$$L_1 = L_S = [S_1, S_2, \ldots, S_m]^T \tag{11.2}$$

is the shear vector at m load locations (x_i, y_i), $i = 1, \ldots, m$,

$$L_2 = L_B = [B_1, B_2, \ldots, B_m]^T \tag{11.3}$$

is the bending moment vector with its ith component located at y_i given by

$$B_i = S_i y_i, \quad i = 1, \ldots, m, \tag{11.4}$$

and

$$L_3 = L_T = [T_1, T_2, \ldots, T_m]^T \tag{11.5}$$

is the torque vector with its ith component located at x_i given by

$$T_i = S_i x_i, \quad i = 1, \ldots, m. \tag{11.6}$$

Let the $m \times n$ influence coefficient matrix M denote the response of the n gauges to the m loads in the calibration process. Specifically, let

$$M = [u_1, u_2, \ldots, u_n] = \begin{bmatrix} M_1 \\ \vdots \\ M_m \end{bmatrix}, \tag{11.7}$$

where each u_i, $i = 1, \ldots, n$, represents the normalized response of the ith gauge to the m loads. Let the $n \times 3$ dependence coefficient matrix b consist of

$$b = [b^{(1)}, b^{(2)}, b^{(3)}], \tag{11.8}$$

where

$$L_i \cong M b^{(i)}, \quad i = 1, 2, 3, \tag{11.9}$$

or, in matrix form,

$$L \cong M b. \tag{11.10}$$

For $i = 1$, the $n \times 1$ vector $b^{(1)}$ yields the dependence of $L_1 = L_S$, the shear vector, to the linear combinations of the influence coefficient vectors $u_1, \ldots, u_n$ of M in (11.7). Similarly, for $i = 2$ and 3, $b^{(2)}$ and $b^{(3)}$ are related to the bending moment vector $L_2 = L_B$ and the torque moment vector $L_3 = L_T$, respectively. In the calibration process, the matrix M as well as L_1, L_2, and L_3 are available. Consider the solution of (11.10). There are two possible cases: $m \geq n$ and $m < n$.

- $m \geq n$: In this case, define

$$M^\dagger \equiv (M^T M)^{-1} M^T \in \Re^{n \times m}, \tag{11.11}$$

such that $M^\dagger M = I_{n\times n}$, the $n \times n$ identity matrix. The normal equation least-squares (NELS) solution to (11.10) is

$$\hat{b} = M^\dagger L. \tag{11.12}$$

During the flight measurement stage, we observe the $1 \times n$-dimensional gauge vector $\tilde{M}$. From (11.10), the predicted 1×3 load vector $\tilde{L}$ is given by

$$\tilde{L} = \tilde{M}\hat{b} = \tilde{M}M^\dagger L = \tilde{M}(M^T M)^{-1} M^T L. \tag{11.13}$$

The first component of $\tilde{L}$ yields the predicted shear,

$$\tilde{S} = \tilde{M}M^\dagger L_1, \tag{11.14}$$

the second and third components of $\tilde{L}$ yield the predicted bending moment $\tilde{B} = \tilde{S}\tilde{y}$ and predicted torque $\tilde{T} = \tilde{S}\tilde{x}$ as given by

$$\tilde{B} = \tilde{S}\tilde{y} = \tilde{M}M^\dagger L_2, \tag{11.15}$$

$$\tilde{T} = \tilde{S}\tilde{x} = \tilde{M}M^\dagger L_3. \tag{11.16}$$

From (11.14)–(11.16), we can solve for $\tilde{y}$ as

$$\tilde{y} = \frac{\tilde{M}M^\dagger L_2}{\tilde{M}M^\dagger L_1}, \tag{11.17}$$

and $\tilde{x}$ as

$$\tilde{x} = \frac{\tilde{M}M^\dagger L_3}{\tilde{M}M^\dagger L_1}. \tag{11.18}$$

Thus, (11.14), (11.17), and (11.18) represent the predicted equivalent net shear, bending moment location, and torque location of the applied load that yielded the measured gauge vector $\tilde{M}$ using the normal equation approach.

- $m \leq n$: In this case, there are many possible solutions to (11.10), since the system is under-determined. One can choose the solution with "minimum energy." Using the Lagrangian multipliers method, one can define

$$f(b) \equiv b^T b + \sum_{i=1}^{m} \lambda_i M_i b = b^T b + \lambda^T M b = b^T b + b^T M^T \lambda, \tag{11.19}$$

where $\lambda = [\lambda_1, \ldots, \lambda_m]^T$ is the vector of Lagrange multipliers. Setting

$$\nabla f(b) = 2b + M^T \lambda = 0, \tag{11.20}$$

one finds

$$\hat{b} = -\frac{1}{2}M^T\lambda, \quad \lambda = -2(MM^T)^{-1}L, \tag{11.21}$$

and the solution $\hat{b}$ is given by

$$\hat{b} = M^T(MM^T)^{-1}L = M^{\ddagger}L, \tag{11.22}$$

if one defines

$$M^{\ddagger} \equiv M^T(MM^T)^{-1}. \tag{11.23}$$

The predicted load, during the flight measurement stage, is given by

$$\tilde{L} = \tilde{M}\hat{b} = \tilde{M}M^{\ddagger}L = \tilde{M}M^T(MM^T)^{-1}L. \tag{11.24}$$

Now consider the use of the SVD technique via Approach 1. Consider a general form of the SVD of the matrix M with rank $p \leq \min\{m,n\}$, as given by

$$M = U_M\Sigma_M V_M^T, \tag{11.25}$$

where $U_M \in \Re^{m\times p}$ and $V_M \in \Re^{n\times p}$ have orthonormal columns and $\Sigma_M = \text{diag}(\sigma_{M1},\ldots,\sigma_{Mp}) \in \Re^{p\times p}$, $\sigma_{M1} \geq \sigma_{M2} \geq \cdots \geq \sigma_{Mp} > 0$, contains the p singular values of M. Then the SVD approach of the LS solution to (11.10) can be written as

$$\hat{b} = M^+L, \tag{11.26}$$

where

$$M^+ = V_M\Sigma_M^{-1}U_M^T \tag{11.27}$$

is the Moore–Penrose pseudo-inverse of M. In the flight measurement stage, we then have

$$\tilde{L} = \tilde{M}\hat{b} = \tilde{M}M^+L. \tag{11.28}$$

Note that the result of (11.28) is valid for the SVD approach for both cases of $m \geq n$ and $m \leq n$. It is most interesting to note that the predicted load vector in (11.28) based on the SVD technique has the same form as the predicted load vector in (11.13) and (11.24) based on the normal equation technique. Indeed, when $m \geq n$ (i.e., the number of loads is greater than or equal to the number of gauges), and when the gauge measurements are quite linearly independent, the pseudo-inverse given by M^+ in (11.27) is equal to the pseudo-inverse given by $M^{\dagger}$ in (11.11). Analogously, when $n \geq m$, and M is full rank, $M^+ = M^{\ddagger}$ in (11.23). Thus, in those cases, either the conventional normal equation or the SVD methods will yield the same predicted load values. Of course, when the measurements are quite linearly dependent, then the SVD approach will be better from the numerical stability point of view.

11.1.2 Approach 2 – Linear Dependence of Gauge Values on Load Values

From a physical cause and effect point of view, it makes sense that the responses of the first gauge to the m loads are given by

$$u_1 = \begin{pmatrix} u_{11} \\ \vdots \\ u_{1m} \end{pmatrix} \cong \begin{pmatrix} s_1 & s_1 y_1 & s_1 x_1 \\ \vdots & \vdots & \vdots \\ s_m & s_1 y_m & s_1 x_m \end{pmatrix} \begin{pmatrix} c_{11} \\ c_{12} \\ c_{13} \end{pmatrix} \tag{11.29}$$

$$= [L_1,\ L_1,\ L_1]c_1 = Lc_1. \tag{11.30}$$

In (11.29), we describe the gauge measurement u_{1i} as a linear combination of $s_i c_{11} + s_i y_i c_{12} + s_i x_i c_{13}$, which depends on the shear, bending moment, and torque. In general, for all n gauges, we have

$$M = [u_1, \ldots, u_n] \cong L[c_1, \ldots, c_n] = Lc, \tag{11.31}$$

where the $3 \times n$ dependence matrix c is denoted by

$$c = [c_1, \ldots, c_n]. \tag{11.32}$$

In the calibration process, as before, M and L are available. In the flight measurement process, as before, we have a measured $\tilde{M}$, given from (11.31) as

$$\tilde{M} = \tilde{L}c. \tag{11.33}$$

In order to solve for the 1×3 predicted load vector

$$\tilde{L} = [\tilde{S}, \tilde{B}, \tilde{T}], \tag{11.34}$$

we need to use (11.31) and (11.33). First, consider the normal equation technique. Again we have two cases, namely $m \geq 3$ and $m \leq 3$, even though this latter case is not likely to happen in practice.

- $m \geq 3$: In this case, we have

$$\hat{c} = L^{\dagger}M = (L^T L)^{-1} L^T M, \tag{11.35}$$

and

$$\tilde{M} = \tilde{L}\hat{c} = \tilde{L}L^{\dagger}M = \tilde{L}(L^T L)^{-1} L^T M, \tag{11.36}$$

where $L^{\dagger}$ is the pseudo-inverse of L defined as

$$L^{\dagger} = (L^T L)^{-1} L^T. \tag{11.37}$$

If $m \leq n$, we can multiply by M^T on both sides of (11.36) and take inverses and, remembering that $L^{\dagger}L = I_{3\times 3}$, we have

$$\tilde{L} = \tilde{M}M^T(MM^T)^{-1}L, \tag{11.38}$$

which corresponds to the NELS solution of (11.36). If $m \geq n$, (11.36) the minimum energy solution for $\tilde{L}L^{\dagger}$ is

$$\tilde{L}L^{\dagger} = \tilde{M}(M^T M)^{-1} M^T, \tag{11.39}$$

hence the NELS solution for $\tilde{L}$ is

$$\tilde{L} = \tilde{M}(M^T M)^{-1} M^T L. \tag{11.40}$$

- $m \leq 3$: If this is the case,

$$\hat{c} = L^{\ddagger} M = L^T (L L^T)^{-1} M \tag{11.41}$$

and

$$\tilde{M} = \tilde{L}\hat{c} = \tilde{L}L^{\ddagger} M. \tag{11.42}$$

Whenever $m \leq n$, the NELS solution yields

$$\tilde{L} = \tilde{M} M^T (M M^T)^{-1} L, \tag{11.43}$$

while when $m \geq n$, the NELS solution yields

$$\tilde{L} = \tilde{M}(M^T M)^{-1} M^T L. \tag{11.44}$$

It is interesting to observe that both Approach 1 and Approach 2 give the same result, as shown by equations (11.13), (11.40), and (11.44), for $m \geq n$ and equations (11.24), (11.38), and (11.43), for $m \leq n$. Now, consider solving for c in (11.31) by using the Moore–Penrose pseudo-inverse of L based on the SVD representation of L. Specifically, consider the SVD of L as given by

$$L = U_L \Sigma_L V_L^T, \tag{11.45}$$

where $\Sigma_L = \text{diag}(\sigma_{L1}, \sigma_{L2}, \sigma_{L3})$. $U_L \in \Re^{m\times 3}$ and $V_L \in \Re^{3\times 3}$ have orthonormal columns. Then the Moore–Penrose pseudo-inverse of L can be written as

$$L^{+} = V_L \Sigma_L^{-1} U_L^T. \tag{11.46}$$

We then have $\hat{c} = L^{+} M$ and

$$\tilde{M} = \tilde{L}\hat{c} = \tilde{L}L^{+} M. \tag{11.47}$$

This, irrespective of m being larger or smaller than n, the SVD solution yields

$$\tilde{L} = \tilde{M} M^{+} L, \tag{11.48}$$

where M^{+} is given by (11.27).

In conclusion, we note that by using the NELS method in both Approach 1 and Approach 2, we can write

$$\tilde{L} = \tilde{M} M^{\dagger} L = \tilde{M}(M^T M)^{-1} M^T L, n \leq m, \tag{11.49}$$

$$\tilde{L} = \tilde{M} M^{\ddagger} L = \tilde{M} M^T (M M^T)^{-1} L, n \geq m. \tag{11.50}$$

Similarly, by using the SVD method in both Approach 1 and Approach 2 for all cases, we have

$$\tilde{L} = \tilde{M}M^{+}L. \tag{11.51}$$

Thus, the most general form of the predicted load using either the NELS or SVD method for either Approach 1 or Approach 2 is given by

$$\tilde{L} = \tilde{M}M^{\natural}L, \tag{11.52}$$

where

$$M^{\natural} = \begin{cases} M^{\dagger}, \text{ for NELS method if } m \geq n, \\ M^{\ddagger}, \text{ for NELS method if } m \leq n, \\ M^{+}, \text{ for SVD method in all cases,} \end{cases} \tag{11.53}$$

and $M^{\dagger}$ is defined by (11.11), $M^{\ddagger}$ is defined by (11.23), and M^{+} is defined by (11.27). From the above discussions, it is clear that there is no fundamental difference between Approach 1 versus Approach 2. The advantages in using the SVD method versus the NELS method are twofold. As is known, the SVD is computationally stable and therefore it is advisable to resort to it whenever the condition number of the problem is high. Moreover, the SVD allows one to tackle rank-deficient problems, whereas the LS method, at least in the form presented here, would not give reliable results. Of course, the NELS method is computationally less costly compared to the SVD method.

11.2 Least-Squares, Total Least-Squares, and Correspondence Analysis

In this section, we first consider the basic properties of the least-squares (LS) estimation of $Ax \approx b$ by showing the equivalency of the minimization of the square of the norm of the residual to the minimization of the square of the norm of the perturbations on b. Simple examples to illustrate the geometric interpretation of the perturbations in LS problems are given. The total least-squares (TLS) estimation approach is used when both the elements of A and b are affected by noises. The TLS approach solves the perturbation problem of $(A + \Delta A)x = b + \Delta b$ under the minimization constraint of $\|\Delta A, \Delta b\|_F$. The formal TLS solution is expressed in terms of the SVD of the composite matrix $[A, b]$. Simple examples and geometric interpretation of TLS estimation are given. Correspondence analysis (CA) is an analytical technique for performing multi-dimensional data reduction, clustering, and displaying by using the essence of the data in an equivalent lower dimensional space. Upon appropriate centering and scaling, we show the equivalency of the CA approach to that of the TLS approach. Examples demonstrating the usefulness of CA are given.

11.2.1 Least-Squares Estimation

The least-squares technique has been used starting from Laplace and Gauss to perform data estimation and curve fitting. In the standard interpretation of the solution of an overdetermined system of equations of the form

$$Ax \approx b, \tag{11.54}$$

where A is a known full rank $m \times n$ matrix with $m \geq n$, x is an $n \times 1$ unknown vector, and b is a known $m \times 1$ vector denoted by

$$A = [\mathbf{a}_1, \mathbf{a}_2, \ldots, \mathbf{a}_n],\ x = [x_1, x_2, \ldots, x_n]^T,\ b = [b_1, b_2, \ldots, b_m]^T.$$

Denote the subspace defined by the range of the matrix A as $R_A = \{y : y = c_1\mathbf{a}_1 + \cdots + c_n\mathbf{a}_n, c_i \in \Re, i = 1, \ldots, n\}$. Then if the given b is not in R_A, then there is no solution $\hat{x}$ such that $A\hat{x} = \mathbf{a}_1\hat{x_1} + \cdots + \mathbf{a}_n\hat{x_n} = b$. This means the expression in (11.54) cannot be satisfied by an equality and we need to minimize the square of the norm of the residual $r = Ax - b$. The LS solution $\hat{x}$ of (11.54) satisfies

$$\min_x \|r\|^2 = \min_x \|Ax - b\|^2 = \|A\hat{x} - b\|^2. \tag{11.55}$$

Since R_A is a subspace of $\Re^m$, denote its complement by $R_A^\perp$. Then every $b \in \Re^m$ has an unique decomposition $b = b_{R_A} + b_{R_A^\perp}$, where $b_{R_A} \in R_A$ and $b_{R_A^\perp} \in R_A^\perp$. Then

$$\min_x \|r\|^2 = \min_x \|A\hat{x} - b\|^2 = \|(A\hat{x} - b_{R_A}) - b_{R_A^\perp}\|^2 = \|A\hat{x} - b_{R_A}\|^2 + \|b_{R_A^\perp}\|^2 = \|b_{R_A^\perp}\|^2. \tag{11.56}$$

The fourth expression of (11.56) follows from the third expression due to the orthogonality of $(A\hat{x} - b_{R_A})$ to $b_{R_A^\perp}$ since $(A\hat{x} - b_{R_A}) \in R_A$ and $b_{R_A^\perp} \in R_A^\perp$. Furthermore, the first term of the fourth expression is zero since $A\hat{x} = b_{R_A}$ yields the LS solution $\hat{x} = (A^TA)^{-1}A^Tb_{R_A} = (A^TA)^{-1}A^Tb$. The last expression follows from the observation that $a_i \perp b_R^\perp, i = 1, \ldots, n$, which is equivalent to $(a_i, b_R^\perp) = a_i^T b_R^\perp = 0, i = 1, \ldots, n$, which in turn is also equivalent to $A^T b_{R_A^\perp}$.

Another way of considering the LS solution of (11.54) is to note that for a given b, in general, b does not belong to R_A. However, it is always possible to find a perturbation vector Δb such that $(b + \Delta b) \in R_A$, in order to convert the approximation of the expression in (11.54) to an exact equality given by

$$A\hat{x} = b + \Delta b. \tag{11.57}$$

By using the unique decomposition of $b = b_{R_A} + b_{R_A^\perp}$ in (11.57), we have $A\hat{x} = b + \Delta b = b_{R_A} + b_{R_A^\perp} + \Delta b$. Since $A\hat{x} \in R_A$ and $b_{R_A} \in R_A$, then $b_{R_A^\perp} + \Delta b = 0$ implies (11.57) is given by

$$\Delta b = -b_{R_A^\perp}. \tag{11.58}$$

The intuitive appeal of this approach is that we assume the observed values of A are exact but the observed values of b represent a noisy perturbation from the true $b_{true} = b + \Delta b = b - b_{R_A^\perp}$. Thus, we seek a perturbation Δb such that $b + \Delta b$ not only satisfies the exact equation of (11.57), but also minimizes the square of the norm of the residual in (11.56). This optimum perturbation vector Δb is then given by (11.58).

Example 11.1 Consider a full rank square matrix A with $m = n$. Then the linear combinations of $\{\mathbf{a}_1, \ldots, \mathbf{a}_n\}$ span $\Re^n$ resulting in $R_A = \Re^n$. Thus, any given $b \in \Re^n$ belongs to R_A and $b_{R_A^\perp} = \Delta b = 0$. Then from (11.56) we have $\|r_{\min}\|^2 = 0$.

Example 11.2 Consider an $m \times n$ matrix A with $m > n = 1$ denoted by $A = \mathbf{a} = [a_1, a_2, \ldots, a_m]^T$. Let the $m \times 1$ vector $b = [b_1 \ldots b_m]^T$ and the scalar solution be denoted by x. Then

$$r = Ax - b = \mathbf{a}x - b = [(a_1 x - b_1), \ldots, (a_{mx} - b_m)]^T$$

and

$$\|r\|^2 = \Sigma_{i=1}^m (a_{ix} - b_i)^2. \tag{11.59}$$

Direct evaluation for the optimum $\hat{x}$ (by either minimizing $\|r\|^2$ in (11.59) or by using the orthogonal principle of $\mathbf{a} \perp (\mathbf{a}\hat{x} - b)$), results in $\mathbf{a}^T(\mathbf{a}\hat{x} - b) = 0$,

$$\hat{x} = \frac{\mathbf{a}^T b}{\|\mathbf{a}\|^2}, \tag{11.60}$$

and

$$\|r_{\min}\|^2 = \Sigma_{i=1}^m (a_i \hat{x} - b_i)^2 = \|b\|^2 - \frac{(\mathbf{a}^T b)^2}{\|\mathbf{a}\|^2}. \tag{11.61}$$

From Fig. 11.4, we note the optimum straight line has a slope $\hat{x}$ given by (11.60) such that the sum of the square of the vertical distances from the set of points (a_i, b_i) to the points $(a_i, a_i\hat{x}), i = 1, \ldots, m$ is minimized and is given by (11.61).

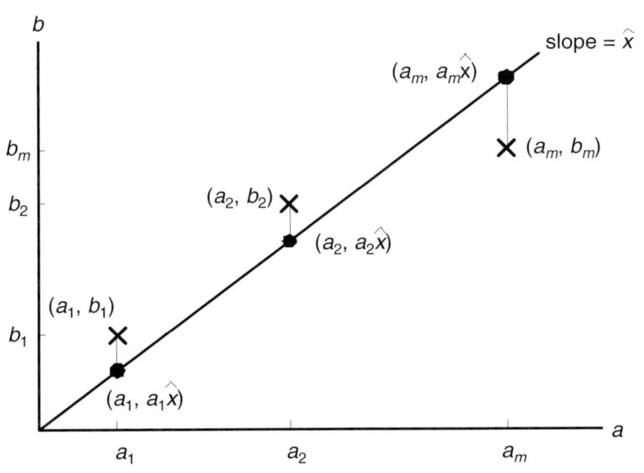

Figure 11.4 LS approximation of an $m \times 1$ vector b by $\hat{x}$ times an $m \times 1$ vector $\mathbf{a}$

11.2.2 Total Least-Squares Estimation

In the LS approach, we assume the elements of the $m \times n$ matrix A are free of error, while all the errors are confined to the vector b. However, in many, if not most, practical estimations, errors occur both in the data matrix A as well as in b. Total least-squares (TLS), on the other hand, makes the assumption that both A and b are affected by noises. The TLS solution for x solves the perturbed problem $(A + \Delta A)x = b + \Delta b$, under the constraint that the Frobenius norm of the perturbation $\|[\Delta A, \Delta b]\|_F$ is minimized. The classical LS solution solves the same perturbed problem, where the matrix A is left unchanged (i.e., $\Delta A = 0$).

Under the assumption that the rows of the compound matrix $[A, b]$ are disturbed by independent and identically distributed (iid) zero-mean white noises, then the TLS solution for x is known to be a strongly consistent estimate of the exact solution, as the number of rows goes to infinity. This property is not shared by the LS solution. In addition, the TLS approach needs an "approximation effort" $\|[\Delta A, \Delta b]\|_F$ which is always smaller than that of the corresponding $\|[\Delta b]\|_F$ for the LS approach. Another feature which makes the TLS approach more attractive than that of the LS in some cases is its smaller sensitivity to noise perturbations.

In order to introduce the concept of TLS, consider the following simple example with dimensions corresponding to those used in Example 11.2.

Example 11.3 Consider an $m \times n$ matrix A with $m > n = 1$ denoted by $A = \mathbf{a} = [a_1 a_2 \ldots a_m]^T$. Let the $m \times 1$ vector $b = [b_1 \ldots b_m]^T$ and let the scalar solution be denoted by x. In the TLS approach, the perturbations on both a_i and b_i have to be minimized. Equivalently, the m pairs of perturbation $(\Delta a_1, \Delta b_1)$, $(\Delta a_2, \Delta b_2), \ldots, (\Delta a_m, \Delta b_m)$ have to satisfy the following equality

$$\begin{pmatrix} a_1 + \Delta a_1 \\ \vdots \\ a_m + \Delta a_m \end{pmatrix} \hat{x} = \begin{pmatrix} b_1 + \Delta b_1 \\ \vdots \\ b_m + \Delta b_m \end{pmatrix}, \tag{11.62}$$

with

$$\left\| \begin{bmatrix} \Delta a_1 & \Delta b_1 \\ \vdots & \vdots \\ \Delta a_m & \Delta b_m \end{bmatrix} \right\|_F = \min_{\{\Delta \tilde{a}_i, \Delta \tilde{b}_i\}} \left\| \begin{bmatrix} \Delta \tilde{a}_1 & \Delta \tilde{b}_1 \\ \vdots & \vdots \\ \Delta \tilde{a}_m & \Delta \tilde{b}_m \end{bmatrix} \right\|_F . \tag{11.63}$$

The TLS solution $\hat{x}$ can be found by minimizing the expression

$$\epsilon^2 = \|[\Delta \tilde{a}, \Delta \tilde{b}]\|_F^2 = \sum_{i=1}^{M} (b_i - a_i x)^2 / (1 + x^2) \tag{11.64}$$

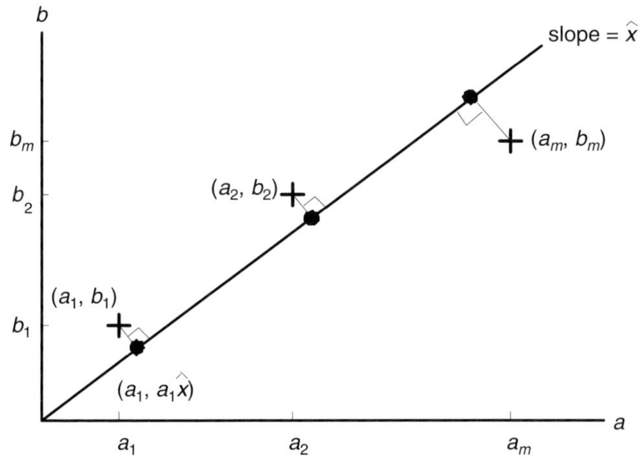

Figure 11.5 TLS approximation of an $m \times 1$ vector b by $\hat{x}$ times an $m \times 1$ vector **a**

with respect to x. Direct solution shows $\hat{x}$ is given by the solution of the second-order equation,

$$\sum_{i=1}^{m} a_i b_i \, \hat{x}^2 + \left(\sum_{i=1}^{m} a_i^2 - \sum_{i=1}^{m} b_i^2 \right) \hat{x} - \sum_{i=1}^{m} a_i b_i = 0,$$

which minimizes ϵ^2. From (11.63), we note the TLS solution minimizes the sum of the square of the distances of the points $(a_i, b_i), i = 1, \ldots, m,$ to their projected nearest points on the interpolating straight line as shown in Fig. 11.5.

In the general case, given an incompatible system

$$AX \approx B,$$

where A is $m \times n$ and has rank J, B is $m \times d$, and X is $n \times d$, the problem is to minimize the approximation error

$$\epsilon_{\text{TLS}}^2 = \min \|[\Delta A, \Delta B]\|_F^2,$$

subject to

$$(A + \Delta A)X = (B + \Delta B),$$

and

$$\text{rank}([A + \Delta A, B + \Delta B]) = J.$$

This is equivalent to computing the best rank-J approximate of the $m \times (n+d)$ composite $[A, B]$ matrix. That is, find the projection on a J-dimensional space which minimizes the energy of the perturbation. Operationally, one is required to compute the SVD of the matrix $[A, B]$ and determine its J largest singular values. Denote this SVD of $[A, B] = U \Sigma V^T$, where the $(n + d) \times (n + d)$ matrix $V = [V_1, V_2]$ has the $(n + d) \times J$ submatrix V_1 having as columns the J right singular vectors corresponding to the J largest singular values, and the $(n+d)\times(n+d-J)$ submatrix V_2 containing the remaining $n+d-J$ right singular vectors. The TLS solution matrix $\hat{X}$ now has to satisfy

$$\begin{pmatrix} \hat{X} \\ -I_{d\times d} \end{pmatrix} = V_2 K, \tag{11.65}$$

for some $(n+d)\times d$ matrix K. In the case where $d=1$ and A is a full rank matrix (i.e., $J=n$), the $(n+1)\times 1$ vector $V_2 = V(:\,,N+1)$ is the $(n+1)$th right singular vector, and $\hat{X}=\hat{x}$ is an $n\times 1$ vector, whose ith component is given by

$$\hat{x}_i = -\frac{V(i,n+1)}{V(n+1,n+1)}. \tag{11.66}$$

Special care has to be taken when $V(N+1,N+1)$ is small or zero.

11.2.3 Correspondence Analysis

In applied statistical and data analysis, there are myriads of analytical, graphical, and intuitive methods for performing multi-dimensional data reduction, clustering, and display. The "correspondence analysis" (CA) technique provides an analytical method of performing data reduction and the displaying of a large centered data matrix A in lower dimensional spaces such as that on the two-dimensional plane. Consider a real-valued $m\times n$ data matrix A denoted by

$$A = [a_{ij}] = \begin{pmatrix} a_{1\cdot} \\ a_{2\cdot} \\ \vdots \\ a_{m\cdot} \end{pmatrix} = [a_{\cdot 1}, a_{\cdot 2}, \cdots, a_{\cdot n}]. \tag{11.67}$$

Clearly, the matrix A can be considered either as m row vectors each of dimension n, or as n column vectors each of dimension m. Each interpretation has its own usefulness for specific applications. When the number of elements mn in A is large, then it is most useful to check if all the observed data are really relevant and in some sense yield useful information. If not, then some data reduction method operating on A may be able to characterize the essence of the original data matrix A, using a smaller number of elements than mn.

Now we want to find a set of new orthonormal coordinates in that the m observed data row vectors represented as m points in $\Re^n$ (or the n observed data column vectors represented as points in $\Re^m$) are shown as distinctly using as few of these coordinates (i.e., J) as possible. That is, what we want to do is to maximize the sum of the square of the distances among all the pairs of the m points (or n points) projected onto the J coordinates after appropriate centering and scalings. The main purpose of using this criterion is to accentuate the differences among the features of the data vectors that are different. Thus, the data vectors that are similar will bunch together while those that are dissimilar will pull apart in the J-dimensional graphical plots. Typically, we use $J=2$ plots.

For simplicity, assume that the matrix A has been "centered," i.e., so that each column has zero-mean. We want the projection coefficients $\{\hat{H}_i^{(j)} = (a_{i\cdot}, v_j^T)\}$ to be as separated as possible with respect to all possible orthonormal bases $\{v_1^T, \ldots, v_n^T\}$. Specifically, define

$$D = \sum_{j=1}^{J} \sum_{i=1}^{m} \sum_{k=1}^{m} (\hat{H}_i^{(j)} - \hat{H}_k^{(j)})^2.$$

This measure is maximum when the projected vectors are as separated as possible. It is possible to prove that, among all orthonormal row bases of dimension J in $\Re^n$, the optimum orthonormal row basis in the sense of maximizing the measure D defined in the above is given by $\{v_1^T, \ldots, v_J^T\}$, the right singular vectors corresponding to the J largest singular values of the SVD of A.

Now we consider two examples of interest.

Example 11.4 In this example, we use the dimensions of Examples 11.2 and 11.3 where a and b are $m \times 1$ vectors. In the CA approach, we want to find a 2×1 unit-norm vector q such that the projections of the m points $p_i = (a_i, b_i)$, $i = 1, \ldots, m$, are as separated as possible. Let us assume that $\sum_{i=1}^{m} a_i = \sum_{i=1}^{m} b_i = 0$, that is, each point cluster is "centered" around the origin. Let p be the $m \times 2$ matrix of points and $h = pq$ be the $m \times 1$ vector of the projections of such points onto the vector q. This maximization problem becomes

$$\max_q \sum_{i=1}^{m} \sum_{j=1}^{m} (h_i - h_j)^2 = 2m \max_q \left(\sum_{i=1}^{m} h_i^2 - m\bar{H}^2 \right),$$

where $\bar{H} = \sum_{i=1}^{m} h_i = 0$ in this case. Since $\sum_{i=1}^{m} h_i^2 = \|h\|_2^2 = \|pq\|_2^2 = \|p\|_F^2 - \|pq^{\perp}\|_2^2$, where $q^T q^{\perp} = 0$, we can see that the problem is equivalent to finding the unit-norm vector $q^{\perp} \triangleq [x \quad -1]^T / \sqrt{1 + x^2}$, which minimizes $\|[a \quad b]\,[x \quad -1]^T / \sqrt{1 + x^2}\|_F^2$. In other words, our purpose is to minimize

$$\epsilon^2 = \sum_{i=1}^{M} (b_i - a_i x)^2 / (1 + x^2),$$

which is the same as the ϵ^2 in (11.64) in the TLS case. Note that for both TLS and CA approaches, we exclude, at least in this example, the possibility that the straight line is vertical or that $q = [0 \quad 1]$.

Example 11.5 Here we consider a real-life NASA generated 18×12 data matrix of load measurements. For this example, we shall first compute the collinearity indices and then relate these observations to the location of these column vectors as points in the $J = 2$ coordinate graph.

The concept of collinearity quantifies the amount of linear dependency for a given column (or row) vector relative to all the other column (or row) vectors of A. The column collinearity index of $A = [a_{\cdot 1}, \ldots, a_{\cdot n}]$, is defined as

$$\kappa_j = \|a_{\cdot j}\| \, \|[A^+]_{j\text{th row}}\| \quad j = 1, \ldots, n,$$

where A^+ is the pseudo-inverse of A.

Consider the 18×12 strain gauge measurement matrix A as given by

$$\begin{pmatrix}
12.5 & -30.8 & 11.6 & 4.2 & 11.8 & 11.9 & 11.3 & 23.4 & 36.9 & 50.3 & 89.4 & 128.7 \\
21.4 & -2.9 & 19.2 & 12.8 & 17.9 & 21.5 & 16.5 & 40.0 & 42.1 & 73.7 & 58.2 & 68.3 \\
30.8 & 9.1 & 23.4 & 23.6 & 20.7 & 36.9 & 28.8 & 67.4 & 28.9 & 38.3 & 40.3 & 27.7 \\
42.5 & 29.7 & 29.1 & 43.6 & 28.2 & 64.4 & 24.6 & 37.2 & 20.9 & 20.4 & 24.8 & 8.9 \\
56.6 & 61.4 & 41.4 & 76.8 & 25.4 & 34.2 & 21.7 & 23.1 & 15.8 & 10.8 & 15.3 & -2.7 \\
77.8 & 126.3 & 41.0 & 54.7 & 24.4 & 21.5 & 19.8 & 12.3 & 8.2 & 2.2 & 3.5 & -15.2 \\
27.0 & -19.2 & 26.3 & 8.5 & 29.2 & 15.3 & 23.6 & 25.8 & 64.1 & 29.6 & 132.8 & 82.1 \\
45.4 & -2.3 & 35.8 & 18.0 & 32.9 & 20.2 & 31.6 & 27.7 & 59.0 & 25.6 & 98.8 & 56.8 \\
62.3 & 14.0 & 45.2 & 24.2 & 37.1 & 21.5 & 37.8 & 27.1 & 45.3 & 21.3 & 72.4 & 34.2 \\
85.8 & 34.3 & 55.7 & 30.0 & 42.8 & 21.2 & 41.1 & 26.5 & 36.4 & 17.0 & 50.2 & 13.9 \\
110.7 & 55.2 & 68.0 & 33.4 & 44.1 & 19.3 & 43.4 & 22.8 & 28.9 & 11.7 & 29.5 & -3.0 \\
131.4 & 82.0 & 66.6 & 44.2 & 39.4 & 21.2 & 37.3 & 19.1 & 13.9 & 6.5 & 5.4 & -18.2 \\
46.3 & -22.6 & 41.9 & 11.9 & 38.1 & 14.7 & 40.1 & 16.6 & 83.4 & 0.9 & 160.6 & 63.0 \\
64.6 & -3.5 & 49.0 & 18.3 & 41.8 & 13.7 & 43.4 & 16.6 & 69.3 & 4.6 & 129.5 & 48.9 \\
138.4 & 51.2 & 80.8 & 19.3 & 53.1 & 9.4 & 52.9 & 20.9 & 30.8 & 10.8 & 35.1 & -2.7 \\
163.3 & 62.6 & 82.7 & 30.7 & 49.4 & 19.0 & 49.1 & 21.9 & 20.0 & 6.8 & 8.7 & -18.9 \\
91.4 & 15.5 & 65.1 & 18.6 & 50.8 & 4.2 & 51.0 & 6.8 & 58.0 & 3.7 & 96.9 & 31.7 \\
114.4 & 35.9 & 74.6 & 18.9 & 53.1 & 1.7 & 54.8 & 14.8 & 45.8 & 10.5 & 68.2 & 16.4
\end{pmatrix}.$$

These data are obtained from a NASA/Ames hypersonic wing test structure (HWTS) load measurement experiment. The 12 columns represent the output responses of 12 strain gauges located in the wing root of the aircraft caused by 18 known input load conditions applied across different locations of the wing structure in the calibration state of the experiment performed on the ground. From these input and output data, the system characterization of the wing structure is obtained. Then during a flight, responses from the gauges are used to estimate the equivalent true loads experienced by the wing. In practice, the number of gauges available during a flight is typically much less than that available during the calibration. Thus, from the data in matrix A, we are motivated to determine redundant gauges that can be eliminated. Direct evaluations of collinearity indices yield

$$\kappa = [24,\ 4.3,\ 52,\ 7.8,\ 34,\ 6.2,\ 29,\ 7.5,\ 32,\ 8.3,\ 32,\ 9.5].$$

Application of the above CA technique on the centered version of the matrix A above, yields

$$S = \operatorname{diag}[0.61,\ 0.31,\ 0.18,\ 0.18,\ 0.083,\ 0.067,\ 0.051,$$
$$0.031,\ 0.011,\ 6.6 \times 10^{-3},\ 5.6 \times 10^{-3},\ 7.2 \times 10^{-17}],$$

while the coefficients of expansion of the 12 column vectors for the two dominant singular vectors are plotted in Fig. 11.6. From these κ values, we note that all the odd numbered column data vectors are fairly collinear. From Fig. 11.6, we note these vectors

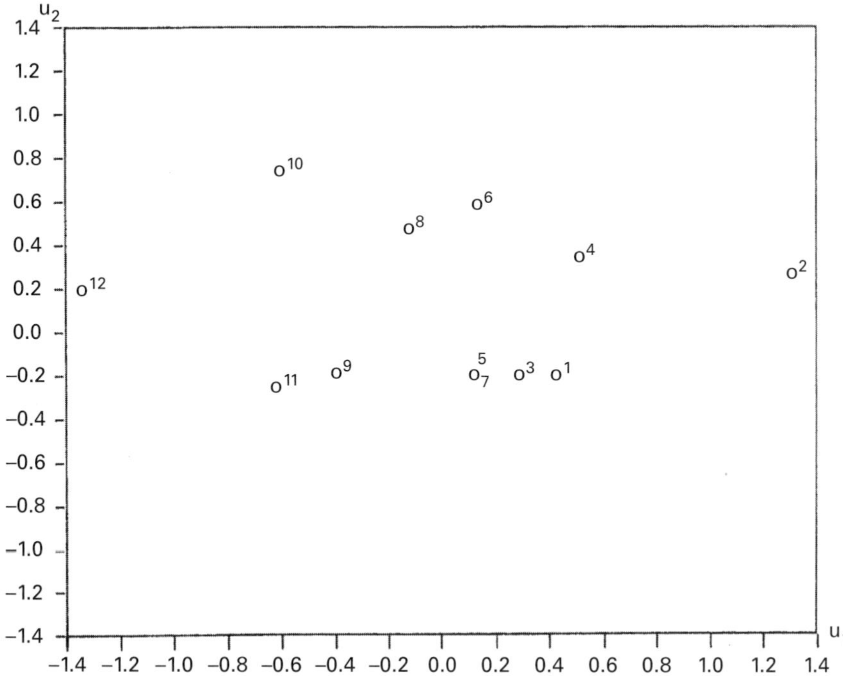

Figure 11.6 $J = 2$ correspondence analysis of an 18×12 load matrix

are bunched closely together. Alternatively, the even-numbered column data vectors are spread loosely about the graph. This observation is consistent with the relatively low values of these collinearity indices. It is interesting to note that all the even numbered column values of κ are from shear strain gauge outputs while the odd number column values are from bending strain gauge outputs. Indeed, physical characterizations of the wing structure do provide justifications for the bending strain gauge outputs to be more correlated than shear strain gauge outputs. From Fig. 11.6, it seems that if we need to eliminate some "nearly redundant" gauges, we may want to drop the gauge corresponding to the third, fifth, or seventh column. Even from this simple example, we can conclude that the CA analysis technique considered here can provide insights and understandings from the data matrix that are generally not possible by detailed human inspections.

11.2.4 Equivalence of TLS and CA

In order to show the equivalence of TLS to CA, let us reformulate the CA problem. Let A be centered such that $\sum_{i=1}^{m} a_{ij} = 0$. Consider the SVD of $A = U\Sigma V^T$. Let $Q = [q^{(1)}, \ldots, q^{(J)}] \in \Re^{n \times J}$ be a matrix with $J \leq N$ orthonormal columns and $H = AQ = (H_i^{(j)})$, with $H_i^{(j)} = a_i.q^{(j)}$, is the matrix which contains the projections of A onto Q. In this context, Q represents the basis-matrix of a J-dimensional subspace. For the purposes of CA, it makes sense to vary Q so to have the projection coefficients $H_i^{(j)}$ as separated as possible, with respect to each basis vector $q^{(j)}$. A possible way to impose this requirement is to perform the maximization of the following quantity over the space spanned by the columns of Q. Thus, max $D = 2m$ max $\|H\|_F^2$. The above maximization can be carried out completely in terms of the SVD of A. In fact,

$$\begin{aligned}\|H\|_F^2 &= \|AQ\|_F^2 = \|U\Sigma V^T Q\|_F^2 \\ &= \|U\Sigma\widetilde{V}^T\|_F^2 = \|U\Sigma[\tilde{v}_{1\cdot}, \ldots, \tilde{v}_{J\cdot}]\|_F^2 \leq \sum_{i=1}^{J} \sigma_i^2.\end{aligned}$$

The equality is obtained when $Q = V_J = [v_{\cdot 1}, \ldots, v_{\cdot J}]$, in which case $\widetilde{V} = Q^T V = [I_J, 0_{J\times(n-J)}]$ and

$$H = AQ = AV_J = U\Sigma V^T V_J = [\sigma_1 u_{\cdot 1}, \ldots, \sigma_J u_{\cdot J}].$$

Now, let $[Q, Q^{\perp}] \in \Re^{n\times n}$ be an orthonormal completion of Q. Then we can write $\|H\|_F^2 = \sum_{k=1}^{n} \sigma_k^2 - \|AQ^{\perp}\|_F^2$. We can therefore infer that the maximization of D is equivalent to the following minimization problem

$$\min \|AQ^{\perp}\|_F^2 = \min \|A - \hat{A}\|_F^2 = \min \|\Delta\hat{A}\|_F^2,$$

where $\hat{A} = HQ^T$ and $\Delta\hat{A} = A - \hat{A}$.

The choice of the J-dimensional subspace implied by the maximization problem stated above has therefore the effect of *minimizing the energy of the perturbation* imposed on the original data matrix A. Such a perturbation reduces the rank of A to rank$(\hat{A}) \leq J$, so that any vector $\underline{X}$ in the orthogonal subspace spanned by $Q^{\perp}$ *exactly* solves the equation $\hat{A}\underline{X} = 0$.

Indeed the TLS problem $\mathcal{A}\mathcal{X} \approx \mathcal{B}$ can be stated as find the minimum norm vector $\underline{X} = [\mathcal{X}^T, -I]^T$ which solves exactly the perturbed equation $\hat{A}\underline{X} = 0$, where $\hat{A} = A - \Delta\hat{A}$ and $A = [\mathcal{A}, \mathcal{B}]$. That is, we want the Frobenius norm of the perturbation applied to the matrix A, $\|\Delta\hat{A}\|_F$, to be minimized. This shows the equivalency of TLS and CA approaches.

11.3 Maximum Entropy Method Spectral Estimation via FBLP System Identification and SVD Approximation

Consider a wide-sense stationary sequence $\{y_m, -\infty < m < \infty\}$. The optimum one-step prediction of y_m by L past values is determined by

$$\min_g \mathrm{E}\left\{|\, y_m - \sum_{l=1}^{L} g_l y_{m-l}\,|^2\right\} = \mathrm{E}\left\{|\, y_m - \sum_{l=1}^{L} \hat{g}_l y_{m-l}\,|^2\right\}, \tag{11.68}$$

where $\hat{g} = (\hat{g}_1, \ldots, \hat{g}_L)^T$ are the optimum coefficients of the prediction estimator. Furthermore, let $\{y_m\}$ be modeled by

$$y_m = \sum_{i=1}^{P} A_i e^{j(2\pi f_i m + \theta_i)} + w_m, \quad m = 1, \ldots, M, \tag{11.69}$$

where $\{w_m\}$ is a zero-mean white sequence, $\{A_i\}$ is a set of p unknown but deterministic real-valued amplitudes, $\{\theta_i\}$ is a set of uniformly distributed and uncorrelated r.v. on $[0, 2\pi)$, and $\{f_i\}$ is a set of unknown but deterministic real-valued frequencies. Then $\{y_m\}$ given by (11.69) is a wide-sense stationary sequence. Given only the observed realization of $\{y_m\}$, the optimum $\{\hat{g}_i\}$ values cannot be obtained (in the sense of the minimum of (11.68), but instead can be obtained under the minimum LS sense of

$$\min_g ||Ag - y|| = ||A\hat{g} - y||, \tag{11.70}$$

where A is a $2(M-L) \times L$ matrix, g is an $L \times 1$ vector, y is an $2(M-L) \times 1$ vector, and $\underline{y}_m$ is an $(M-L) \times 1$ vector. Specifically,

$$\overset{2(M-L)\times L}{A} = \begin{pmatrix} \underline{y}_L & \underline{y}_{L-1} & \cdots & \underline{y}_1 \\ \underline{y}_2^* & \underline{y}_3^* & \cdots & \underline{y}_{L+1}^* \end{pmatrix}, \tag{11.71}$$

$$\overset{L\times 1}{g} = [g_1, \ldots, g_L]^T, \tag{11.72}$$

$$\overset{2(M-L)\times 1}{y} = [\underline{y}_{L+1}^T, y_M^T, y_1^{\mathrm{H}}, y_2^{\mathrm{H}}, \ldots, y_{M-L}^{\mathrm{H}}]^T, \tag{11.73}$$

$$\overset{(M-L)\times 1}{\underline{y}_m} = [y_m, y_{m+1}, \ldots, y_{M-L+m-1}]^T, \ m = 1, \ldots, L+1. \tag{11.74}$$

From (11.70), the optimum $\hat{g} = [\hat{g}_1, \ldots, \hat{g}_L]^T$ are chosen under the LS sense to approximate the $(M-L)$ forward one-step predictions of

$$\begin{aligned} \sum_{i=1}^{L} \hat{g}_i y_{L+1-i} &\approx y_{L+1}, \\ \sum_{i=1}^{L} \hat{g}_i y_{L+2-i} &\approx y_{L+2}, \\ &\vdots \\ \sum_{i=1}^{L} \hat{g}_i y_{M-i} &\approx y_M, \end{aligned} \tag{11.75}$$

as well as the $(M - L)$ backward one-step predictions of

$$\begin{aligned}
\sum_{i=1}^{L} \hat{g}_i \; y_{i+1}^* &\approx y_1^*, \\
\sum_{i=1}^{L} \hat{g}_i \; y_{i+2}^* &\approx y_2^*, \\
&\vdots \\
\sum_{i=1}^{L} \hat{g}_i \; y_{M-L+i}^* &\approx y_{M-L}.
\end{aligned} \tag{11.76}$$

Thus, the LS estimation under (11.70) is called a forward and backward linear prediction (FBLP) approach for identifying the set of optimum system coefficients in $\hat{g}$.

A linear one-step prediction error of y_m by L past values is denoted by

$$e_m = y_m - \sum_{i-1}^{L} g_i \; y_{m-i}. \tag{11.77}$$

In the z-transform domain,

$$E(z) = Y(z) - \sum_{i=1}^{L} g_i \; Y(z) z^{-i} = Y(z) \left[1 - \sum_{i=1}^{L} g_i \; z^{-i} \right]. \tag{11.78}$$

The transfer function of the optimum prediction error filter is given by

$$H(z) = \frac{E(z)}{Y(z)} = 1 - \sum_{i=1}^{L} \hat{g}_i \; z^{-i}. \tag{11.79}$$

A block diagram of $H(z)$ is given in Fig. 11.7.

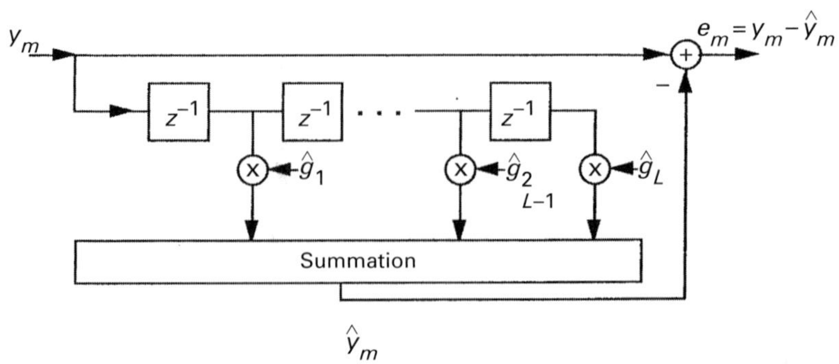

Figure 11.7 Optimum prediction-error filter $H(z)$

Now, we want to show that $|H(z)|$ in (11.79) takes small values for $z = e^{j2\pi f}$ with $f \in \{f_1, f_2, \ldots, f_p\}$. First, take the case of a single complex sinusoid (i.e., $p = 1$) with no noise. Then the LS approximation of (11.75) shows

$$\sum_{i=1}^{L} \hat{g}_i \, y_{m-i} \approx y_m, \quad m = L+1, \ldots, M, \tag{11.80}$$

and by using the form of (11.77), we have

$$\begin{aligned} 0 \approx y_m - \sum_{i=1}^{L} \hat{g}_i \, y_{m-i} &= A_1 e^{j(2\pi f_1 m + \theta_1)} \left[1 - \sum_{i=1}^{L} \hat{g}_i \, e^{-j2\pi f_1 i} \right] \\ &= A_1 e^{j(2\pi f_1 m + \theta_1)} H(z)|_{z=e^{j2\pi f_1}}. \end{aligned} \tag{11.81}$$

Thus, in this case, $|H(e^{j2\pi f_1})| \approx 0$.

Consider p complex sinusoids in the absence of noise. Then (11.81) generalizes to

$$\begin{aligned} 0 \approx y_m - \sum_{i=1}^{L} \hat{g}_i \, y_{m-i} &= A_1 e^{j(2\pi f_1 m + \theta_1)} H(e^{j2\pi f_1}) + A_2 e^{j(2\pi f_2 m + \theta_2)} H(e^{j2\pi f_2}) \\ &\quad + \cdots + A_p e^{j(2\pi f_p m + \theta_p)} H(e^{j2\pi f_p}). \end{aligned} \tag{11.82}$$

For any realization of $\{\theta_1, \ldots, \theta_p\}$, the p vectors of $\{A_i e^{j(2\pi f_i m + \theta_i)}, \; m = 1, \ldots, M, \; i = 1, \ldots, p\}$ are generally linearly independent, then (11.82) implies

$$0 \approx H(e^{j2\pi f_i}), \quad i = 1, \ldots, p. \tag{11.83}$$

In the presence of white noise in (11.69), (11.83) is no longer valid. Nevertheless, $|H(e^{j2\pi f})|$ has small values for $f \simeq f_i$, $i = 1, \ldots, p$.

The maximum entropy method (MEM) uses the autoregressive (AR) model to estimate the power spectral density of $\{y_m\}$ in (11.69) with regard to the frequency locations $\{f_i\}$ by considering

$$S(e^{j2\pi f}) = \frac{c}{|H(z)|^2_{z=e^{j2\pi f}}} = \frac{c}{|1 - \sum_{i=1}^{L} \hat{g}_i \, e^{-j2\pi f_i}|^2}, \quad 0 \le f < 1. \tag{11.84}$$

Clearly, from (11.83), if $f \simeq f_i$, $i = 1, \ldots, p$, then $|H(e^{j2\pi f})|$ is small and thus $S(e^{j2\pi f})$ is large. Thus, by setting up a threshold T_0, those values of f such that

$$S(e^{j2\pi f}) > T_0 \tag{11.85}$$

can be used to estimate f_i, $i = 1, \ldots, p$.

In practice, there are various possible difficulties with the MEM approach for spectral estimation. One basic problem is the proper selection of the order L of the FIR prediction filter as well as the number of terms M of the data needed to determine good LS approximation for $\hat{g}$. The solution to (11.70) can be written in various forms. If the matrix A is full rank and well-conditioned, then $\hat{g}$ can be expressed in terms of the normal equations,

$$\hat{g} = (A^{\mathrm{H}} A)^{-1} A^{\mathrm{H}} y, \quad 2(M - L) > L.$$

If the matrix A is rank deficient or ill-conditioned, then one needs to compute a low-rank approximation of A. A possible way is to compute the SVD of A, $A = \sum_{i=1}^{L} \sigma_i u_i v_i^{\mathrm{H}}$, determine its (numerical) rank, r, and define the r-rank approximation matrix

$$A_r = \sum_{i=1}^{r} \sigma_i u_i v_i^{\mathrm{H}}.$$

The LS solution is given by

$$\hat{g} = A^{+} y, \quad A^{+} = \sum_{i=1}^{r} \sigma_i^{-1} v_i u_i^{\mathrm{H}}.$$

In reality, there is no difference in the nature of the entries of the matrix A and the entries of vector y, and both A and y are affected by perturbations of the same entity. Therefore, it is perfectly justified to consider optimizing the choice of the coefficients $\{\hat{g}_i\}$ under a total least-squares (TLS) criterion. The solution then is the minimum norm solution to the following minimization problem

$$\begin{aligned} &\text{minimize} \quad \|[\Delta A, \Delta y]\|_F, \\ &\text{subject to} \quad (A + \Delta A)\hat{g} = y + \Delta y. \end{aligned}$$

In order to compute the values of the coefficients $\{\hat{g}_i\}$, one has to compute the SVD of the composite matrix $[A, y]$, determine its rank, $r \leq L$, and impose that the vector $[\hat{g}^T, 1]^T$ lies in the numerical null-space of the r-rank approximate of $[A, y]$. Another basic problem is the precise evaluation of T_0 for the proper operation of (11.85).

11.4 Reduced Rank FIR Filter Approximation

Consider a pth order FIR filter transfer function denoted by

$$H(z) = \sum_{i=0}^{p-1} h_i z^{-i}, \tag{11.86}$$

where $\{h_0,\ h_1, \ldots,\ h_{p-1}\}$ are the impulse responses of the filter. A typical realization of this filter given in Fig. 11.8 shows p multiplications are needed. Now, assume the order is a composite number given by $p = mn$. Clearly, the p summations of (11.86) can be regrouped as summation of n grouped terms with m terms in each group, as shown in (11.87).

$$\begin{aligned} H(z) = {} & (h_0 + h_1 z^{-1} + \cdots + h_{m-1} z^{-(m-1)}) + (h_m z^{-m} + h_{m+1} z^{-(m+1)} \\ & + \cdots + h_{2m-1} z^{-(2m-1)}) + \cdots + (h_{(n-1)m} z^{-(n-1)m} + h_{(n-1)m+1} z^{-((n-1)m+1)} \\ & + \cdots + h_{nm-1} z^{-(nm-1)}). \end{aligned} \tag{11.87}$$

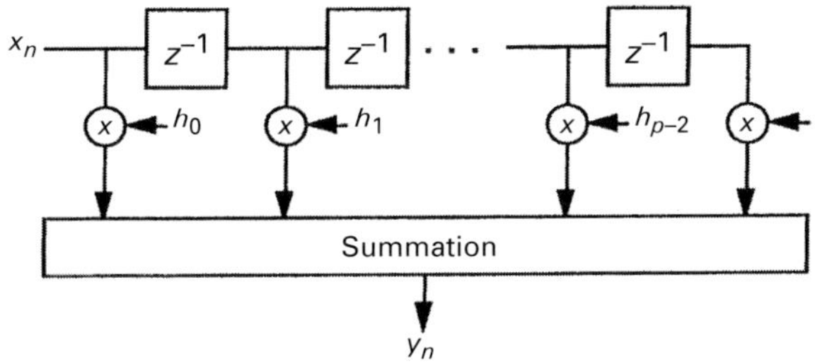

Figure 11.8 A realization of a pth order FIR filter

By factoring the expression of $1, z^{-m}, \ldots,$ and $z^{-(n-1)m}$ from each of the m terms respectively in (11.87), we have

$$\begin{aligned} H(z) = (h_0 + h_1 z^{-1} + \cdots + h_{m-1} z^{-(m-1)}) + z^{-m}(h_m + h_{m+1} z^{-1} + \cdots + \\ h_{2m-1} z^{-(m-1)}) + \cdots + z^{-(n-1)m}(h_{(n-1)m} + h_{(n-1)m+1} z^{-1} \\ + \cdots + h_{nm-1} z^{-(m-1)}). \end{aligned} \tag{11.88}$$

Denote

$$a = [1,\ z^{-1},\ \ldots,\ z^{-(m-1)}]^T, \tag{11.89}$$

$$H_0 = \begin{bmatrix} h_0 \\ h_1 \\ \vdots \\ h_{m-1} \end{bmatrix}, H_1 = \begin{bmatrix} h_m \\ h_{m+1} \\ \vdots \\ h_{2m-1} \end{bmatrix}, \ldots, H_{n-1} = \begin{bmatrix} h_{(n-1)m} \\ h_{(n-1)m+1} \\ \vdots \\ h_{nm-1} \end{bmatrix}, \tag{11.90}$$

$$b = [1, z^{-m},\ \ldots,\ z^{-(n-1)m}]^T. \tag{11.91}$$

By observing that each grouped term in (11.88) is an inner product of the form $a'H_i$, then $H(z)$ can be expressed as

$$H(z) = a^T H_0 + a^T H_1 z^{-m} + \cdots + a^T H_{n-1} z^{-(n-1)m} \tag{11.92}$$

$$= a^T (H_0 + H_1 z^{-m} + H_{n-1} z^{-(n-1)m}) \tag{11.93}$$

$$= a^T [H_0,\ H_1,\ \ldots,\ H_{n-1}] b \tag{11.94}$$

$$= a^T H b. \tag{11.95}$$

The grouped term in (11.93) is also expressible as an inner product in the form of (11.94) and (11.95), where H is an $m \times n$ matrix of the sequence of filter impulse responses $\{h_0,\ \ldots,\ h_{nm-1}\}$ given by

$$H = [H_0,\ H_1,\ \ldots,\ H_{n-1}]. \tag{11.96}$$

Let the SVD of the $m \times n$ matrix H be denoted by

$$H = USV^T = [u_1, u_2, \ldots, u_n][\text{diag}[s_1, s_2, \ldots, s_n]] \begin{bmatrix} v_1^T \\ v_2^T \\ \vdots \\ v_n^T \end{bmatrix}$$

$$= \sum_{i-1}^{n} s_i u_i v_i^T. \tag{11.97}$$

Then by using (11.97) in (11.95), we have

$$H(z) = \sum_{i=1}^{n} s_i a^T u_i v_i^T b \tag{11.98}$$

$$= \sum_{i=1}^{n} F_i(z) G_i(z), \tag{11.99}$$

where

$$F_i(z) = s_i[1, z^{-1}, \ldots, z^{-(m-1)}]u_i$$

$$= [1, z^{-1}, \ldots, z^{-(m-1)}] \begin{bmatrix} u_{i,1}s_i \\ u_{i,2}s_i \\ \\ u_{i,m}s_i \end{bmatrix}$$

$$= f_{i,0} + f_{i,1}z^{-1} + \cdots + f_{i,m-1}z^{-(m-1)} \tag{11.100}$$

$$G_i(z) = v_i^T \begin{bmatrix} 1 \\ z^{-m} \\ \vdots \\ z^{-(n-1)m} \end{bmatrix} = [v_{i,1}, v_{i,2}, \ldots, v_{i,n}] \begin{bmatrix} 1 \\ z^{-m} \\ \vdots \\ z^{-(n-1)m} \end{bmatrix}$$

$$= g_{i,0} + g_{i,1}z^{-m} + \cdots + g_{i,n-1}z^{-(n-1)m}. \tag{11.101}$$

From the SVD of H, all the s_i, u_i, and v_i are known. Thus, each $F_i(z)$ can be considered as an mth order FIR filter needing m multiplications and each $G_i(z)$ is an nth order FIR filter needing n multiplications.

Now, suppose $H(z)$ is approximated by rank r matrix $H_r(z)$. Then (11.99) becomes

$$H_r(z) = \sum_{i=1}^{r} F_i(z) G_i(z). \tag{11.102}$$

A realization of (11.102) in Fig. 11.9 shows r parallel branches consisting of the cascading of $F_i(z)$ and $G_i(z)$. Since each branch needs $(m+n)$ multiplications, the total number of multiplication M_r for the rank r approximated $H_r(z)$ becomes

$$M_r = r(m+n). \tag{11.103}$$

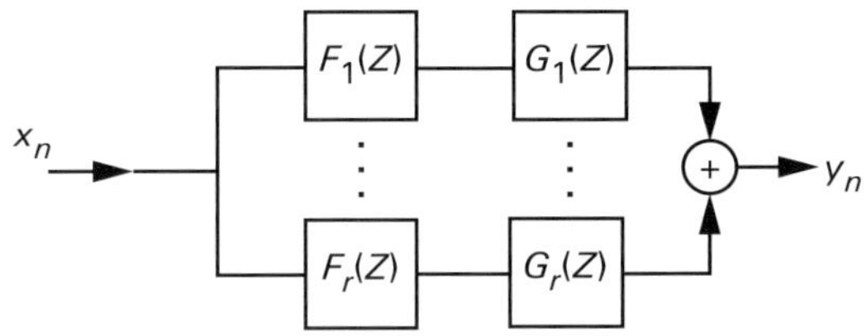

Figure 11.9 A realization of the rth order approximation of the FIR filter $H(z)$

Indeed, if $M_r < p = mn$, then the reduced rank FIR realization of $H_r(z)$ in Fig. 11.9 provides a more hardware efficient realization of the original realization in Fig. 11.8.

Example 11.6 Consider $h_i = c^{-i}$, $i = 0, \ldots, p-1$. Then (11.90) yields

$$H = [H_0, \ldots, H_{n-1}] = \begin{bmatrix} 1 & c^{-m} & \cdots & c^{-(n-1)m} \\ c^{-1} & c^{-(m+1)} & & c^{-((n-1)m+1)} \\ \vdots & \vdots & & \vdots \\ c^{-(m-1)} & c^{-(2m-1)} & & c^{-(nm-1)} \end{bmatrix}. \tag{11.104}$$

Clearly, since the ith column, $i = 2, \ldots, n$ is just $c^{-m(i-1)}$ times the first column, then H is of rank $r = 1$. Thus

$$H = s_1 \, u_1 \, v_1^T, \tag{11.105}$$

and

$$H(z) = F_1(z)G_1(z). \tag{11.106}$$

The number of multiplications $M_1 = m + n$ needed in (11.106) is generally less than that of $p = mn$ for m and n greater than 2.

11.5 Applications of SVD to Speech Processing

The SVD can be successfully used in a number of applications where it is required to give characterization of the data. The SVD can be used to extract underlying features of a number of signals. Unfortunately, most signals have a non-stationary behavior, which would require a repeated computation of the SVD. A full singular value decomposition is a particularly burdensome task, especially for large matrices. A better approach is to limit the computation to the quantities of interest, which are typically the singular values and/or the numerical rank, and the right singular vectors, or a subset thereof. In addition, much research has been lately devoted to the devisement of real-time SVD updating procedures. By these approaches, the information regarding the SVD computed at earlier times is retained and properly updated to reflect the variations in the data stream. The availability of real-time non-stationarity tracking and SVD updating can be exploited for the simultaneous extraction of different useful parameters from the arriving data. In certain speech-processing applications, for instance, it may be required to simultaneously discriminate voiced from unvoiced segments, estimate the best-fitting AR filter parameters, and determine the glottal closure epochs. In the following, we consider two applications.

Example 11.7 We demonstrate the real-time tracking of the SVD of the data matrix of speech signals. Speech utterances can be segmented into *voiced* and *unvoiced* sounds. Voiced speech is characterized by pronounced resonances, which are missing in unvoiced speech spectra. A widely used speech production model is composed of an excitation of periodic pulses for voiced speech or white noise for unvoiced speech onto a vocal tract filter. The contributions of the glottal flow, the vocal tract, and the radiation at the lips can be modeled by an all-pole (autoregressive, AR) time-varying linear filter. This filter is responsible for the short-time spectral envelope of the speech signal, while the fine structure is determined by the periodic or non-periodic characteristics of the excitation. Although an all-pole filter does not efficiently model spectral dips, it can approximate the spectral balance contributions of zeros very accurately. The order of the model (i.e., the number of coefficients in the all-pole filter) is determined by numerous factors, such as vocal tract length, the coupling of the nasal cavities, the place of the excitation, the nature of the glottal flow. The time-varying transfer function has the following z-transform

$$H(z) = G \Big/ \left(1 - \sum_{i=1}^{p} a_{iz}^{-i}\right),$$

where G is a gain factor, p is the model order, and $\{a_i\}_{i=1}^{p}$ are the filter's AR coefficients. The characteristics of speech vary in time according to the rate at which the articulators change. This rate, sometimes referred to as *articulator rate*, or *phoneme rate*, is relatively low, and approximately equal to 100 Hz. This means that the speech waveform $y(n)$ can be analyzed over periods of 10 ms, and its characteristics can be considered almost constant over these intervals. Construct an $L \times J$ Toeplitz data matrix

$$Y_m = \begin{pmatrix} y(m-L+1) & y(m-L) & \cdots & y(m-J-L+2) \\ \vdots & \vdots & & \vdots \\ y(m-1) & y(m-2) & \cdots & y(m-J) \\ y(m) & y(m-1) & \cdots & y(m-J+1) \end{pmatrix},$$

from an observed speech waveform $y(m)$ corresponding to a given utterance. The SVD of Y_m is dependent on both the coefficients of the all-pole filter and on the correlation properties of the excitation. The rank of this matrix is related to the model order p.

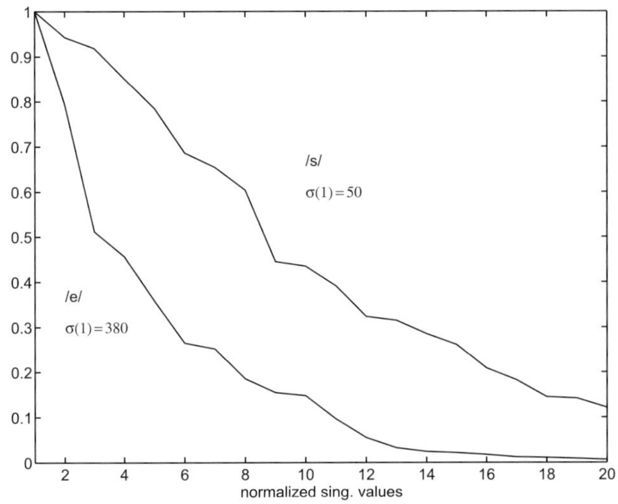

Figure 11.10 Normalized singular value distributions corresponding to sounds /e/ and /s/

When the excitation is periodic, then the singular value distribution of Y_m tends to display a low-rank structure, and the smallest singular values are very close to zero. On the other hand, when the vocal tract filter is excited by random white noise, then the singular value distribution spreads out. It is often the case that voiced sounds are associated to higher energy than unvoiced. The previous considerations are verified by Fig. 11.10. In this figure, we show the distribution of the singular values corresponding to 20×20 data matrices associated to two segments of the utterance "test." The first one corresponds to the sound /e/, while the second corresponds to the sound /s/. The distributions are normalized to the first singular value, which is equal to the norm of the data matrix and is therefore indicative of the energy in the given time window. Note that the singular value distribution for the sound /e/ falls off more rapidly.

Based on the above considerations, the detection of voiced/unvoiced speech segments can be provided by measuring the percentage of data energy, observed within a given window, that falls in the subspace spanned by the right singular vectors corresponding to the smallest singular values (the "noise subspace"). In this way, we can detect the boundary between voiced and unvoiced sounds, a task which is known to be difficult in practice. If the sound is voiced, the ratio between the energy measured in the "noise subspace" and the total data energy falls below a threshold value, whereas it remains above it when the sound is unvoiced. These results are shown in Fig. 11.11, where the waveform associated to the utterance "test" and the described non-stationarity indicator are plotted against time. When the voiced sound /e/ exists over a time unit of about 1,500 to over 3,000, there is a corresponding dip in the indicator for this interval.

Example 11.8 We show how the computed SVD can be used for adaptive parameter estimation. Consider the running SVD of the data matrix, and compute an estimate of the all-pole filter coefficients. For an AR filter model,

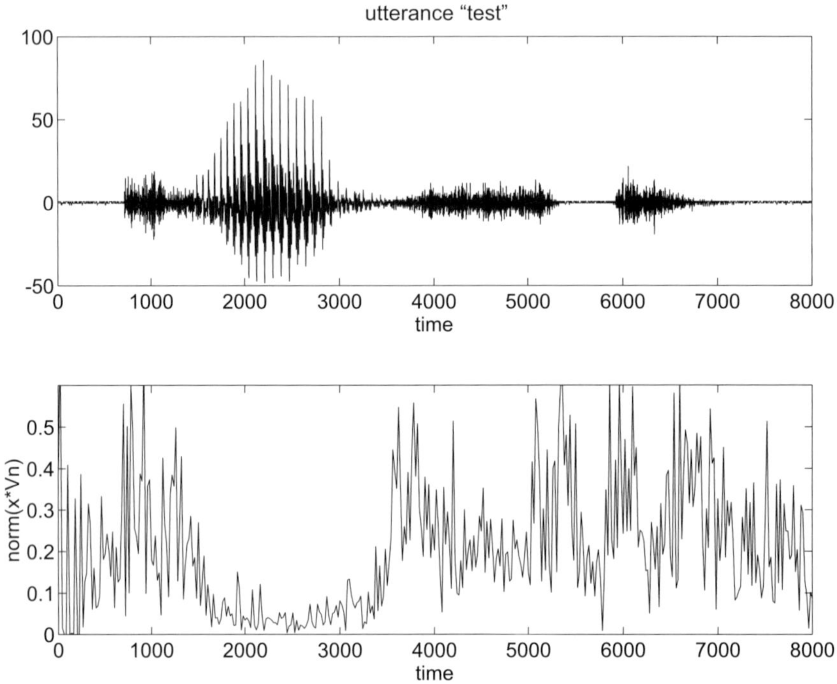

Figure 11.11 Time waveform for the word "test" and non-stationarity indicator as function of time

$$y(m) = Gu(m) + a_1 y(m-1) + \cdots + a_{py}(m-p),$$

for all time instants m, where $u(m)$ are unit variance zero-mean excitation samples. Given the data matrix, Y_m, defined earlier, when $J > p$, and the $J \times 1$ vector $a \equiv (1, a_1, \ldots, a_p, 0, \ldots, 0)^T$, we have that

$$Y_m a = G u_m,$$

where $u_m \equiv (u(m-L+1), u(m-L+2), \ldots, u(m))$. In order to estimate the AR coefficients, a possible strategy is to find the vector $\hat{a}$ that minimizes the norm of Y_{ma}. If the SVD of Y_m is given as $Y_m = U_m \Sigma_m V_m^{\mathrm{H}}$, then we have

$$\hat{a} = \arg \min_{x, x(1)=1} \|Y_{mx}\| = \arg \min_{x, x(1)=1} \|\Sigma_m V_m^{\mathrm{H}} x\|.$$

The vector $\hat{z}$ which minimizes the norm of Σ_{mz} is given by $\hat{z} = e_n^T \equiv (0, 0, \ldots, 0, 1)^T$, and the norm is equal to the smallest singular value. It follows that $\hat{a} = v_{\min}/v_{\min}(1)$, where $v_{\min}$ is the right singular vector corresponding to the smallest singular value. In conclusion, the desired estimated AR coefficients are given by the singular vector corresponding to the smallest singular value. In Fig. 11.12, we show the coefficient values of a order-9 AR filter estimated adaptively by use of the proposed SVD updating algorithm. The coefficients shown were computed between times 2160 and 2165

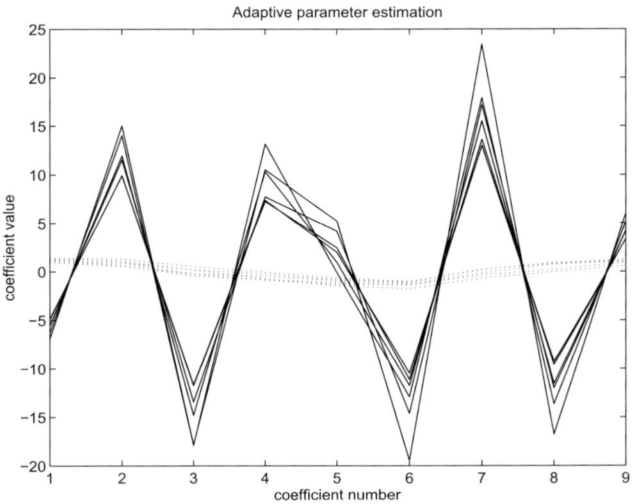

Figure 11.12 AR coefficients of order-9 filter computed using the SVD updating algorithm. Solid lines: times 2160 to 2165, sound /e/ between glottal closures. Dotted lines: times 4360 to 4365, sound /s/

(corresponding to sound /e/, between glottal closures), and between times 4360 and 4365 (corresponding to sound /s/).

It is well-known that linear prediction models fit best during speech segments of less than one pitch period, between instants of glottal closure, or "epochs." At glottal closure, the excitation is present in the data, with the consequence that the linear prediction model does not fit the data properly and the prediction error is large. The large deviation between actual and predicted data around a glottal closure instant, due to the abrupt change in glottal flow, has been used for epoch estimation by many authors. In the absence of excitation, the linear filter model of order p imposes a linear dependence among the columns of Y_m. Consequently, the determinant of the matrix $Y_m^{\mathrm{H}} Y_m$ increases sharply when the speech segment covered by Y_m contains an excitation. Similarly, when the analysis window contains a glottal pulse, the LS residual energy, $\|Y_m \hat{a}\|^2$ increases rapidly. It can be shown that both the determinant of $Y_m^{\mathrm{H}} Y_m$ and the LS residual energy are related to the singular values of Y_m. In particular, the singular values of Y_m tend to have sharp peaks corresponding to the instants of glottal closure. Any algorithm for SVD updating can therefore be successfully employed for glottal closure detection, by keeping track of the behavior of the individual singular values. In Fig. 11.13, we show the behavior of three computed singular values, namely $\sigma_{\max}$, $\sigma_{\min}$, and σ_5, corresponding to the sounds /e/ and /s/ of the utterance "test." From this figure, we note the clear periodic pattern of the singular values during a voiced segment, with pulses at the glottal closure instants. No particular pattern is discernible for the unvoiced sound. For comparison, we also show the running Frobenius norm of Y_m (which is the sum of the squared singular values) for the two cases. Note that the smaller singular values have sharper peaks around the epochs than both the larger singular values and the Frobenius norm.

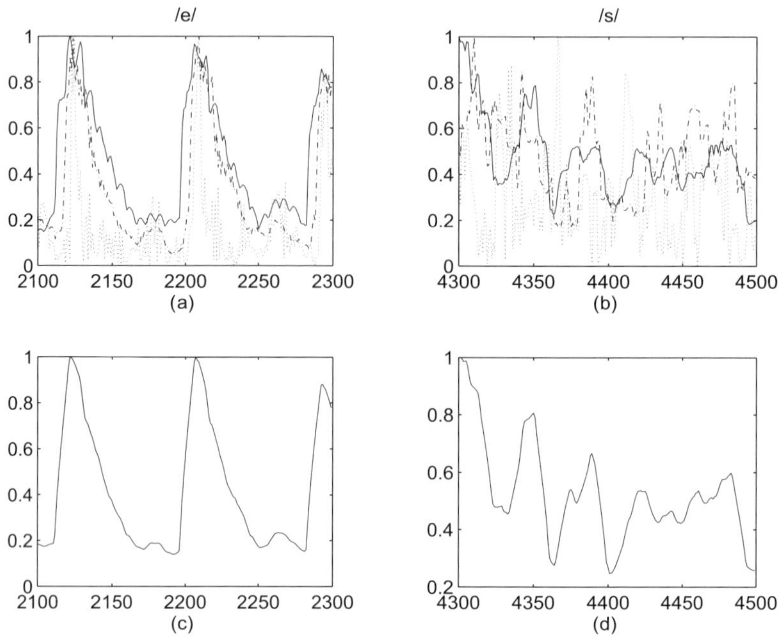

Figure 11.13 Normalized singular values ($\sigma_{\max}$ (solid), σ_5 (dash-dot), and $\sigma_{\min}$ (dotted)) and Frobenius norm of windowed data matrices. (a) and (b) normal. sing. values for segment corresponding to /e/ and /s/; (c) and (d) normal; Frobenius norm for segment corresponding to /e/ and /s/

11.6 Applications of SVD to DOA

Consider an array of K sensors, on which M narrowband sources impinge from angles $\theta_1, \ldots, \theta_M$. The outputs of the sensors can be collected in an $N \times K$ matrix X, where N is the length of the observation interval, or snap-shots. Subspace oriented methods for DOA, such as MUSIC allow the estimation of the DOA, by computing the "signal" and "noise subspaces" of the $K \times K$ sample correlation matrix $R_X = X^H X$. If the eigenvalue decomposition of R_X is given by

$$R_X = E \Lambda E^H,$$

where $\Lambda = \text{diag}(\lambda_1, \ldots, \lambda_K)$ is diagonal, and $E = [E_s, E_n]$ is a unitary $K \times K$ matrix. In the model for MUSIC, we assume the noise is white with power σ_n^2, then

$$\lambda_1 \geq \lambda_2 \geq \cdots \geq \lambda_M \gg\approx \sigma_n^2 \approx \lambda_{M+1} \approx \cdots \approx \lambda_K.$$

Then if we denote $a(\theta)$ as the generic steering vector, and with the $K \times (K - M)$ noise subspace matrix E_n consisting of the K-dimensional column eigenvectors corresponding to the $K - M$ smallest eigenvalues, the maxima of the function

$$S(\theta) = \frac{1}{a(\theta)^H E_n E_n^H a(\theta)},$$

are attained approximately near $\theta = \theta_1, \ldots, \theta = \theta_M$.

It is well known that the SVD of X,

$$X = U \Lambda V^H,$$

where U is an $N \times K$ matrix and V is a $K \times K$ matrix both with orthonormal column vectors, and $\Lambda = \text{diag}(\sigma_1, \ldots, \sigma_K)$ is a diagonal matrix, is related to the eigenvalue decomposition of R_X in the following manner

$$\Lambda = \Sigma^2, \quad E = [E_s, E_n] = V = [V_s, V_n].$$

The function $S(\theta)$ can be computed from the SVD of X, as follows

$$S(\theta) = \frac{1}{\|a(\theta)^H V_n\|^2},$$

where $V_n = E_n$ is the rightmost $K \times (M - K)$ submatrix of V and E.

It is often the case that the DOAs vary with time, and thus information in the newly observed data need to be included and the older data need to be de-emphasized. In the sliding window approach, the newly observed data are appended at the bottom of the matrix X and the oldest data at the top of X are dropped, so that the updated matrix X remains of constant size $N \times K$. In the exponential window approach, as the new data are updated, the older data are forgotten in an exponentially decaying manner. In both approaches, the SVD of the updated and windowed data need not be computed from scratch at every instant. In a recursive SVD updating algorithm, the information regarding the previously computed SVD is retained and used as a starting point for the calculation of the current computation. Additional schemes can be added for the estimation of the numerical rank, which is related to the number of signals, M, currently impinging on the sensor array.

11.7 Conclusion

In Section 11.1, we applied least-squares solution of a linear system of equations and SVD methods to an aircraft load measurement problem. Section 11.2 compared the least-squares, total least-squares, and correspondence analysis methods. Section 11.3 considered the rank reduction property of SVD to the forward–backward linear prediction estimation problem as applied to the maximum entropy spectral estimation problem. Section 11.4 used the rank reduction property of SVD to perform complexity reduction for a class of FIR filter problem. Section 11.5 utilized the SVD technique on the subspaces of the voiced and unvoiced speech sounds for segmentation purposes.

11.8 References

The materials presented in Section 11.1 were taken from our own past work with NASA Dryden Flight Research Laboratory's aircraft measured data [1]. The TLS problem was first proposed by Van Huffel [2] the correspondence analysis method was originated by J.P. Benzecri [3] with more detailed explanations given in Greenacre [4]. The materials in Section 11.2.3 appeared in [5]. The equivalency of the correspondence analysis and TLS of Section 11.2.4 was considered in [6]. Detailed materials in Section 11.3 can be found in [7]. The use of SVD for reduced rank FIR filter design was proposed by Mitra for 1D and by [8] for 2D FIR filters in Section 11.4. The materials in Section 11.5 appeared in [9]. Many papers have dealt with the DOA problem using MUSIC and SVD methods. A collection of papers on this problem appeared in [10].

[1] K. Yao, "Equivalency of Normal Equation and SVD Techniques for Linear Modeling of Approaches 1 and 2," UCLA-Dryden Memo, 2004.

[2] S. Van Huffel and J. Vandewalle, *The Total Least Squares Algorithm*, Kluwer, 1991.

[3] J.P. Benzecri, "L'Analyse des Donnees," *L'Analyse des Correspondences*, Dunod, 1973.

[4] M. Greenacre, *Theory and Applications of Correspondence Analysis*, Academic Press, 1983.

[5] K. Yao, "SVD-Based Data Reduction and Classification Techniques," in *Advances in Statistical Signal Processing, Vol. 2*, ed. H.V. Poor and J.B. Thomas, JAI Press, 1993, pp. 411–428.

[6] K. Yao, F. Lorenzelli, and J. Kong, "The Equivalence of Total Least Squares and Correspondence Analysis," in *Proceedings of the IEEE Conference on ICASSP*, 1992, pp. 585–588.

[7] D.W. Tufts and R. Kumaresan, "Estimation of Frequencies of Multiple Sinusoids: Making Linear Prediction Perform like Maximum Likelihood," *Proceedings of the IEEE*, 1982, pp. 975–989.

[8] A. Antoniou and W.-S. Lu, "Design of Two-dimensional Digital Filters by Using the Singular Value Decomposition," *IEEE Trans. Circuits Systems*, 1987, pp. 1191–1198.

[9] F. Lorenzelli and K. Yao, "Jacobi SVD Algorithms for Tracking of Nonstationary Signals," *IEEE ICASSP*, 1995, pp. 3183–3186.

[10] T.E. Tuncer and B. Friedlander, *Classical and Modern Direction-of-Arrival Estimation*, Elsevier, 2009.

11.9 Exercises

1. Suppose we are given two pairs of measured data $[a1, b1] = [0.98, 0.99]$ and $[a2, b2] = [2.01, 2.03]$. We assume the true model satisfies $b = ax$, where $b = [b1, b2]^T$ and $a = [a1, a2]^T$. (a) Find under the least-squares criterion the optimum x_{LS}. (b) Find under the total least-squares criterion the optimum x_{TLS}.
2. Consider two pairs of points (a1,b1) = (1,1.11) and (a2,b2) = (2,2.15).

a. Use the LS criterion to find the optimum scalar x_{LS} such that

$$\min_{x}\left\{\left\|\begin{bmatrix} a_1 \\ a_2 \end{bmatrix} x - \begin{bmatrix} b_1 \\ b_2 \end{bmatrix}\right\|^2\right\} = \left\|\begin{bmatrix} a_1 \\ a_2 \end{bmatrix} x_{LS} - \begin{bmatrix} b_1 \\ b_2 \end{bmatrix}\right\|^2 = \varepsilon_{LS}^2.$$

b. Find the minimum ε_{LS}^2.

c. Use the TLS criterion to find the optimum scalar x_{TLS} for these data. Hint: You have two ways of solving this problem. Either way will be OK.

d. For this example, how does ε_{LS}^2 compare to ε_{TLS}^2? In general, what can one say about ε_{LS}^2 compared to ε_{TLS}^2?

3. Consider a 6×5 data matrix

$$X = \begin{bmatrix} 2 & 2 & 0 & 0 & 1 \\ 0 & 0.1 & 0 & 0 & 2 \\ 0 & 0 & 1 & 1 & 3 \\ 0 & 0 & 0 & 0.1 & 4 \\ 0 & 0 & 0 & 0 & 5 \\ 1 & 1 & 1 & 1 & 6 \end{bmatrix}.$$

The five column collinearity indices are given by

$$k = \begin{bmatrix} 24 & 24 & 17 & 18 & 1.8 \end{bmatrix}.$$

Plot the five column vectors of X with respect to the two dominant singular vectors. Show the first two vectors are similar and the third and fourth vectors are also similar.

4. Show the TLS estimation method and the correspondence approach method are equivalent.

5. Consider the following annual rainfall data from 8 cities around the world. Use the correspondence analysis approach to find similarities and differences among these cities.

	Santa Monica	Westwood	Moscow	San Francisco	Tokyo Meguro-ku	Kobe	Ulaanbaatar	Shanghai
Jan	8.58	5.4	3.99	10.65	4.86	3.89	0.19	7.75
Feb	8.92	5.56	3.25	9.96	6.02	5.42	0.19	5.06
Mar	5.73	4.67	2.72	7.45	11.45	9.08	0.28	9.43
Apr	1.82	2.56	2.42	3.34	13.03	12.14	0.64	7.17
May	0.63	2.45	3.92	1.19	12.8	14.21	1.46	9.33
Jun	0.24	1.65	5.93	0.33	16.49	18.96	3.67	19.62
Jul	0.15	0.95	5.56	0.03	16.15	14.58	4.65	11.66
Aug	0.21	0.81	5.64	0.12	15.51	10	4.61	22.46
Sep	0.69	1.24	4.61	0.6	20.85	17.14	2.26	7.07
Oct	1.44	2.46	4.99	2.44	16.31	10.6	0.53	5.86
Nov	2.8	4.32	3.82	5.84	9.25	6.47	0.48	4.91
Dec	4.87	5.43	3.57	9.33	3.96	3.98	0.35	4.85

6. Consider a simplified fading wireless communication model. Suppose a transmitter sends a signal vector $[1, 2, 1]^T$. The direct path has a delay such that the transmitted vector arrives at the receiver at times of $10, 11,$ and 12 with an attenuation of x_1. The reflected path arrives at times of $12, 13,$ and 14 with an attenuation of x_2. The received data at times $10, 11, 12, 13,$ and 14 are given by $[4, 7, 8, 6, 3]^T$ which contains the two transmitted signal vectors with attenuations plus negligible random noise.

   a. Formulate this problem as an LS estimation problem of the form $Ax = b$. What are $A, x,$ and b for this problem? Hint: A can be taken as a 5×2 matrix.
   b. Use the normal equation pseudo-inverse method to find the solution denoted by x_0.
   c. Use the Moore–Penrose pseudo-inverse method to find the solution denoted by x_0.
   d. Why are the two solutions for x_0 essentially the same?

12 Quantization, Saturation, and Scaling

In a digital signal processing system, binary words of finite length are used for storage and computations. On the other hand, continuous-time and continuous-amplitude signals are first sampled in time to obtain a sequence of discrete-time and and continuous-amplitude signals. After passing this sequence of data through a quantizer with a finite number of levels coded in some binary format, we obtain the binary words appropriate for digital processing. A quantizer is an analog-to-digital converter (ADC) that outputs one level for all input continuous-amplitude signals with values inside each step size. Continuous-amplitude signals need in principle an infinite number of steps of arbitrarily small size in the quantizer for an error-free representation. Thus, every quantizer with a finite number of steps introduces a quantization noise in the continuous-amplitude to discrete-amplitude conversation process. In Section 12.1, we introduce the truncation and roundoff quantization error models. By using a sufficient number of bits to represent the quantizer levels, the quantization error can be reduced to a tolerable amount suitable for a specified application. In practice, we must know completely (or with some high probability of certainty) the range of the input continuous-time signals. Then the dynamic range of the input of the quantizer given by the quantizer step size times the number of levels must be set appropriately relative to the range of the input continuous-amplitude signals. In Section 12.2, we consider the resulting saturation distortions when the range of the input continuous-signals is greater than the dynamic range of the digital processing system for some simple cases. In Section 12.3, we consider the internal state values of a second-order recursive digital filter. Near the transition region of any recursive digital filter, these state values can be quite large. If the range of these state variables is greater than the dynamic range of the digital processing system, then distortions similar to those considered in Section 12.2 can occur. Both time-domain and frequency-domain techniques are used to estimate the range of the state variables in the filter. In Section 12.4, we consider the combined uniform quantization and saturation error models for Gaussian input signals. For a fixed number of steps in the quantizer, the objective is to select the step size sufficiently small so that the quantization error is under control, but at the same time the input dynamic range of the quantizer is sufficiently large (with high probability) relative to the input random signals so that distortions occur infrequently. Numerical results for Gaussian input signals are presented to illustrate optimum parameter selections for minimizing the combined total quantization and distortion errors. In Section 12.5, we consider the power spectral density of the distortion sequence due to saturation for a narrowband Gaussian random sequence. Some numerical results are presented to illustrate the concept.

12.1 Processing Noise Model

Consider a digital signal processing system using binary registers or words of length B bits to store and process its data. Then each word contains 2^B possible contents and can be represented as non-negative integers from 0 to $2^B - 1$. Upon the multiplication of two such B bit words, the result may need to be represented by a $2B$ bit word. After repeated multiplications, the results may require an arbitrarily large number of bits for exact representation. In practice, we often want to reduce a B bit word by N bits for storage and subsequent processing. For example, after multiplying two B bit words, we set $N = B$ to retain a B bit word. This bit reduction generally leads to processing errors in digital signal processing.

12.1.1 Truncation and Roundoff Errors

Consider the allowable contents of an original B bit binary word represented by non-negative integers given by

$$\mathcal{S}_B = \{0, 1, \ldots, 2^B - 1\}. \tag{12.1}$$

Suppose we want to use a $(B - N)$ bit word to represent the contents in (12.1). This implies that we want to reduce the allowable size of $\mathcal{S}_B$ by a factor of 2^N. There are many possible ways of mapping each group of 2^N values in $\mathcal{S}_B$ to a single value. Truncating N bits is one way to achieve this reduction.

Truncation of a B bit word by N bits can be considered as achieved in two stages. First, each

$$y = \ell 2^N + i, i = 0, \ldots, 2^N - 1, \ell = 0, \ldots, 2^{B-N} - 1, \tag{12.2}$$

in $\mathcal{S}_B$ is arranged in $2^{(B-N)}$ groups, each with 2^N elements given by

$$\mathcal{S}_B = \Big\{\{0, 1, \ldots, 2^N - 1\}, \{2^N, 2^N + 1, \ldots, 2^{N+1} - 1\}, \ldots, \{2^B - 2^N, \ldots, 2^B - 1\}\Big\}. \tag{12.3}$$

Then truncation of y by N bits of elements in (12.3) can be expressed as a mapping of the form

$$\mathcal{F}_1(\ell 2^N + k) = \ell 2^N, \quad \ell = 0, 1, \ldots, 2^{(B-N)} - 1, \quad k = 0, 1, \ldots, 2^N - 1. \tag{12.4}$$

We note, while there are only $2^{(B-N)}$ values after the mapping in (12.4), the values still belong to $\mathcal{S}_B$ and thus need B bits for representation. Therefore, we need a second mapping

$$\mathcal{F}_2(\ell 2^N) = \ell, \quad \ell = 0, 1, \ldots, 2^{(B-N)} - 1. \tag{12.5}$$

Now, the 2^{B-N} values after the mapping in (12.5) are elements of

$$\mathcal{S}_{B-N} = \{0, 1, \ldots, 2^{(B-N)} - 1\}, \tag{12.6}$$

which can be represented directly by a $(B - N)$ bit word. In summary, the truncation operation, $\mathcal{F}_t(\cdot)$, is a mapping from y in $\mathcal{S}_B$ represented by (12.2) to ℓ in $\mathcal{S}_{B-N}$ represented by (12.5) such that

$$\mathcal{F}_t(\ell 2^N + k) = \mathcal{F}_2(\mathcal{F}_1(\ell 2^N + k)) = \ell, \quad \ell = 0, 1, \ldots, 2^{(B-N)} - 1, \quad k = 0, 1, \ldots, 2^N - 1. \tag{12.7}$$

Example 12.1 Let $B = 5$ and $N = 2$. Then the mapping under $\mathcal{F}_1(\cdot)$ and $\mathcal{F}_2(\cdot)$ can be expressed as

$$\begin{array}{rcccc} & \mathcal{F}_1(\cdot) & & \mathcal{F}_2(\cdot) & \\ \{0,1,2,3\} & \longrightarrow & 0 & \longrightarrow & 0 \\ \{4,5,6,7\} & \longrightarrow & 4 & \longrightarrow & 1 \\ \{8,9,10,11\} & \longrightarrow & 8 & \longrightarrow & 2 \\ \vdots & & \vdots & & \vdots \\ \{28,29,30,31\} & \longrightarrow & 28 & \longrightarrow & 7 \end{array}$$

Thus, while $\{0, 4, \ldots, 28\} \subset \mathcal{S}_5$, $\{0, 1, \ldots, 7\} \subset \mathcal{S}_3$.

In general, any integer $y \in \mathcal{S}_B$ can also be represented by

$$y = y_{B-1} \cdot 2^{B-1} + y_{B-2} \cdot 2^{B-2} + \cdots + y_1 \cdot 2 + y_0, \tag{12.8}$$

where y_j, $j = 0, 1, \ldots, B - 1$, can take the values of 0 or 1. The mapping $\mathcal{F}_1(\cdot)$ in (12.4) is equivalent to setting $y_0, y_1, \ldots, y_{N-1}$ to zero in (12.8), while the mapping $\mathcal{F}_2(\cdot)$ in (12.5) is equivalent to multiplying the expression in (12.8) by 2^{-N}. Thus, the truncation operation of any $y \in \mathcal{S}_B$ given by (12.8) becomes

$$\ell = y_{B-1} \cdot 2^{B-N-1} + y_{B-2} \cdot 2^{B-N-2} + \cdots + y_{N+1} \cdot 2 + y_N, \tag{12.9}$$

which indeed is an element of $\mathcal{S}_{B-N}$.

Example 12.2 Consider the integer 11 in Example 12.1. From (12.8), $y_4 = 0, y_3 = 1, y_2 = 0, y_1 = 1$, and $y_0 = 1$. Setting y_0 and y_1 to zero, which is equivalent to applying $\mathcal{F}_1(\cdot)$ only, we obtain

$$8 = 0 \cdot 2^4 + 1 \cdot 2^3 + 0 \cdot 2^2. \tag{12.10}$$

Then after multiplying (12.10) by 2^{-2}, we also have

$$2 = 0 \cdot 2^2 + 1 \cdot 2^1 + 0, \tag{12.11}$$

which agrees with the result in Example 12.1.

Now, consider the mean and variance associated with the N bit truncation operation of (12.4) and (12.5). From the final $(B - N)$ bit word point of view, the truncation error is given by

$$\epsilon_{\ell,k} = \ell - (\ell 2^N + k)2^{-N} = -k2^{-N}, \qquad \ell = 0, 1, \ldots, 2^{(B-N)} - 1, \qquad k = 0, 1, \ldots, 2^N - 1. \tag{12.12}$$

If we assume that the original 2^B values in $\mathcal{S}_B$ are equally likely, the mean truncation error is given by

$$\begin{aligned} \mu_1 = E\{\epsilon_{\ell,k}\} &= 2^{-B} \sum_{\ell=0}^{2^{B-N}-1} \sum_{k=0}^{2^N-1} (-k2^{-N}) \\ &= -\frac{1 - 2^{-N}}{2} \simeq -\frac{1}{2}. \end{aligned} \tag{12.13}$$

The second moment of the truncation error is given by

$$\mu_2 = E\{\epsilon_{\ell,k}^2\} = \frac{(1 - 2^{-N})(1 - 2^{-N-1})}{3}. \tag{12.14}$$

The variance of the truncation error is

$$\sigma_t^2 = \mu_2 - \mu_1^2 = \frac{1 - 2^{-2N}}{12} \simeq \frac{1}{12}. \tag{12.15}$$

A rounding operation on a binary word can be expressed as adding 2^{N-1} in modulo 2^B followed by truncation. Specifically, define

$$\mathcal{F}_0(y) = (y + 2^{N-1}) \pmod{2^B}, \qquad y \in \mathcal{S}_B. \tag{12.16}$$

Then the rounding of a B bit word to a $(B - N)$ bit word can be expressed as

$$\mathcal{F}_r(y) = \mathcal{F}_2(\mathcal{F}_1(\mathcal{F}_0(y))), \qquad y \in \mathcal{S}_B, \tag{12.17}$$

where $\mathcal{F}_0$ is given by (12.16), $\mathcal{F}_1$ is given by (12.4), and $\mathcal{F}_2$ is given by (12.5).

Example 12.3 Consider the $B = 5$ and $N = 2$ case as treated in Examples 12.1 and 12.2. The rounding of this 5 bit word to a 3 bit word can be expressed as

$$\begin{array}{lclclcl} & \mathcal{F}_0(\cdot) & & \mathcal{F}_1(\cdot) & & \mathcal{F}_2(\cdot) & \\ \{0,1\} & \longrightarrow & \{2,3\} & \longrightarrow & 0 & \longrightarrow & 0 \\ \{2,3,4,5\} & \longrightarrow & \{4,5,6,7\} & \longrightarrow & 4 & \longrightarrow & 1 \\ \{6,7,8,9\} & \longrightarrow & \{8,9,10,11\} & \longrightarrow & 8 & \longrightarrow & 2 \\ \vdots & & \vdots & & \vdots & & \vdots \\ \{26,27,28,29\} & \longrightarrow & \{28,29,30,31\} & \longrightarrow & 28 & \longrightarrow & 7 \\ \{30,31\} & \longrightarrow & \{0,1\} & \longrightarrow & 0 & \longrightarrow & 0 \end{array}$$

Similar to (12.13)–(12.15), the mean, second moment, and variance of the roundoff error are given by

$$\mu_1 = 2^{-N-1} \simeq 0, \tag{12.18}$$

$$\mu_2 = \frac{2^{-2N}(2^{2N-1}+1)}{6} \simeq \frac{1}{12}, \tag{12.19}$$

$$\sigma_r^2 = \mu_2 - \mu_1^2 = \frac{1-2^{-2N}}{12} \simeq \frac{1}{12}. \tag{12.20}$$

12.2 Distortion Due to Saturation

In the previous section, we considered only the distortions caused by approximating a real number by a nearby discrete-valued number. While these ill effects are relevant to the design of the digital system, generally they are not as crucial as those of the distortions caused by saturation.

Consider a simple memoryless continuous-time system that can only respond linearly to inputs in $[-V, V]$. The input and output relationship is given by Fig. 12.1 where

$$y(x) = \begin{cases} V, & x > V \\ x, & -V \le x \le V \\ -V, & x < -V \end{cases}.$$

Since $y(x)$ saturates for $x > V$ and $x < -V$, the system becomes non-linear for large positive and large negative inputs. Consider an input

$$x(t) = A\sin(2\pi f t), \qquad -\frac{T}{2} \le t < \frac{T}{2}, \tag{12.21}$$

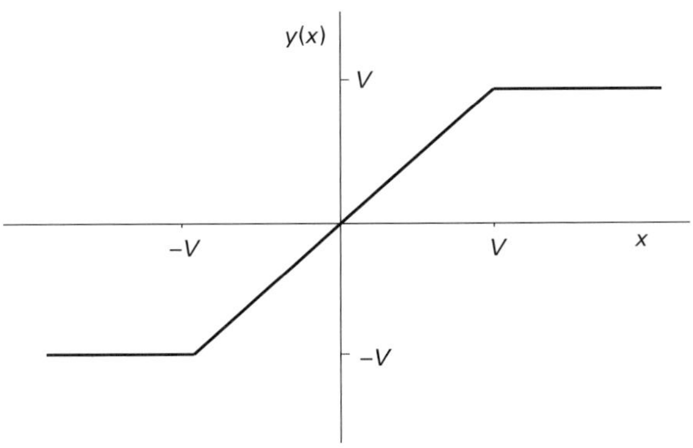

Figure 12.1 A saturation model

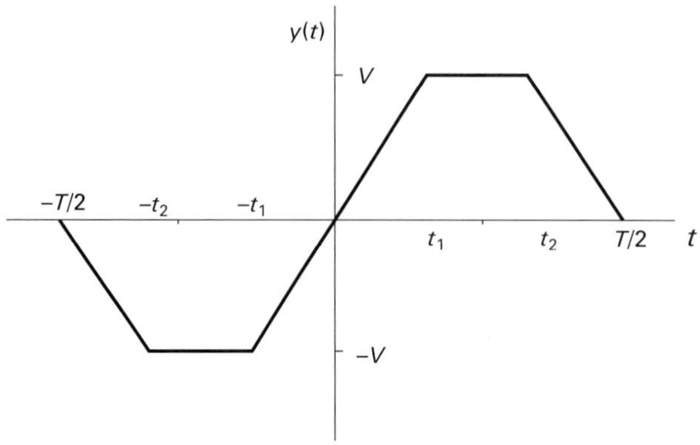

Figure 12.2 Saturation of a sinusoidal input

with $f = \frac{1}{T}$ and $A > V$. Then

$$y(t) = \begin{cases} V, & t \in [t_1, t_2] \\ A\sin(2\pi ft), & t \in [-\frac{T}{2}, -t_2) \cup (-t_1, t_1) \cup (t_2, \frac{T}{2}), \\ -V, & t \in [-t_2, -t_1] \end{cases} \tag{12.22}$$

where

$$\begin{aligned} t_1 &= \frac{\theta}{2\pi f} \\ t_2 &= \frac{T}{2} - t_1 = \frac{1}{2f}\left(1 - \frac{\theta}{\pi}\right), \\ \theta &= 2\pi f t_1 = \sin^{-1} K, \\ \text{and} \quad K &= \frac{V}{A}, \end{aligned}$$

as shown in Fig. 12.2.

Expanding $y(t)$ in Fourier series, we have

$$y(t) = \sum_{m=1}^{\infty} b_m \sin(2\pi mft), \qquad -\frac{T}{2} \le t < \frac{T}{2}, \tag{12.23}$$

where

$$b_m = \frac{4}{T}\int_0^{\frac{T}{2}} y(t)\sin(2\pi mft)dt, \qquad m = 1, 2, \ldots,$$

since $y(t)$ is an odd function of t. Using (12.22), (12.23) becomes

$$y(t) = \sum_{m=1,\, m \text{ odd}}^{\infty} b_m \sin(2\pi mft), \qquad -\frac{T}{2} \le t < \frac{T}{2}. \tag{12.24}$$

Denote the ratio of the magnitude of harmonic distortions $\beta(m)$ as

$$\beta(m) = |\frac{b_m}{b_1}|, m = 3,5,7,\ldots \tag{12.25}$$

Clearly, $\beta(m)$ depends on m and K. In general, $\beta(3)$ is the dominant term although it is not a monotonically decreasing function of odd m. If we plot $\beta(3)$ as a function of K, Fig. 12.3 shows that at $K = 0.776$, the third harmonic is only -3.55 dB down from the magnitude of the fundamental frequency. Indeed, even for a very small amount of saturation, there is an appreciable amount of third-order harmonic distortion. In Fig. 12.4, $\beta(m)$ versus m is plotted for $K = 0.776$. After the third-order harmonic, the next dominant terms are the fifth and the eleventh order harmonics,

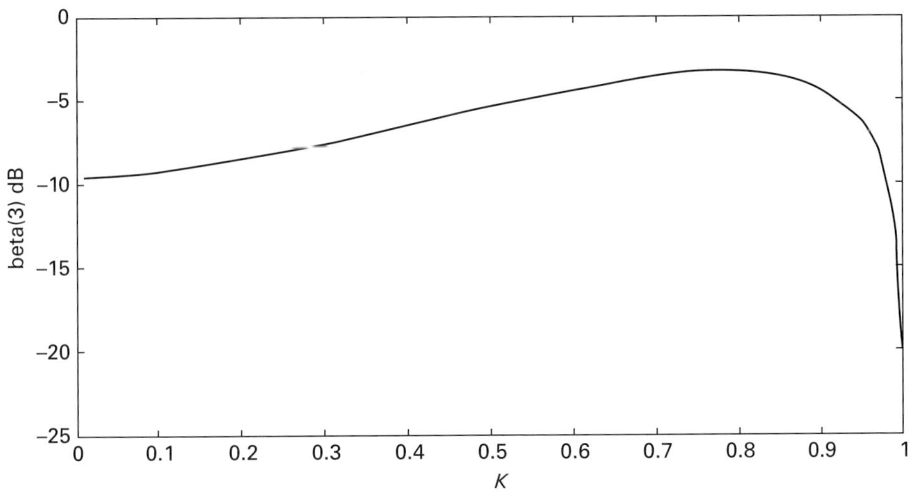

Figure 12.3 The magnitude of the third harmonic as a function of K

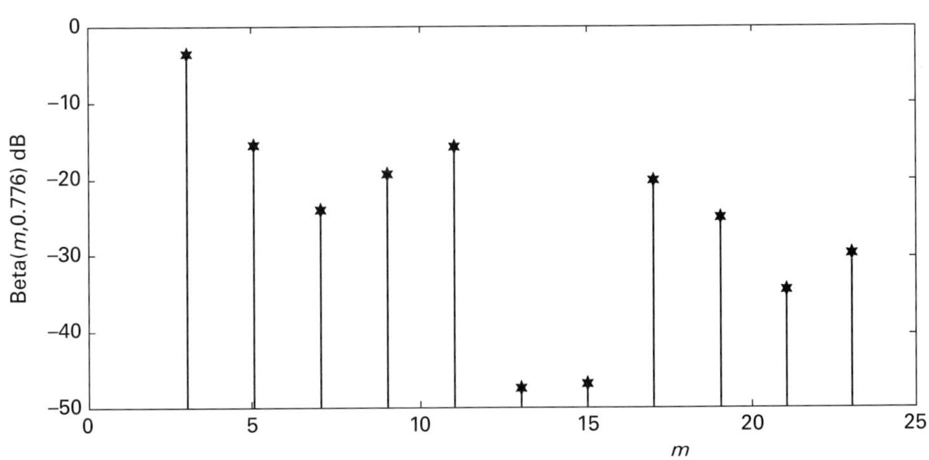

Figure 12.4 Plot of $\beta(m)$ versus m for $K = 0.776$

both at about -15 dB from the magnitude of the fundamental sinusoid. Of course, whether these or other harmonics are of significance depends on the specific system requirements.

12.3 Dynamic Range in a Digital Filter

Unlike the conventional analog filters, digital filters are the computer realization of some mathematical algorithms. While the use of digital filtering produces definite advantages, the digital system does create new types of machine error which may not exist in the analog system. These errors introduce non-linearity which should be controlled appropriately.

There are two basic kinds of errors in a digital filter: coefficient quantization error and arithmetic processing error. Both of them are essentially unavoidable for any fixed processor word length. While the processing error can be modeled statistically, coefficient quantization error can be controlled by the system designer. There are numerous discrete programming techniques to design a filter with discrete coefficients. On the other hand, processor error can only be reduced by rearranging the hardware configuration and the form of implementation.

If the processor word length is allowed to increase, these two types of errors can be reduced simultaneously. However, the corresponding processing time will be increased. This tradeoff is especially critical in the design of recursive filters using fixed point arithmetic. To keep the probability of internal overflow within a reasonable limit, the dynamic range of some intermediate summing points must be evaluated.

There are two ways to analyze the performance of a digital filter: frequency- and time-domain analysis. If we use $z = \exp(i\theta)$, $-\pi \leq \theta < \pi$ to determine the spectral content of the digital filter, we implicitly find the steady state performance of the filter and ignore the transient effect. The transient analysis of the digital filter can be accomplished by using the state space time-domain approach.

12.3.1 Dynamic Range of a Second-Order Recursive Digital Filter

In order not to saturate a digital filter operating under a finite number of bits, we must determine the largest possible values of all the variables in the digital filter. In a recursive (i.e., IIR) digital filter, due to the feedbacks, for certain values of the feedback coefficients and input frequencies, there can be quite large voltage growth at certain nodes of the filter. We can use both time-domain as well as frequency-domain analysis to analyze this problem.

Consider a second-order digital filter expressed in the standard canonical direct form II in Fig. 12.5. Now consider the peak magnitude response of $x(k)$. From Fig. 12.5, initially with $K = 1$,

$$x(k) = u(k) + a_1 x(k-1) + a_2 x(k-2), \quad -\infty < k < \infty. \tag{12.26}$$

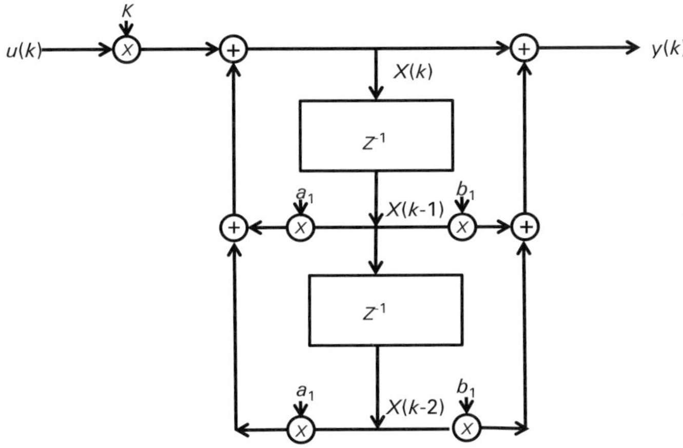

Figure 12.5 A second-order canonical direct form II digital filter

Upon taking the $\mathcal{Z}$-transform of (12.26) and solving for $U(z)$, we obtain

$$U(z) = X(z) - a_1 z^{-1} X(z) - a_2 z^{-2} X(z). \tag{12.27}$$

Then the transfer function $F(z)$ from the input $U(z)$ to the node of $X(z)$ is given by

$$F(z) = \frac{X(z)}{U(z)} = \frac{1}{1 - a_1 z^{-1} - a_2 z^{-2}}. \tag{12.28}$$

Example 12.4 Consider a second-order low-pass elliptic digital filter of 0.1 dB ripple over the passband of $[0, \theta_p]$ and 16 dB rejection over the rejection band of $[\theta_r, \pi]$, where $\theta_p = 0.00480\pi$ and $\theta_r = 0.0143\pi$. Then the filter coefficients are given by

$$a_1 = 1.93877, \quad a_2 = -0.941354, \quad b_1 = -1.984214, \quad b_2 = 1.$$

Direct evaluation of $F(e^{i\theta})$ in (12.28) for these coefficients yields

$$\max_{0 \le \theta < 2\pi} |F(e^{i\theta})| = 408 \approx 52.2\,\text{dB}. \tag{12.29}$$

From (12.29) and Fig. 12.5, we see that registers at $\{x(k), x(k-1), x(k-2)\}$ must be able to handle 52.2 dB larger values than the input in order to avoid saturation. The number of bits needed to prevent this saturation is

$$b = \frac{52.2\,\text{dB}}{6\,\text{dB/bit}} = 8.6\,\text{bits} \approx 9\,\text{bits}.$$

Of course, the actual number of bits needed must be b plus the number of bits needed for proper sign and quantization noise control. From this example, it is clear that we need to find an explicit procedure to determine various node voltages in the filter.

Time-Domain Approach

We can use the time-domain technique to determine the voltages at the internal states and the output of the filter. By definition, the two states $x_1(k)$ and $x_2(k)$ of this filter are given by

$$\mathbf{x}(k) = \begin{bmatrix} x_1(k) \\ x_2(k) \end{bmatrix} = \begin{bmatrix} x(k-1) \\ x(k-2) \end{bmatrix}. \tag{12.30}$$

The state equation and output equation of this dynamical system are given by

$$\mathbf{x}(k+1) = \mathbf{A}\mathbf{x}(k) + \mathbf{B}u(k), \tag{12.31}$$

$$y(k) = \mathbf{C}\mathbf{x}(k) + \mathbf{D}u(k), \tag{12.32}$$

where

$$\mathbf{A} = \begin{bmatrix} a_1 & a_2 \\ 1 & 0 \end{bmatrix}, \tag{12.33}$$

$$\mathbf{B} = \begin{bmatrix} 1 \\ 0 \end{bmatrix}, \tag{12.34}$$

$$\mathbf{C} = \begin{bmatrix} b_1 + a_1 & b_2 + a_2 \end{bmatrix}, \tag{12.35}$$

$$\mathbf{D} = 1. \tag{12.36}$$

The output at time N, $y(N)$, is given by

$$y(N) = \begin{cases} \mathbf{C}\mathbf{x}(0) + u(0), & N = 0, \\ \mathbf{C}\Phi(N)\mathbf{x}(0) + \sum_{i=0}^{N-1} \mathbf{C}\Phi(N-i-1)\mathbf{B}u(i) + \mathbf{D}u(N), & N \geq 0, \end{cases} \tag{12.37}$$

where the fundamental matrix is given by

$$\begin{aligned} \Phi(k) &= \begin{bmatrix} \Phi_{11}(k) & \Phi_{12}(k) \\ \Phi_{21}(k) & \Phi_{22}(k) \end{bmatrix} \\ &= \mathbf{A}^k \\ &= \mathcal{Z}^{-1}\left\{z(z\mathbf{I} - \mathbf{A})^{-1}\right\}, \quad k = 0, 1, \ldots \end{aligned} \tag{12.38}$$

In other words, the i, jth element of $\Phi(k)$ is the i, jth element of the matrix associated with z^{-k} in the curly bracket on the right-hand side of (12.38). Direct evaluation shows

$$\begin{aligned} \Phi_{11}(k) &= \left(\frac{R^{k+1}}{d}\right) \sin(k+1)\theta, \\ \Phi_{21}(k) &= \left(\frac{R^k}{d}\right) \sin k\theta, \\ \Phi_{12}(k) &= a_2 \Phi_{21}(k), \\ \Phi_{22}(k) &= \Phi_{11}(k) - a_1 \Phi_{21}(k), \end{aligned} \tag{12.39}$$

where

$$R = \sqrt{-a_2},$$
$$\theta = \tan^{-1}\left(\frac{2d}{a_1}\right),$$
$$d = \sqrt{-a_2 - \left(\frac{a_1^2}{4}\right)}. \tag{12.40}$$

Now, by using $\Phi_{ij}(k)$ given by (12.39) and (12.40) in (12.37), we can obtain the output $y(N)$ of the filter. In particular, for the zero initial condition of $\mathbf{x}(0) = \mathbf{0}$, then $y(0) = u(0)$, and for $N = 1, 2, \ldots,$

$$y(N) = u(N) + \left(\frac{R^N}{d}\right)\sum_{i=0}^{N-1} R^{-i}u(i)\cdot$$
$$\Big[(b_1 + a_1)\sin(N-i)\theta + (b_2 + a_2)R^{-1}\sin(N-i-1)\theta\Big]. \tag{12.41}$$

Furthermore, the states as defined in (12.30) are given by

$$\mathbf{x}(N) = \Phi(N)\mathbf{x}(0) + \sum_{i=0}^{N-1} \Phi(N-i-1)\mathbf{B}u(i), \quad N = 1, 2, \ldots \tag{12.42}$$

In particular, with $\mathbf{x}(0) = \mathbf{0}$, for $N = 1, 2, \ldots,$

$$x_1(N) = \frac{R^N}{d}\sum_{i=0}^{N-1} R^{-i}u(i)\sin(N-i)\theta,$$
$$x_2(N) = \frac{R^{N-1}}{d}\sum_{i=0}^{N-1} R^{-i}u(i)\sin(N-i-1)\theta\,. \tag{12.43}$$

While the expressions in (12.41) and (12.43) for the output and state voltages are explicit, the computations involved for large values of N may be objectionable. In particular, in order to determine the peak magnitude of the state, we need to evaluate (12.43) for all possible input frequencies of $u(k)$. Thus, the computational effort needed based on the time-domain approach may not be too practical.

Frequency-Domain Approach

Now, consider the frequency-domain approach using a scaling constant K in Fig. 12.5 such that the maximum $|x(k)|$ is constrained to be less than or equal to one for all values of K. Then the transfer function $F(z)$ of (12.28) becomes

$$F(z) = \frac{X(z)}{U(z)} = \frac{K}{1 - a_1 z^{-1} - a_2 z^{-2}}, \tag{12.44}$$

and

$$X(z) = F(z)U(z). \tag{12.45}$$

By applying the inverse $\mathcal{Z}$-transform on (12.45), we have

$$x(k) = \frac{1}{2\pi}\int_0^{2\pi} U(e^{i\theta})F(e^{i\theta})e^{ik\theta}d\theta, \quad -\infty < k < \infty. \tag{12.46}$$

With the $|x(k)| \leq 1$ constraint imposed on (12.46), we have

$$|x(k)| = \left|\frac{1}{2\pi}\int_0^{2\pi} U(e^{i\theta})F(e^{i\theta})e^{ik\theta}d\theta\right| \leq 1, \quad -\infty < k < \infty. \tag{12.47}$$

Consider the Holder inequality of

$$\left|\frac{1}{2\pi}\int_0^{2\pi} F(e^{i\theta})G(e^{i\theta})d\theta\right| \leq ||F(e^{i\theta})||_p \cdot ||G(e^{i\theta})||_q, \quad \frac{1}{p} + \frac{1}{q} = 1, \tag{12.48}$$

where $F(\cdot) \in \mathcal{L}_p$ and $G(\cdot) \in \mathcal{L}_q$ class of functions on $[0, 2\pi)$. The $\mathcal{L}_p$ norm of $F(\cdot)$ is defined by

$$||F(e^{i\theta})||_p = \left\{\frac{1}{2\pi}\int_0^{2\pi} |F(e^{i\theta})|^p d\theta\right\}^{\frac{1}{p}}, \quad 1 \leq p \leq \infty. \tag{12.49}$$

Specifically, for $p = 1, 2$, and ∞, we have

$$\mathcal{L}_1 \text{ norm} : ||F(e^{i\theta})||_1 = \frac{1}{2\pi}\int_0^{2\pi} |F(e^{i\theta})|d\theta, \tag{12.50}$$

$$\mathcal{L}_2 \text{ norm} : ||F(e^{i\theta})||_2 = \left\{\frac{1}{2\pi}\int_0^{2\pi} |F(e^{i\theta})|^2 d\theta|\right\}^{\frac{1}{2}}, \tag{12.51}$$

$$\mathcal{L}_\infty \text{ norm} : ||F(e^{i\theta})||_\infty = \max_{0 \leq \theta < 2\pi} |F(e^{i\theta})|. \tag{12.52}$$

From (12.48), with $G(e^{i\theta}) = U(e^{i\theta})e^{ik\theta}$ and $F(e^{i\theta})$ given by (12.44), then (12.47) yields

$$|x(k)| \leq ||F(e^{i\theta})||_p \cdot ||G(e^{i\theta})||_q = 1, \quad \frac{1}{p} + \frac{1}{q} = 1. \tag{12.53}$$

Among all possible positive p and q satisfying

$$\frac{1}{p} + \frac{1}{q} = 1,$$

we want to use p and q such that $||F(e^{i\theta})||_p$ and $||G(e^{i\theta})||_q$ can be evaluated readily. Specifically, consider $p = \infty$ and $q = 1$. For a sinusoidal input of θ_0 radian frequency,

$$u(k) = \sin(\theta_0 k) = -\sin(\theta_0 k), \quad -\infty < k < \infty. \tag{12.54}$$

Then

$$
\begin{aligned}
U(e^{i\theta}) &= \sum_{k=-\infty}^{\infty} u(k)e^{-ik\theta} = \sum_{k=0}^{\infty} \sin(\theta_0)ke^{-ik\theta} - \sum_{k=1}^{\infty} \sin(\theta_0)ke^{ik\theta} \\
&= \sum_{k=1}^{\infty} \sin(\theta_0)k \left[e^{-ik\theta} - e^{ik\theta}\right] \\
&= -2i \sum_{k=1}^{\infty} \sin(\theta_0)k \sin(\theta k) \\
&= -i \sum_{k=1}^{\infty} [\cos(\theta - \theta_0)k - \cos(\theta + \theta_0)k], 0 \le \theta < 2\pi. \qquad (12.55)
\end{aligned}
$$

From operational calculus, we know

$$
\delta(\theta - \theta_0) = \frac{1}{2\pi}\left[1 + 2\sum_{k=1}^{\infty} \cos(\theta - \theta_0)k\right], \quad 0 \le \theta < 2\pi, \qquad (12.56)
$$

$$
\delta(\theta + \theta_0) = \frac{1}{2\pi}\left[1 + 2\sum_{k=1}^{\infty} \cos(\theta + \theta_0)k\right], \quad 0 \le \theta < 2\pi. \qquad (12.57)
$$

From (12.56) and (12.57), (12.55) is given by

$$
U(e^{i\theta}) = -i\pi\,[\delta(\theta - \theta_0) - \delta(\theta + \theta_0)], \quad 0 \le \theta, \theta_0 < 2\pi. \qquad (12.58)
$$

By using (12.50), we have

$$
\begin{aligned}
||G(e^{i\theta})||_1 &= ||U(e^{i\theta})e^{i\theta k}||_1 \\
&= \frac{1}{2}\left[\int_0^{2\pi} \delta(\theta - \theta_0)d\theta + \int_0^{2\pi} \delta(\theta + \theta_0)d\theta\right] = 1. \qquad (12.59)
\end{aligned}
$$

From (12.44) and (12.52), we have

$$
\begin{aligned}
||F(e^{i\theta})||_\infty &= \max_{0 \le \theta < 2\pi} \left|\frac{K}{1 - a_1e^{-i\theta} - a_2e^{-i2\theta}}\right| \\
&= \frac{K}{\min_{0 \le \theta < 2\pi} D(\theta)}, \qquad (12.60)
\end{aligned}
$$

where

$$
\begin{aligned}
D^2(\theta) &= |1 - a_1e^{-i\theta} - a_2e^{-i2\theta}|^2 \\
&= 1 + a_1^2 + a_2^2 + 2a_1(a_2 - 1)\cos\theta - 2a_2\cos(2\theta), \quad 0 \le \theta < 2\pi. \qquad (12.61)
\end{aligned}
$$

From a variational argument, taking the derivative of (12.61) with respect to θ and setting it to zero, we have

$$\begin{aligned}\frac{dD^2(\theta)}{d\theta} &= -2a_1(a_2-1)\sin(\theta)+4a_2\sin(2\theta)\\ &= -2a_1(a_2-1)\sin(\theta)+8a_2\sin(\theta)\cos(\theta)\\ &= 2\sin(\theta)(4a_2\cos(\theta)-a_1(a_2-1))\\ &= 0. \end{aligned} \tag{12.62}$$

Thus, possible candidates for $\theta_{\min}$ are given by

$$\theta_{\min} = \begin{cases} 0, \\ \cos^{-1}\left(\frac{a_1(a_2-1)}{4a_2}\right). \end{cases} \tag{12.63}$$

By using (12.59), (12.60), and (12.63) in (12.53), we obtain

$$K = D(\theta_{\min}). \tag{12.64}$$

Thus, by using a scaling constant of $K = D(\theta_{\min})$ with a unit amplitude sinusoidal input of (12.54), $|x(k)| \le 1$ for all k. However, in the absence of this scaling factor

$$x_{\max} = \max_{0\le\theta<2\pi} |x(k)| = \frac{1}{K} = \frac{1}{D(\theta_{\min})}. \tag{12.65}$$

Example 12.5 Consider the coefficients a_1 and a_2 of the second-order elliptic digital filter of Example 12.3. Direct evaluation of (12.63) yields $\theta_{\min} = 0.028949$ with $K = D(\theta_{\min}) = 2.4286 \times 10^{-3}$ or $\frac{1}{K} = 407.688$. A plot of $|X(e^{i\theta})|$ in units of dB for a unit amplitude sinusoidal input versus θ is given in Fig. 12.6. An expanded version of this plot in Fig. 12.7 shows indeed

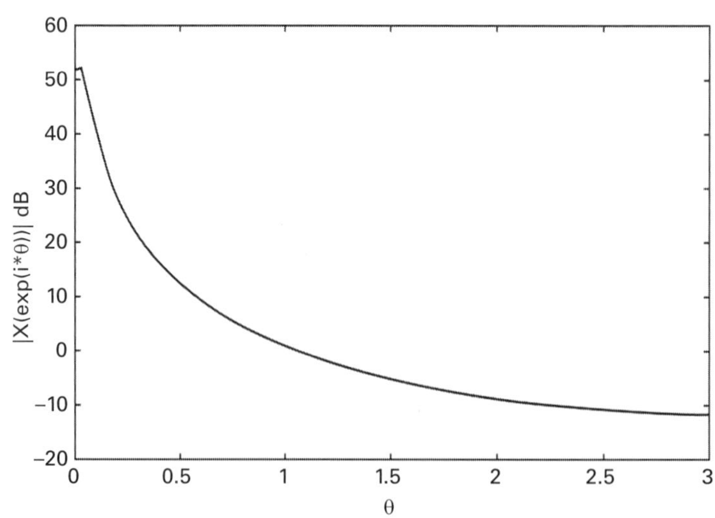

Figure 12.6 Plot of $|X(exp(i\theta))|$ dB with unit amplitude sinusoidal input

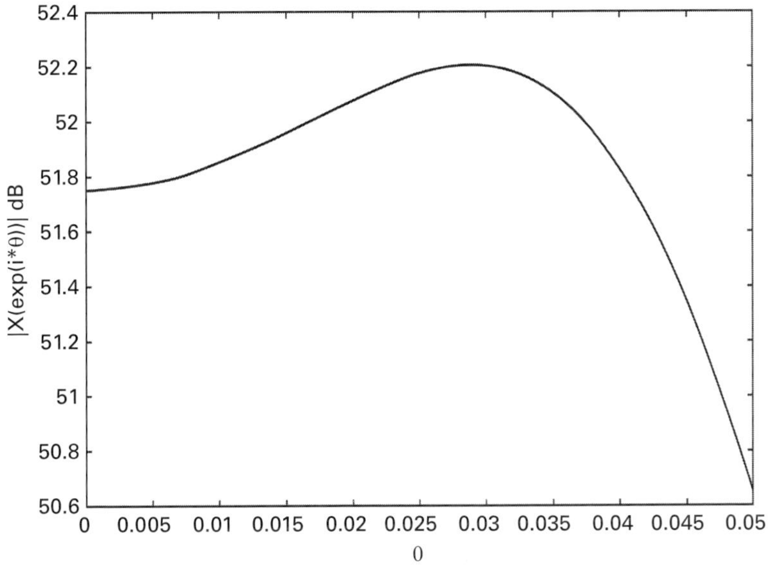

Figure 12.7 Plot of $|X(exp(i\theta))|$ dB with unit amplitude sinusoidal input

$$|X(e^{i\theta_{\min}})| = \max_{0\leq\theta<2\pi} |X(e^{i\theta})| = 407.688 = 52.2\,\text{dB} \tag{12.66}$$

with $\theta_{\min} = 0.028949$.

12.4 Combined Uniform Quantization and Saturation Error Model

In Fig. 12.1, it is clear that if the real-valued input data x are in the interval $(-V, V)$, then the output of the A/D contains only the quantization error. On the other hand, if the data x are outside the interval $(-V, V)$, then the output of the A/D has saturation error. In Section 12.4.1, we consider the case where x is the sum of a Gaussian signal and a Gaussian noise. Then, on some statistical basis, x will be inside as well as outside the interval $(-V, V)$. It is then important to scale the input data relative to V such that the total mean-square A/D error consisting of the quantization error and saturation error is minimized. We shall also see that this criterion is equivalent to maximizing the output SNR for any fixed input SNR.

12.4.1 Optimum Scaling for Gaussian Signal and Gaussian Noise

Consider the input data x to be modeled by a Gaussian signal v plus a Gaussian noise n as given by

$$x = v + n. \tag{12.67}$$

Both random variables are assumed to be statistically independent, have zero-means, and variances σ_v^2 and σ_n^2, respectively. Then the total input power is given by

$$\sigma_x^2 = \sigma_v^2 + \sigma_n^2, \tag{12.68}$$

and the input SNR is given by

$$(SNR)_{\text{in}} = \frac{\sigma_v^2}{\sigma_n^2}. \tag{12.69}$$

Since x is also a Gaussian r.v., then for any non-zero σ_x^2, x can take values on $(-\infty, \infty)$ and there is always some probability of saturation. Specifically, let the saturation limit V be related to σ_x by

$$V = k\sigma_x. \tag{12.70}$$

Now, we want to find the optimum loading parameter k to minimize the total MS A/D error given in (12.71).

$$\sigma_T^2 = E\left\{e^2\right\} = \int_{-\infty}^{\infty} e^2(x)\, p(x)\, dx = \sigma_q^2 + \sigma_{\text{sat}}^2. \tag{12.71}$$

In (12.71), $e(x)$ is the A/D error, and $p(x)$ is a Gaussian probability density with a zero-mean and a variance σ_x^2. First, consider the MS quantization error σ_q^2 given by (12.72).

$$\begin{aligned}\sigma_q^2 &= \int_{-V}^{V} e^2(x)\, p(x)\, dx = 2\int_{0}^{k\sigma_x} e^2(x)\, p(x)\, dx \\ &\cong 2\,\frac{\Delta^2}{12}\left[\frac{1}{2} - Q(k)\right].\end{aligned} \tag{12.72}$$

In (12.72), Δ, the quantization spacing, is related to V, σ_x, and B as follows

$$M\Delta = V = k\sigma_x, \qquad M = 2^{B-1} - 1, \tag{12.73}$$

and $Q(x)$ is the complementary Gaussian distribution of a zero-mean unit variance r.v. In (12.72), we used several approximations to get to the final reasonably simple expression for σ_q^2. In particular, if σ_x^2 is fixed, then from (12.70) and (12.73), the normalized MS quantization error in (12.72) can be expressed as

$$\frac{\sigma_q^2}{\sigma_x^2} = \frac{k^2}{6M^2}\left[\frac{1}{2} - Q(k)\right]. \tag{12.74}$$

The MS saturation noise is given by

$$\sigma_{\text{sat}}^2 = \int_{-\infty}^{-V} (x+V)^2\, p(x)\, dx + \int_{V}^{\infty} (x-V)^2\, p(x)\, dx. \tag{12.75}$$

12.4.2 Introduction

Consider a real-valued signal modeled as a zero-mean Gaussian r.v. X of variance σ^2. This signal is quantized by an M levels uniform quantizer having a dynamic range of

$[-V, V]$. We want to determine the total A/D mean-square error distortion due to the quantization and saturation effects. Denote the loading factor k by

$$k = \frac{V}{\sigma} = \frac{\text{Total dynamic range}}{2\sigma}. \quad (12.76)$$

For a fixed σ, a smaller value of k implies that there is a greater fraction of the time the signal may hit the saturation level at $\pm V$. Thus, the saturation error will decrease with increasing k. On the other hand, for a fixed σ and a fixed number of quantization levels M in the interval $[-k\sigma, k\sigma]$, the quantization error will increase with increasing k. Since the total A/D mean-square error distortion is the sum of both of these errors, we may expect this distortion to have a minimum as a function of k. The actual value of k that achieves the minimum would also depend on the values of σ and M.

Define the total A/D mean-square error σ_T^2 by

$$\sigma_T^2 = E\{e^2\} = \int_{-\infty}^{\infty} e^2(x)p(x)dx = \sigma_q^2 + \sigma_{\text{sat}}^2, \quad (12.77)$$

where $e(x)$ is the A/D error (to be described more precisely shortly), $p(x)$ is the pdf of a zero-mean and variance σ^2 Gaussian r.v., σ_q^2 is the quantization MS error, and σ_{sat}^2 is the saturation MS error.

12.4.3 MS Quantization Error

First, consider the MS quantization error σ_q^2 given by

$$\begin{aligned}
\sigma_q^2 &= \int_{-V}^{V} e^2(x)p(x)dx = \int_{-k\sigma}^{k\sigma} e^2(x)p(x)dx \\
&\approx \sum_{m=-(M-1)/2}^{(M-1)/2} \int_{(m-0.5)\Delta}^{(m+0.5)\Delta} (x - m\Delta)^2 p(x)dx \\
&\approx \sum_{m=-(M-1)/2}^{(M-1)/2} p(m\Delta) \int_{-\Delta/2}^{\Delta/2} x^2(x)dx = (\Delta^2/12) \sum_{m=-(M-1)/2}^{(M-1)/2} p(m\Delta)\Delta \\
&\approx (\Delta^2/12) \int_{-k\sigma}^{k\sigma} p(x)dx = 2(\Delta^2/12)[0.5 - Q(k)], \quad (12.78)
\end{aligned}$$

where M is the total number of quantization levels with the middle level set equal to zero and is related to the number of bits B by

$$M = 2^B - 1, \quad (12.79)$$

the uniform quantizer step size is given by $\Delta = (2V)/(M - 1)$, and $Q(x)$ is the complementary Gaussian distribution of a zero-mean unit variance r.v. The normalized MS quantization error then can be expressed by

$$\frac{\sigma_q^2}{\sigma^2} = \frac{2k^2(0.5 - Q(k))}{3(M-1)^2}. \quad (12.80)$$

12.4.4 MS Saturation Error

The MS saturation error σ_{sat}^2 is given by

$$\sigma_{\text{sat}}^2 = \int_{-\infty}^{-V} (x+V)^2 p(x)dx + \int_{V}^{\infty} (x-V)^2 p(x)dx = 2\sigma^2[I_1 + I_2 + I_3], \quad (12.81)$$

where $I_1,\ I_2$, and I_3 are given by

$$I_1 = Q(k) + k(1/2\pi)^{0.5}\exp(-k^2/2), \quad (12.82)$$

$$I_2 = (-2k/(2\pi)^{0.5})\exp(-k^2/2), \quad (12.83)$$

$$I_3 = k^2 Q(k). \quad (12.84)$$

Thus, the normalized MS saturation noise can be expressed as

$$\frac{\sigma_{\text{sat}}^2}{\sigma^2} = 2[(1+k^2)Q(k) - (k/(2\pi)^{0.5})\exp(-k^2/2)]. \quad (12.85)$$

12.4.5 Normalized Total A/D MS Error Distortion

By combining the normalized MS quantization and saturation errors in (12.80) and (12.85), the normalized total A/D MS error is given by

$$g(k,M) = \frac{\sigma_T^2}{\sigma^2} = Q(k)\left[2 + 2k^2(1 - \frac{1}{3(M-1)^2})\right] + \frac{k^2}{3(M-1)^2} - \frac{2k}{(2\pi)^{0.5}}\exp(-k^2/2). \quad (12.86)$$

In Fig. 12.8, we plot $g(k)$ versus k for values of $B = 4, 6, 8, 10$, and 12 bits corresponding to $M = 15, 63, 255, 1023$, and 4095 levels, respectively. From these curves we note that for each value of B, there is an optimum loading parameter k that minimizes the normalized total A/D MS error. In general, the values of $g(k)$ vary relatively slowly as k varies about the optimum k. From a system point of view, this is desirable, since in practice the input power σ^2 may change with time and it is not always possible to operate at this optimum k.

12.5 Total A/D MS Error Distortion for Fixed *V*

In (12.86) of Section 12.4.5, we obtained the total A/D MS error normalized by the variance of the input signal. In that expression, we fixed σ and vary k and thus also vary $V = k\sigma$ for given $M = 2^{B-1}$ levels. In practice, we may be more interested in the case of fixing the dynamic range parameter V and the number of levels M, and let σ vary as a function of time under various scenarios of interest. Thus, by using $k = V/\sigma$, we can express the total A/D MS error from (12.86) simply as

$$\sigma_T^2 = \sigma^2 g\left(\frac{V}{\sigma}, M\right). \quad (12.87)$$

By using (12.87), we generate Fig. 12.9 showing the log of the total A/D MS error versus σ for a fixed $V = 5$ and a different number of bits $B = 4, 6, 8, 10$, and 12.

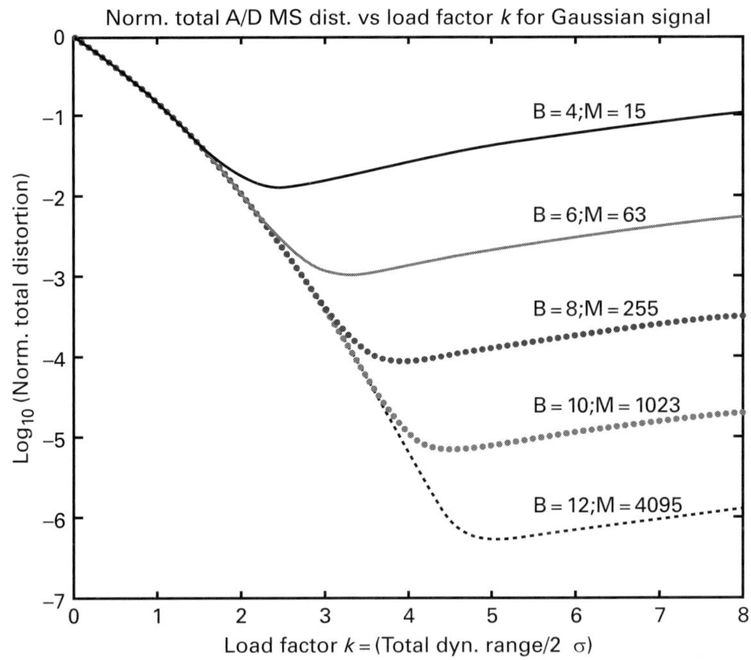

Figure 12.8 $g(k)$ vs k for $B = 4, 6, 8, 10$, and 12 bits

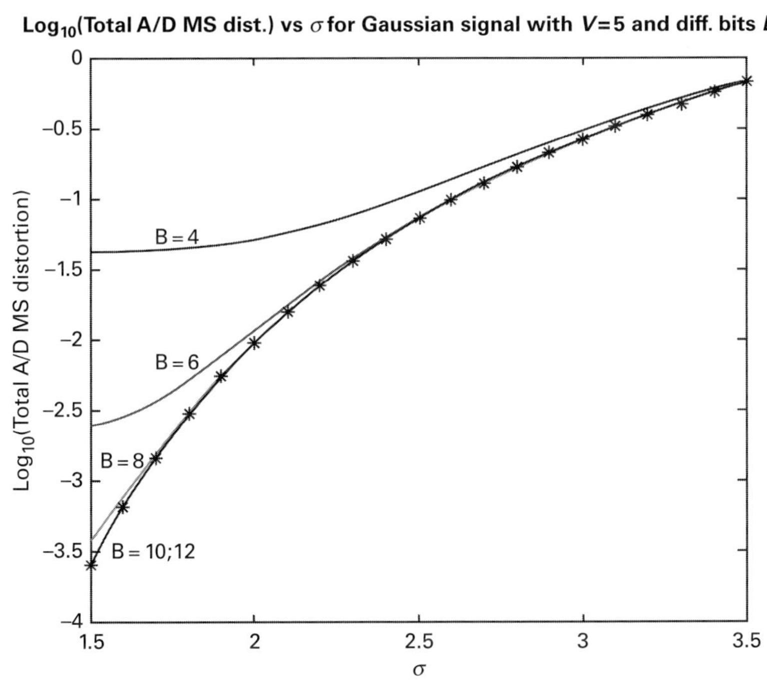

Figure 12.9 Total A/D MS error for different B values

We note that the increase of the log of the total A/D MS error for increasing σ is due to the degradation of the saturation effect. For a fixed V, increasing σ is equivalent to decreasing k. From Fig. 12.8, we know that as k decreases the total A/D MS error increases rapidly. In Fig. 12.9, we see that for large σ most of the contribution to the total A/D MS error is due to saturation, and thus increasing the number of bits B is not relevant in reducing the total A/D MS error. Thus, the $B = 6, 8, 10$, and 12 curves essentially have merged together. In practice, as σ changes, the total A/D MS error can be evaluated using the single heavy curve in Fig. 12.9 for all values of B greater than 4.

12.6 Distortion Spectra of Narrowband Gaussian Waveforms

Consider an IF waveform modeled as a narrowband Gaussian random process at the input of the two channel coherent digital processor. After coherent demodulation with $\cos(2\pi f_0 t)$ and $\sin(2\pi f_0 t)$, where f_0 is the IF frequency, the in-phase and quadrature waveforms are each sampled and passed through an A/D device. Let the saturation effect of the A/D device be modeled by that of a linear memoryless saturating amplifier given by Fig. 12.1 as considered in Section 12.2. Due to the non-linear effect of saturation, distortions exist at the output of the device. While various mean-square error criteria can be used to measure the extent of this distortion, for many digital signal processing applications, the distortion spectral density is the most useful method of characterizing this disturbance. In general, the evaluation of this spectral distortion is quite complicated and tedious. Specifically, consider the narrowband Gaussian random process after coherent demodulation and sampling has a Gaussian shaped spectral density given by

$$S(f) = \frac{\sigma^2}{B_1\sqrt{\pi}}\, e^{-(f-f_I)^2/B_1^2}, \qquad 0 \le f < f_s, \tag{12.88}$$

where the sampling frequency is given by $f_s = 1/T$, σ^2 is the input waveform power, the offset frequency f_I is the difference between f_c the central frequency of the symmetrical input waveform spectrum and the IF frequency f_0, and B_1 is related to the two-sided 3 dB bandwidth B_0 by

$$B_1 = \frac{B_0}{2\sqrt{\ln 2}}, \tag{12.89}$$

$$f_I = f_c - f_0. \tag{12.90}$$

Then the normalized total spectral density at the output of the A/D device, $S_T(f)$, consists of the sum of the input spectral density $S(f)$ and the distortion spectral density $S_0(f)$, and is given by

$$S_T(f) = S(f) + S_0(f) \tag{12.91}$$

$$= \frac{4V^2T}{\pi} \sum_{\substack{k=1 \\ k \text{ odd}}}^{\infty} B(k)\left\{1 + 2\sum_{n=1}^{\infty}\left[\left(\cos^k 2\pi f_I nT\right)\left(\cos 2\pi f nT\right) + \left(\sin^k 2\pi f_I nT\right)\left(\sin 2\pi f nT\right)\right]e^{-k\pi B_1 nT}\right\}, \quad 0 \le f < f_s, \tag{12.92}$$

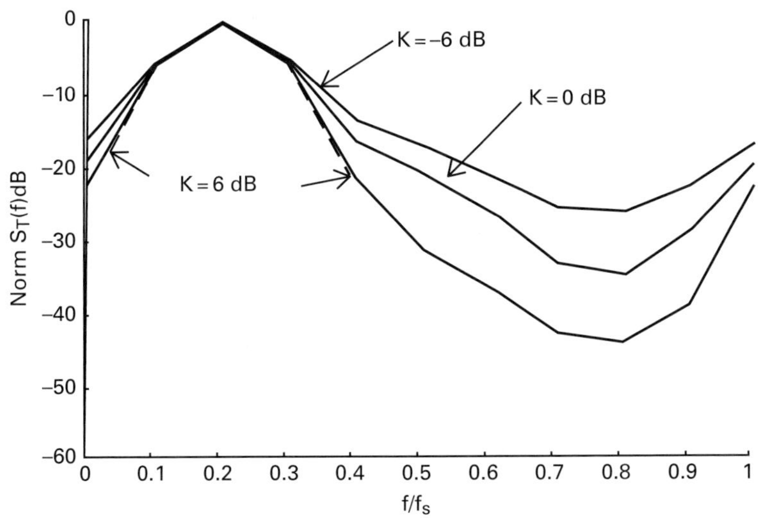

Figure 12.10 Normalized total spectral density $S_T(f)$ vs f/f_s for three $K = V/\sigma$ values

where

$$B(1) = \left[1 + \sum_{n=1}^{\infty} \frac{(-1)^n}{(2n+1)n} \left(\frac{V}{\sqrt{2}\sigma}\right)^{2n}\right]^2, \tag{12.93}$$

$$B(k) = 4\left(\frac{\sigma}{V}\right)^2 H^2(k)\, e^{-(V/\sigma)^2}, \quad k \geq 3, \tag{12.94}$$

$$H(k) = \frac{V}{\sigma}\ \frac{H(k-1)}{\sqrt{k}} - (k-3)\,\frac{H(k-2)}{\sqrt{k(k-1)}}, \quad k \geq 4, \tag{12.95}$$

$$H(2) = 2^{-1.5}, \qquad H(3) = \frac{1}{\sqrt{24}}\ \frac{V}{\sigma}. \tag{12.96}$$

In Fig. 12.10, $S_T(f)$ is plotted versus f/f_s, for $B_0 = 0.15$ with $f_I/f_s = 0.2$ and $V/\sigma = -6$, 0, and 6 dB. In all cases, the effect of the saturation caused the total output spectral density to be considerably broader than the input spectral density. Furthermore, this broadness increases as the saturation effect increases with smaller values of (V/σ). The results in this section allow one to evaluate the disturbance effect of a strong signal or interferer due to A/D saturation across the entire spectrum region and in particular with respect to some other bands of possible interest.

12.7 Conclusion

Digital processing became possible with the availability of digital computers and dedicated digital processors. While digital processing has the advantage and flexibility of digital programming, it also has intrinsic distortions due to quantization errors from the finite resolution of the original analog data as well as saturation errors from the finite dynamic ranges of digital processors. In Section 12.1, we introduced the truncation and roundoff quantization error models. In Section 12.2, we considered the resulting saturation distortions when the range of the input continuous-signals is greater than the dynamic range of the digital processing system. In Section 12.3, we considered the internal state values of a second-order recursive digital filter, where when near the transition region of any recursive digital filter, these state values can be quite large. Both time-domain and frequency-domain techniques are used to estimate the range of the state variables in the filter. In Section 12.4, we considered the combined uniform quantization and saturation error models for Gaussian input signals. In Section 12.6, we considered the power spectral density of the distortion sequence due to saturation for a narrowband Gaussian random sequence.

12.8 References

Materials in Sections 12.1 and 12.2 can be found in many digital signal processing books such as [1]. Materials on digital filtering quantization and dynamic range issues in Section 12.3 can be found in [1] and [2]. The material on the use of Holder inequality for estimating the maximum magnitude of a second-order digital filter appears to be original. Materials in Sections 12.4–12.6 can be found in [3] and [4].

[1] A.V. Openheim and R.W. Schafer, *Digital Signal Processing*, Prentice-Hall, 1975.

[2] R.A. Roberts and C.T. Mullis, *Digital Signal Processing*, McGraw-Hill, 1967, Chap. 9.

[3] G.A. Gray and G.W. Zeoli, "Quantization and Saturation Noise Due to A/D Conversion," *IEEE Trans. Aerospace and Electronic Systems*, 1971, pp. 222–223.

[4] G.W. Zeoli, "IF versus Video Limiting for Two-Channel Coherent Signal Processors," *IEEE Trans. on Information Theory*, 1971, pp. 579–585.

12.9 Exercises

1. Duplicate the results of Fig. 12.3.
2. Duplicate the results of Fig. 12.6 and Fig. 12.7 using the frequency-domain approach of Section 12.3.1.
3. Duplicate the results of Fig. 12.8.
4. Duplicate the results of Fig. 12.9.

13 Introduction to Systolic Arrays

13.1 Introduction to Systolic Arrays

The term *systolic array* was coined by H.T. Kung and C.E. Leisserson [1] and H.T. Kung [2] around 1980 to denote a particular simple form of parallel processing. Some of the basic properties associated with a systolic array (SA) are

- There is only one (or few) kind of processing element (PE) arranged in a regular array of one (or two) dimensions.
- There are only adjacent (or nearby) communications and connections.
- All the PEs operate in a synchronous (rhythmic) manner.
- There is minimal control of the PEs.

Consider a conventional processing system as shown in Fig. 13.1, where the processing performance is bounded by the memory access limitation. In this model, for each access to and from the memory, the PE can perform μ MOPS. In Fig. 13.2, by arranging N such PE in a one-dimensional systolic array (SA), each memory access can achieve $N\mu$ MOPS. This increase of a factor of N in the processing capability was achieved without demanding faster hardware capability on the memory, bus lines, PE, etc. Of course, we need to use N instead of only one such PE. The ability to use a one-dimensional or even two-dimensional SA as shown in Fig. 13.3 depends on the nature of the algorithm and its reorganization for systolic operation. Due to the regular structure and local communication properties, an SA is ideally suited for VLSI implementation. There is much interest at the theoretical level for using SA for high performance real-time signal processing applications.

13.1.1 Correlation

Since correlation is one of the simplest and most basic operations in signal processing, we consider several heuristic one-dimensional SA implementations in this section. Let us denote the weights by $\{w_1, w_2, \ldots, w_N\}$, the inputs by $\{x_1, x_2, \ldots, x_M\}$, and the outputs by $\{y_1, y_2, \ldots\}$. In general

$$y_i = w_1 x_i + w_2 x_{i+1} + \cdots + w_N x_{i+N-1} = \sum_{j=1}^{N} w_j x_{i+j-1}.$$

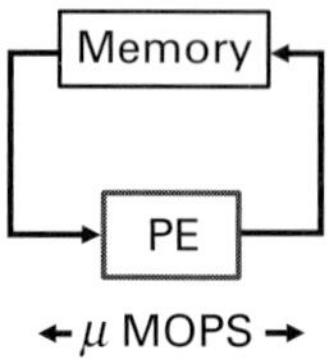

Figure 13.1 Single PE

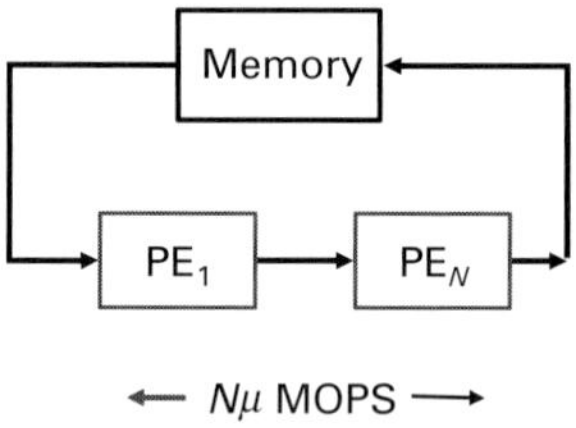

Figure 13.2 One-dimensional systolic array

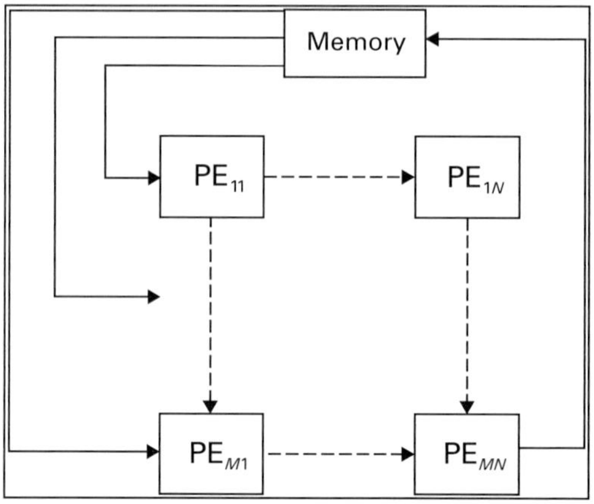

Figure 13.3 Two-dimensional systolic array

For $N = 3$ and $M > N$, we have

$$\begin{aligned} y_1 &= w_1x_1 + w_2x_2 + w_3x_3 \\ y_2 &= \qquad\quad w_1x_2 + w_2x_3 + w_3x_4 \\ y_3 &= \qquad\qquad\qquad w_1x_3 + w_2x_4 + w_3x_5 \\ &\vdots \end{aligned}$$

Consider Design B1 in Fig. 13.4. Here, the weights w_i are preloaded to the cell and stay throughout the computation. Partial results y_i move systolically from cell to cell. Starting at the third iteration, $y_1, y_2, \ldots,$ are output from the right-most cell at the rate

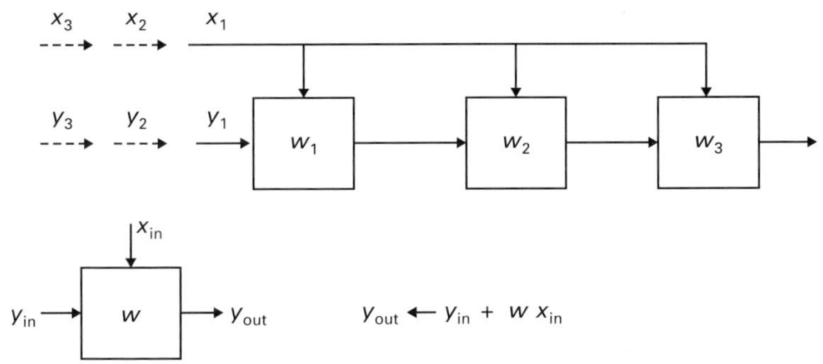

Figure 13.4 Design B1

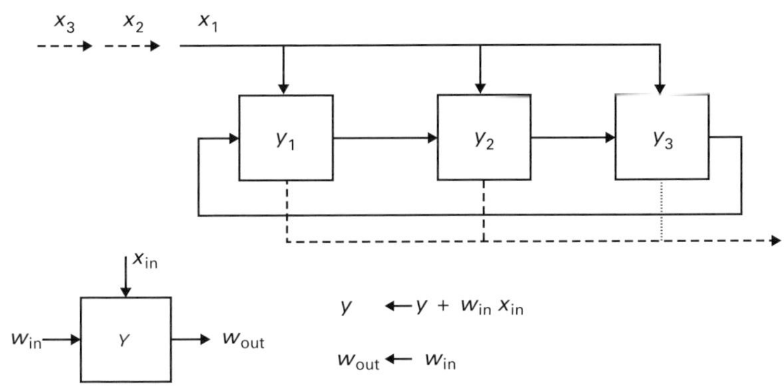

Figure 13.5 Design B2

of one output per iteration. For each iteration, an x_i is broadcast to all the cells, and a y_i, initialized to zero, enters the left-most cell. The broadcast data x_i are marked with an arrow, $^{\downarrow}$. Indeed, by comparison we see y_1, y_2, and y_3 output at iteration $T = 3, 4,$ and 5 agree with those given previously.

In Design B2, shown in Fig. 13.5, each input x_i is again broadcast to each cell, each y_i stays at each cell to accumulate terms, while the weights w_i circulate around the cells in the array. A tag bit is associated with w_1 to reset the contents of the accumulator, while a tag bit is associated with w_3 to output the contents of the accumulator after the first two iterations. Data movements in Design B2 are shown in Table 13.2. We note, resets occur at cell 1 at iteration 1, cell 2 at iteration 2, cell 3 at iteration 3, cell 1 at iteration 4, etc. Similarly, output y_1 occurs from cell 1 at iteration 3, y_2 from cell 2 at iteration 4, y_3 from cell 3 at iteration 5, etc.

In Design F, as shown in Fig. 13.6, the weights w_i stay in each cell, the input data x_i move systolically one cell to the right at each iteration, and results from the cells are

Table 13.1 Data Movement in Design B1

Iteration	Cell 1	Cell 2	Cell 3
$T = 1$	$\overset{\downarrow}{x_1}\ w_1$	$\overset{\downarrow}{x_1}\ w_2$	$\overset{\downarrow}{x_1}\ w_3$
$T = 2$	$\overset{\downarrow}{x_2}\ w_1$	$x_1w_1 + \overset{\downarrow}{x_2}\ w_2$	$x_1w_2 + \overset{\downarrow}{x_2}\ w_3$
$T = 3$	$\overset{\downarrow}{x_3}\ w_1$	$x_2w_1 + \overset{\downarrow}{x_3}\ w_2$	$x_1w_1 + x_2w_2 + \overset{\downarrow}{x_3}\ w_3 \rightarrow y_1$
$T = 4$	$\overset{\downarrow}{x_4}\ w_1$	$x_3w_1 + \overset{\downarrow}{x_4}\ w_2$	$x_2w_1 + x_3w_2 + \overset{\downarrow}{x_4}\ w_3 \rightarrow y_2$
$T = 5$	$\overset{\downarrow}{x_5}\ w_1$	$x_4w_1 + \overset{\downarrow}{x_5}\ w_2$	$x_3w_1 + x_4w_2 + \overset{\downarrow}{x_5}\ w_3 \rightarrow y_3$

Table 13.2 Data Movement in Design B2

Iteration	Cell 1	Cell 2	Cell 3
$T = 1$	$0 + w_1\ \overset{\downarrow}{x_1}$	$w_3\ \overset{\downarrow}{x_1}$	$w_2\ \overset{\downarrow}{x_1}$
$T = 2$	$w_1x_1 + w_2\ \overset{\downarrow}{x_2}$	$0 + w_1\ \overset{\downarrow}{x_2}$	$w_2x_1 + w_3\ \overset{\downarrow}{x_2}$
$T = 3$	$w_1x_1 + w_2x_2 + w_3\ \overset{\downarrow}{x_3} \rightarrow y_1$	$w_1x_2 + w_2\ \overset{\downarrow}{x_3}$	$0 + w_1\ \overset{\downarrow}{x_3}$
$T = 4$	$0 + w_1\ \overset{\downarrow}{x_4}$	$w_1x_2 + w_2x_3 + w_3\ \overset{\downarrow}{x_4} \rightarrow y_2$	$w_1x_3 + w_2\ \overset{\downarrow}{x_4}$
$T = 5$	$w_1x_4 + w_2\ \overset{\downarrow}{x_5}$	$0 + w_1\ \overset{\downarrow}{x_5}$	$w_1x_3 + w_2x_4 + w_3\ \overset{\downarrow}{x_5} \rightarrow y_3$

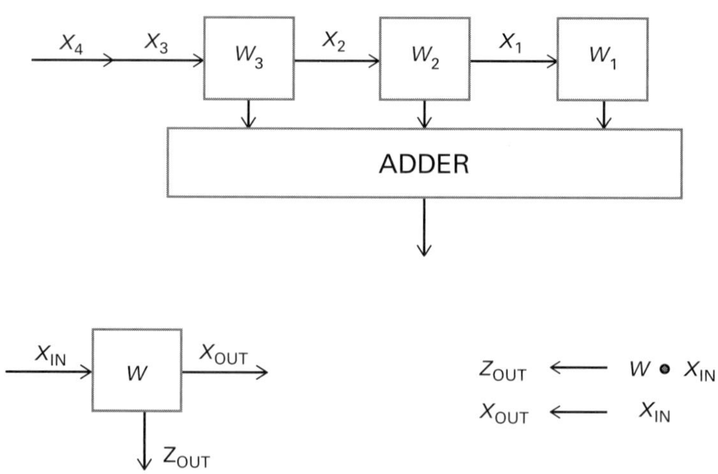

Figure 13.6 Design F

fanned out and summed by an adder to form the outputs y_i. The data movements of Design F are shown in Table 13.3 and we observe the outputs $y_1, y_2, y_3, \ldots$, appear at iterations $3, 4, 5, \ldots$

Table 13.3 Data Movements in Design F

Iteration	Cell 1		Cell 2		Cell 3		y_i
$T = 1$	x_1w_3	+	0	+	0		
$T = 2$	x_2w_3	+	x_1w_2	+	0	$\rightarrow$	0
$T = 3$	x_3w_3	+	x_2w_2	+	x_1w_1	$\rightarrow$	y_1
$T = 4$	x_4w_3	+	x_3w_2	+	x_2w_1	$\rightarrow$	y_2
$T = 5$	x_5w_3	+	x_4w_2	+	x_3w_1	$\rightarrow$	y_3

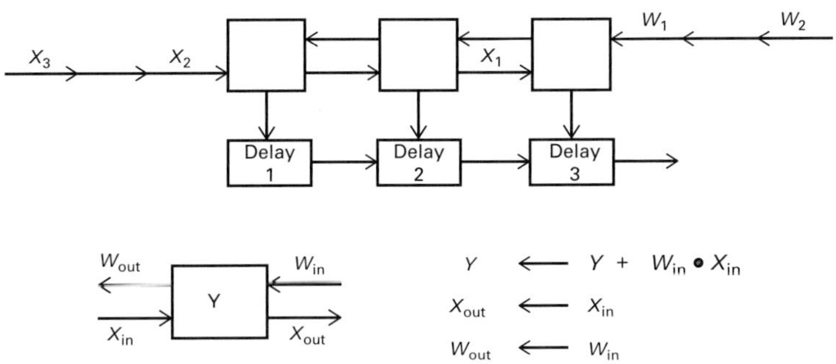

Figure 13.7 Design R1

In Design R1, as shown in Fig. 13.7, each datum x_i and weight w_i move systematically in opposite directions. When an x_i meets a w_i, they are multiplied and the product is accumulated to the y in the cell. Consecutive x_i and w_i are separated by *blank* markers in order to produce the proper timing operations. As in Design B2, a tag bit is associated with w_3 to trigger an output from each cell. Each such output is then fed into an array of delay cells. In Table 13.4, data movements in the three processing cells and the three delay cells for R1 are given. We note, the outputs $y_1, y_2, y_3, y_4, \ldots$, appear at iterations $6, 8, 10, 12, \ldots$ Indeed, the output appears at every other time unit just as in the data x_i and the weights w_i are also being input at every other time unit.

13.1.2 DFT

Consider a given data sequence $\{x_0, x_1, \ldots, x_{N-1}\}$. Its DFT sequence is given by $\{X(0), X(1), \ldots, X(N-1)\}$ with

$$X(k) = \sum_{n=0}^{N-1} x_{nw}^{nk}, \quad k = 0, \ldots, N-1, \tag{13.1}$$

where $w = e^{-i2\pi/N}$. Direct implementation of (13.1) needs N^2 multiplications and $(N^2 - N)$ additions. FFT implementation needs $(N/2)\log_2 N$ multiplications and $N \log_2 N$ additions.

Table 13.4 Data Movements in Design F

Iteration	Cell 1	Cell 2	Cell 3	Delay 1	Delay 2	Delay 3	y_i
$T = 1$	$x_2 \rightarrow$		$x_1 \rightarrow \leftarrow w_1$				
			$x_1 w_1$				
$T = 2$		$x_2 \rightarrow \leftarrow w_1$					
		$x_2 w_1$	$x_1 w_1$				
$T = 3$	$x_3 \rightarrow \leftarrow w_1$		$x_2 \rightarrow \leftarrow w_2$				
	$x_3 w_1$	$x_2 w_1$	$x_1 w_1 + x_2 w_2$				
$T = 4$		$x_3 \rightarrow \leftarrow w_2$					
	$x_3 w_1$		$x_1 w_1 + x_2 w_2$				
$T = 5$	$x_4 \rightarrow \leftarrow w_2$		$x_3 \rightarrow \leftarrow w_3$				
	$x_3 w_1 + x_4 w_2$	$x_2 w_1 + x_3 w_2$	$x_1 w_1 + x_2 w_2 + x_3 w_3$ - - - - - - →	- - - - - - →	- - - - - - →	y_1'	
$T = 6$		$x_4 \rightarrow \leftarrow w_3$	0				
	$x_3 w_1 + x_4 w_2$	$x_2 w_1 + x_3 w_2 + x_4 w_3$ - - - - - - →	- - - - - - →	- - - - - - →	y_2'		$y_1 = x_1 w_1 + x_2 w_2 + x_3 w_3$
$T = 7$	$x_5 \rightarrow \leftarrow w_3$	0	$x_4 \rightarrow \leftarrow w_1$				
	$x_3 w_1 + x_4 w_2 + x_5 w_3$ - - - - - - →	- - - - - - →	- - - $x_4 w_1$ - - - - - - →	y_3'		y_2'	
$T = 8$		$x_5 \rightarrow \leftarrow w_1$					
	0	$x_5 w_1$	$x_4 w_1$		y_3'		$y_2 = x_2 w_1 + x_3 w_2 + x_4 w_3$
$T = 9$	$x_6 \rightarrow \leftarrow w_1$		$x_5 \rightarrow \leftarrow w_2$				
			$x_4 w_1 + x_5 w_2$			y_3'	
$T = 10$		$x_6 \rightarrow \leftarrow w_2$					
	$x_6 w_1$	$x_5 w_1 + x_6 w_2$	$x_4 w_1 + x_5 w_2$				$y_3 = x_3 w_1 + x_4 w_2 + x_5 w_3$

Consider (13.1) from an evaluation of the polynomial

$$f(z) = x_{N-1}z^{N-1} + x_{N-2}z^{N-2} + \cdots + x_1 z + x_0, \tag{13.2}$$

with $z = 1, w, w^2, \ldots, w^{N-1}$.

Example 13.1 Consider the Horner method of evaluating (13.2) for $N = 5$.

$$\begin{aligned}
X(0) &= f(1) = (((x_4 \cdot 1 + x_3) \cdot 1 + x_2) \cdot 1 + x_1) \cdot 1 + x_0,\\
X(1) &= f(w) = (((x_4 \cdot w + x_3) \cdot w + x_2) \cdot w + x_1) \cdot w + x_0,\\
X(2) &= f(w^2) = (((x_4 \cdot w^2 + x_3) \cdot w^2 + x_2) \cdot w^2 + x_1) \cdot w^2 + x_0,\\
X(3) &= f(w^3) = (((x_4 \cdot w^3 + x_3) \cdot w^3 + x_2) \cdot w^3 + x_1) \cdot w^3 + x_0,\\
X(4) &= f(w^4) = (((x_4 \cdot w^4 + x_3) \cdot w^4 + x_2) \cdot w^4 + x_1) \cdot w^4 + x_0.
\end{aligned}$$

In general, Horner's method is given by

$$\begin{aligned}
X(0) &= f(1) = (((x_{N-1} \cdot 1 + x_{N-2}) \cdot 1 + \cdots + x_1) \cdot 1 + x_0,\\
X(1) &= f(w) = (((x_{N-1} \cdot w + x_{N-2}) \cdot w + \cdots + x_1) \cdot w + x_0,\\
&\ \ \vdots\\
X(N-1) &= f(w^{N-1}) = (((x_{N-1} \cdot w^{N-1} + x_{N-2}) \cdot w^{N-1} + \cdots + x_1) \cdot w^{N-1} + x_0.
\end{aligned}$$

Now, consider a systolic implementation of (13.1), based on a linear array. In Fig. 13.8, we consider a basic cell consisting of a delay unit modeled by $a_{\text{out}} \leftarrow a_{\text{in}}$ and a multiply-and-add unit modeled by $b_{\text{out}} \leftarrow b_{\text{in}} \cdot a_{\text{in}} + c_i$. In Fig. 13.9, a linear array of $(N-1)$ cells has the weights w^i enter from the left at the top line, the last piece of data x_{N-1} enters from the left at the bottom line, while the other $\{x_0, x_1, \ldots, x_{N-2}\}$ are preloaded to the cells.

In general, there is a latency of $(N-2)$ iterations before the useful DFT output appears at the output of Fig. 13.9. Afterwards, there is an output every iteration. The total number of multiplications is N^2. While the number of arithmetic operations for the proposed systolic DFT is worse than for a conventional FFT, it has a much simpler architectural structure. If an FFT uses one single processor, it needs $(N/2)\log_2 N$ units of time, while the systolic DFT needs only $2(N-1)$ units of time.

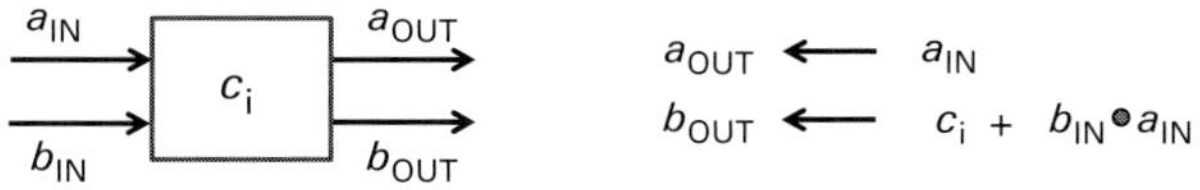

Figure 13.8 Basic systolic DFT cell

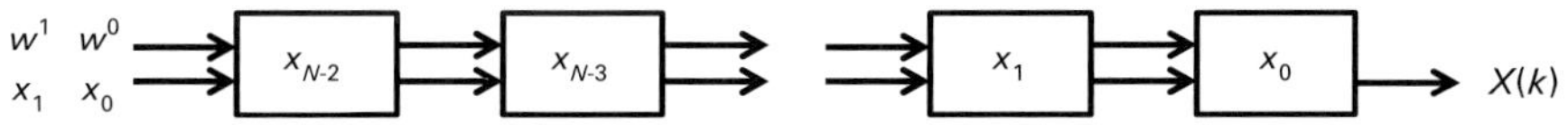

Figure 13.9 Linear systolic array for DFT

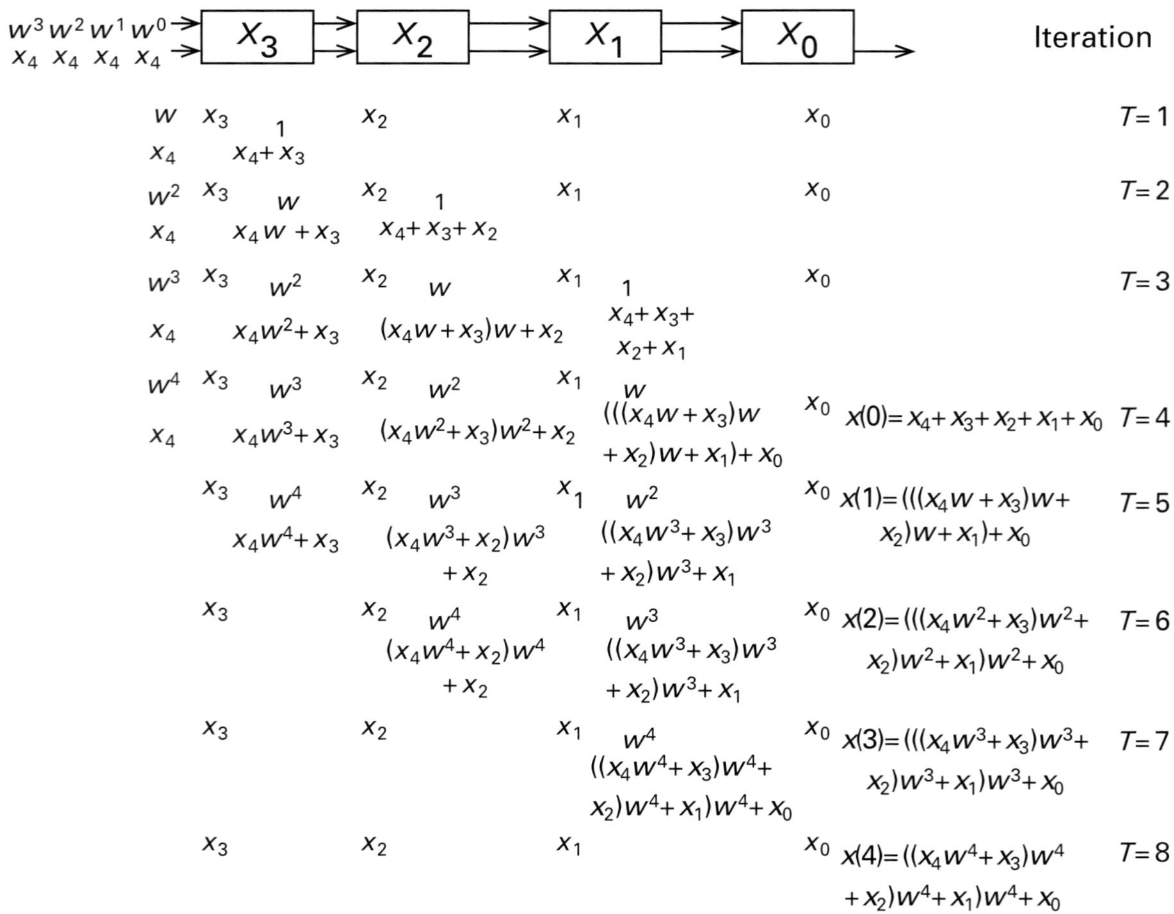

Figure 13.10 Data movements of a systolic DFT

13.1.3 Matrix–Vector Multiplication

Consider the multiplication of a 4×4 matrix A denoted by

$$A = \begin{bmatrix} a_{11} & a_{12} & a_{13} & a_{14} \\ a_{21} & a_{22} & a_{23} & a_{24} \\ a_{31} & a_{32} & a_{33} & a_{34} \\ a_{41} & a_{42} & a_{43} & a_{44} \end{bmatrix},$$

with the 4×1 vector $x = [x_1, \ldots, x_4]^T$ to yield

$$y = \begin{bmatrix} y_1 \\ y_2 \\ y_3 \\ y_4 \end{bmatrix} = \begin{bmatrix} a_{11}x_1 + a_{12}x_2 + a_{13}x_3 + a_{14}x_4 \\ a_{21}x_1 + a_{22}x_2 + a_{23}x_3 + a_{24}x_4 \\ a_{31}x_1 + a_{32}x_2 + a_{33}x_3 + a_{34}x_4 \\ a_{41}x_1 + a_{42}x_2 + a_{43}x_3 + a_{44}x_4 \end{bmatrix}$$

using a linear systolic array of 7 processing cells. The input of each processing cell is shown on the left-hand side of and the output of the processing cell is shown on the right-hand side of Fig. 13.11. Let the input of the $\{a_{ij}\}$ components be described in Fig. 13.12.

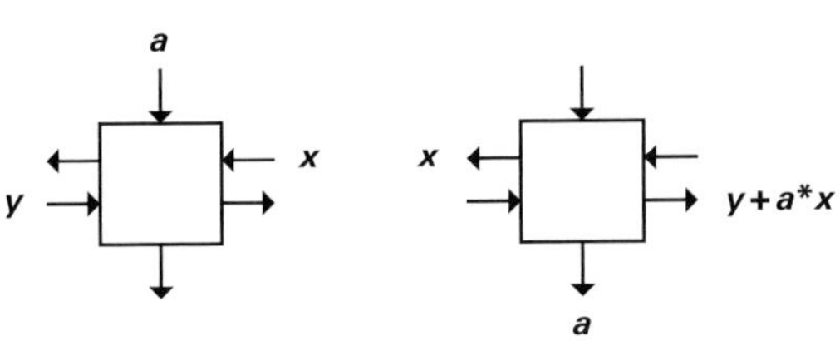

Figure 13.11 Processing cell for linear systolic matrix–vector multiplication

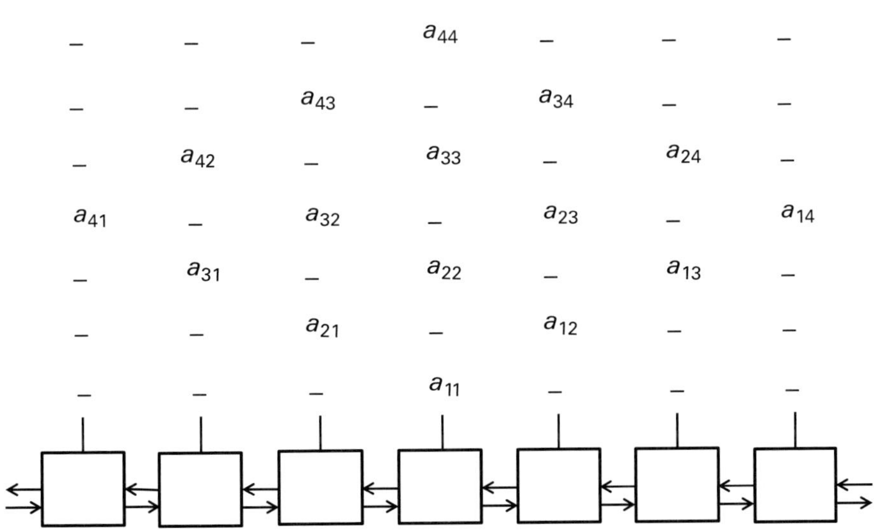

Figure 13.12 Input of the a_{ij} elements of the matrix A into the matrix–vector linear systolic array

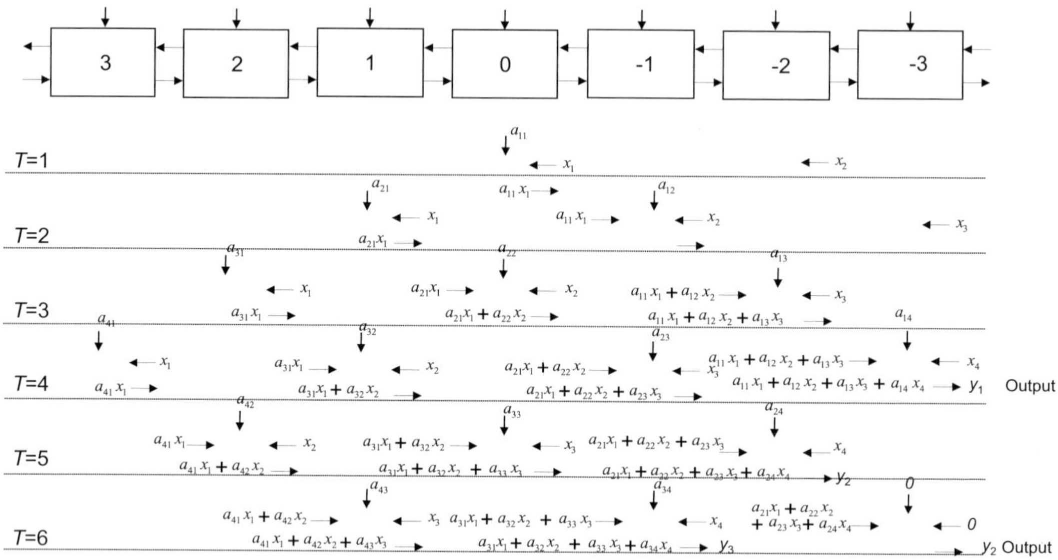

Figure 13.13 Movement of data and associated timing values of the matrix–vector multiplication on a linear systolic array

In Fig. 13.13, we show the movements of the data and associated timing values in the linear systolic array for performing the matrix–vector multiplication based upon the informal and intuitive SA design approach. In Fig. 13.13, denote as time $T = 1$, when x_1 and a_{11} have reached cell 0 and x_2 has reached cell -2. We note the a_{ij} wavefront (i.e., each element on the wavefront having the same time value) entering the linear array from the top consists of the 45 degree diagonal of elements in the A matrix. That is, the first wavefront consists of a_{11}, the second wavefront consists of $\{a_{21}, a_{12}\}$, the third wavefront consists of $\{a_{31}, a_{22}, a_{13}\}$, etc. The x_i values enter from the right and move to the left, at every other time instant. For example, at cell 0, x_1 reaches it at time $T = 1$, while x_2 reaches it at time $T = 3$, and x_3 reaches it at time $T = 5$, and x_4 reaches it at time $T = 7$. The output y_1 is formed at cell -3, and is output at time $T = 4$ from cell -3; while y_2 is formed at cell -2 at time $T = 5$, but is output at time $T = 6$ from cell -3; while y_3 is formed at cell -1 at time $T = 6$, but is output at time $T = 8$ from cell -3; and finally y_4 is formed at cell 0 at time $T = 7$, but is output at time $T = 10$ from cell -3. The names of these cells at this point are arbitrary, but are taken to be consistent with the names used in the following approach.

13.1.4 Matrix–Matrix Multiplication

Consider the matrix operation of

$$\underbrace{D}_{m\times p} = \underbrace{A}_{m\times n} \times \underbrace{B}_{n\times p} + \underbrace{C}_{m\times p},$$

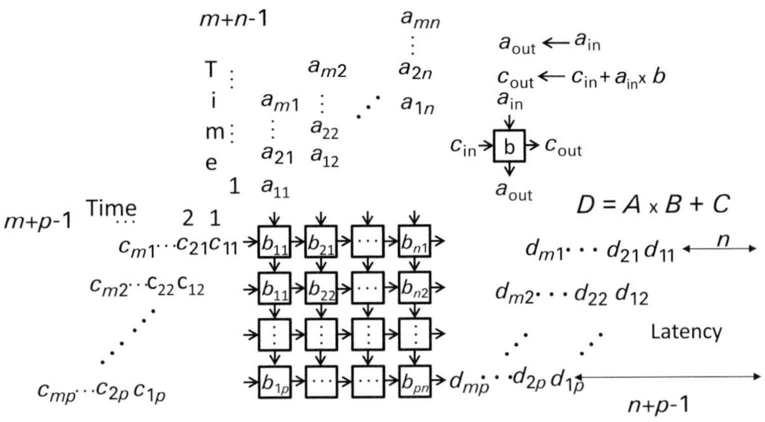

Figure 13.14 Systolic matrix multiplication of $A \times B + C$ on $p \times n$ processors

where $A = (a_{ij})$ is an $m \times n$ matrix, $B = (b_{ij})$ is an $n \times p$ matrix, and $C = (c_{ij})$ and $D = (d_{ij})$ are $m \times p$ matrices. The basic cell in Fig. 13.14 performs a multiply and add operation as the a_{in} data move down the array with the output moving to the right. The rectangular array is of size $p \times n$. We assume that the b_{ij} data are preloaded into the array. There is a latency period of n iterations and then the d_{ij} appear at the outputs to the right of the array. The last d_{ij} appears at iteration $(n + p + m - 1)$.

13.2 Conclusion

In the first section of this chapter, we introduced the concept of *systolic array computation* based on four simple assumptions of: using only one (or few) kind of processing element; adjacent (or nearby) communications and connections; all PE operate in a synchronous manner; there is minimal control of the PE. These ideas were introduced by H.T. Kung and C.E. Leisserson [1] and H.T. Kung [2] and illustrated in some details for the simple signal processing algorithms of correlation, matrix–vector and matrix–matrix multiplications, and DFT evaluations. While all the operations of these algorithms considered in this chapter were made at the simple and heuristic levels, in the next chapter, various more formal characterizations of systolic operations will be considered.

13.3 References

The term systolic array was coined by H.T. Kung in [1] [2] with illustrations to correlation, matrix–vector and matrix–matrix multiplications, and DFT evaluations.

[1] H.T. Kung and C.E. Leisserson, “Systolic Arrays for VLSI,” in *Introduction to VLSI Systems*, ed. C.A. Mead and L.A. Conway, Addison-Wesley, 1980, Section 8.3.

[2] H.T. Kung, “Why Systolic Architecture?” *IEEE Computer*, 1982, pp. 37–46.

13.4 Exercises

1. Consider the matrix–matrix multiplication problem of $C = AB$, where A is $m \times n$, and B is $n \times q$. We can denote the *ijk* form of implementing $c_{ij} = c_{ij} + a_{ik}b_{kj}$, by

$$\begin{aligned} \text{for } i &= 1 \text{ to } m, \\ \text{for } j &= 1 \text{ to } q, \\ \text{for } k &= 1 \text{ to } n\,. \end{aligned}$$

   a. Enumerate the five other possible forms of performing the matrix–matrix multiplication by: the *jik* form, the *ikj* form, the *jki* form, the *kij* form, and the *kji* form. See J.M. Ortega, *Introduction to Parallel and Vector Solution Linear Systems,* Plenum, 1988, pp. 54–55.

   b. Why are all six of these forms of matrix–matrix multiplication not suitable for possible systolic implementation?

2. Consider a matrix–vector multiplication problem. Draw a dependence graph for the matrix–vector problem with global communication capability. Draw a dependence graph for the matrix–vector problem with only local communication capability.

3. In VLSI and systolic implementations, a square-root operation is considered to be not desirable. However, in least-squares solutions of a linear system of equations based on the Givens transformation in the QR approach, square-root operations are needed. Formulate a square-root Givens transformation initially advocated in W.M. Gentleman, "Least Squares Computations by Given Transformations Without Square Roots," *J. Int. Maths. Applies,* 1973, pp. 329–336, and later with more detail in S.F. Hsieh, K.J.R. Liu, and K. Yao, "A Unified Square-Root-Free Approach for QRD-Based Recursive Least Squares Estimation," *IEEE Trans. on Signal Processing,* 1993, pp. 1405–1409.

4. Fault tolerance is a desirable feature in VLSI and systolic designs. Fault tolerance can be achieved using hardware or through system algorithms. Least-squares solutions of linear system of equations can be obtained using QR systolic algorithms. The residuals of the solutions can be obtained using triangular systolic arrays; see S.G. McWhirter, "Recursive Least-Squares Minimization using a Systolic Array," *Proceedings SPIE*, 1983. Demonstrate the method proposed in K.J.R. Liu and K. Yao, "Grace Fully Degradable Real-Time Algorithm Based Fault-Tolerant Method for QR Recursive Least-Squares Systolic Array," in *Systolic Array Processors, Proceedings 1989 Internation Conference on Systolic Arrays*, eds. J.V. McCanny, J.G. McWhiter, and E. Swartzlander, Prentice Hall, 1989, pp. 401–410, to achieve fault-tolerance computing in a triangular systolic array.

14 Systolic Array Design by Dependence Graph Mapping

14.1 Introduction

Systolic array designs can be obtained based on ad hoc or heuristic approaches as initially proposed by H.T. Kung and C.E. Leisserson [1]. Later, the semi-formal space–time transformation approach was proposed by P.R. Cappello and K. Steiglitz [2] to perform systolic array designs for certain classes of relatively simple algorithms that included convolutions, correlations, matrix–matrix multiplications. More formal procedures for the systematic design of systolic arrays were proposed by D.I. Moldovan and J.A.B. Fortes [3], P. Quinton [4], S.K. Rao [5], S.Y. Kung [6], A. Darte [7], J. Bu, and others. All these more formal procedures are collectively referred to as dependence graph mapping techniques for systolic array design.

14.2 Single Assignment Algorithm

All the algorithms under consideration are called *single assignment algorithms*. Each variable has a unique value during the evaluation of the algorithm. Those variables with multiple values can be converted to single values by vectorizing the variables through the introduction of new indices.

Example 14.1 Consider the matrix multiplication of $C = AB$, where $A = [a_{ik}]$ is $N_1 \times N_3$, $B = [b_{kj}]$ is $N_3 \times N_2$, and $C = [c_{ij}]$ is $N_1 \times N_2$. A formal algorithm is given by

Conventional Matrix–Matrix Multiplication Algorithm

For $i = 1$ to N_1
  For $j = 1$ to N_2
    $c_{ij} = 0$
    For $k = 1$ to N_3

$$c_{ij} = c_{ij} + a_{ik}b_{kj}. \tag{14.1}$$

In this approach, c_{ij} has multiple values for $k = 1, \ldots, N_3$. We can modify this to have single values by replacing it with the variable c_{ijk}. Then (14.1) has the recurrence equation form of

$$c_{ijk} = c_{ij(k-1)} + a_{ik}b_{kj},\ c_{ij0} = 0, \quad c_{ij} = c_{ijN_3}. \tag{14.2}$$

We assume all algorithm variables are indexed variables with V variable names, denoted by the generic names of $X_m, 1 \leq m \leq V$. In Example 14.1, $V = 3$ with $X_1 = c$, $X_2 = a$, and $X_3 = b$. For each variable name, the domain of the index vectors is a subset in an S-dimensional space. This subset is called the algorithm's *index space* and S is its dimension. For most iterative signal processing problems, time is usually one of the index space coordinates. For the above matrix–matrix multiplication, we need to propagate a_{ik} across the j variables as well as b_{kj} over the i variables in order to perform (14.1) or (14.2). These a_{ik} and b_{kj} are propagating variables since they involve no computations but need to be made available at various stages of the computation. For Example 14.1, clearly $S = 3$ and the index space is given by $\mathbf{S}_0 = \{(i, j, k) : 1 \leq i \leq N_1,\ 1 \leq j \leq N_2,\ 1 \leq k \leq N_3\}$. Furthermore, the initializations of the new variables are given by

$$a(i,0,k) = a_{ik}, \quad b(0,j,k) = b_{kj}, \quad c(i,j,0) = 0, \quad c_{ij} = c(i,j,N_3).$$

Then the algorithm in (14.1) can be rewritten as:

New Matrix–Matrix Multiplication Algorithm

$$\begin{aligned} &\text{For } (i,j,k) \in \mathbf{S}_0, \\ &\quad a(i,j,k) = a(i,j-1,k), \\ &\quad b(i,j,k) = b(i-1,j,k), \\ &c(i,j,k) = c(i,j,k-1) + a(i,j,k)b(i,j,k). \end{aligned} \tag{14.3}$$

We note, (14.3) has the recurrence equation form of (14.2) as noted above.

In general, a point (or node) in the index space is called an *index point*. We denote $X_m(I)$ as the variable X_m defined at the index point I. A dependence graph mapping is a representation of a single assignment algorithm, where the dependencies among the variables are represented by directed arcs among the nodes. A basic property of the class of algorithms of interest is that of *shift invariance*.

14.3 Shift-Invariant Algorithms

An algorithm is shift-invariant if the dependence graph is regular. That is if $X(I)$ depends on $Y(J)$, then $X(I + K)$ depends on $Y(J + K)$ for all I, J, $I + K$, and $J + K$ in the index space. Three well known shift invariance algorithms include:

1. Uniform recurrence equations (URE) (Quinton (1984) and others)

$$X_1(I) = F_1(X_1(I - D_1), \ldots, X_V(I - D_V)), \tag{14.4}$$
$$X_i(I) = X_i(I - D_i), \quad 2 \le i \le V. \tag{14.5}$$

There is computation only in $F_1(\cdot)$ of (14.4) and propagations in all the other variables in (14.5). Clearly, the new matrix–matrix multiplication algorithm of (14.3) is a URE algorithm. Here $V = 3$, $F_1(\cdot)$ of (14.4) is given by (14.3), with $X_1(\cdot) = c(\cdot)$, $X_2(\cdot) = a(\cdot)$, $X_3(\cdot) = b(\cdot)$, $I = [i, j, k]^T$, $D_1 = [0, 0, 1]^T$, $D_2 = [0, 1, 0]^T$, and $D_3 = [1, 0, 0]^T$.

2. Generalized uniform recurrence equations (GURE) (Quinton (1984))

$$X_m(I) = F_m(X_{m_{1(m)}}(I - D_{m_1}), \ldots, X_{m_{k(m)}}(I - D_{m_{k(m)}})), \quad 1 \le m \le V, \tag{14.6}$$

where $m_{1(m)}, \ldots, m_{k(m)}$ belong to $\{1, \ldots, m\}$. In (14.6), we can have computations in all V functions of $F_m(\cdot)$. The number of independent variables, $m_{k(m)}$ depends on each m. The shift index dependency, $I - D_{m_i}$, is fixed for each X_{m_i}.

3. Regular iterative algorithm (RIA) (Rao (1985))

$$X_m(I) = F_m(X_{m_1}(I - D_{m_1,m}), \ldots, X_{m_{k(m)}}(I - D_{m_{k(m)},m})), \ 1 \le m \le V. \tag{14.7}$$

In (14.7), the shift index dependency, $I - D_{m_i,m}$, is not fixed but is a function of m_i and m.

14.4 Scheduling and Allocation Functions

It is assumed that each processor of the systolic array has all the necessary computational modules to compute $F_m(\cdot)$. For URE, we need only one such module, but for GURE and RIA, we need V modules. The time required for the computation of $F_m(\cdot)$ (i.e., for the evaluation of $X_m(\cdot)$) is denoted by h_m and the minimum time between such computations is denoted by τ_m. However, most researchers limit themselves (and so do we) to unit valued h_m and τ_m. This implies that the time slots have been rescaled in order to define as unity the time needed by the "longest" computation. These limitations do not restrict the scope of our study too much, especially when we deal with systolic structures, which are, by their very nature, fully synchronized.

The design of a processor array to perform the algorithm requires spatial and temporal assignments. Each $X_m(I)$ must be assigned to a processor at each integral time slot. The processor allocation function, $A(I)$, assigns all variables with the index I to the processors in the array. The scheduling function, $S_m(I)$, assigns the start of the computation for the variable $X_m(I)$. The simplest form of scheduling and processor allocation is based on the projection of the multi-dimensional dependence graph onto the lower dimensional processor array. Variables represented by nodes in the dependence graph are mapped to processors which perform the computations. The directed arcs of the dependence graphs are transformed to physical communication links in the processor array.

The essence of the allocation function $A(\cdot)$ is thus to return for every index value $I \in \mathbf{S}_0$ a vector which indicates the processor in charge of the computation (a point in a lower dimensional space). Analogously, the scheduling function $S_m(\cdot)$ provides the relative start of the execution for the computation indexed by I. Observe that these two functions cannot be chosen independently: two computations assigned to the same processor cannot be scheduled for the same time (compatibility constraint). While in principle $A(\cdot)$ and $S_m(\cdot)$ can be any function, we shall consider only *affine* functions, in the sense that

$$A(I) = A^T I$$
$$S_m(I) = \lambda^T I + \gamma_m,$$

A being a suitable matrix, λ a vector, and γ_m a constant.

The dependence graph of an algorithm can be interpreted as a lattice embedded in a multi-dimensional integral space (i.e., a proper bounded subset of $\mathbf{Z}^S$, where $\mathbf{Z}$ is the set of relative integers), enclosed in a convex polyhedron. A simplifying assumption is to consider only "dense" lattices, in the sense that all the integral points in it correspond to actual computations. The whole procedure of mapping an algorithm onto a systolic-type processor can thus be rephrased using the theory of integral matrices. The mapping involved in the systolic design, as mentioned earlier, consists in fact of two conceptually different, even if interdependent, operations: a space transformation and a time transformation. The former actually *projects* the dependence graph onto a lower dimensional structure (which can be mapped one-to-one onto the physical array), while the latter gives the start of the execution of each computation.

For simplicity, suppose we want to project the S-dimensional space onto an $(S-1)$-dimensional processor space. The more general problem of projecting the dependence graph onto an $(S-p)$-dimensional space ($p \geq 1$) can be expressed using similar, but more involved, notation. Instead of talking about allocation functions, we will refer to the vector u which is orthogonal to the processor space (i.e. the space onto which we project, and for this reason we name such vector *projection vector*) as shown in Fig. 14.1.

Assume that we have chosen both the projection and the scheduling vectors (u and λ respectively). For normalization purposes, it is customary to choose them as coprime vectors, that is such that the greatest common divisor of their components be one, and with their first non-zero element positive.

There are a number of constraints which have to be satisfied by u and λ. Assume nodes I and J are located along a direction parallel to the projection vector u (i.e., $J = I + \alpha u$, $\alpha \in \mathbf{Z}$). Then the computations associated with the two nodes will be projected onto the same processor. Consequently, for reasons of compatibility, one has to make sure that they be performed at different times. In particular, since h_m is the time required for the evaluation of $F_m(\cdot)$, either we have to start the computation of node J after the completion of the computation of node I or vice versa. That is, we must satisfy both inequalities of:

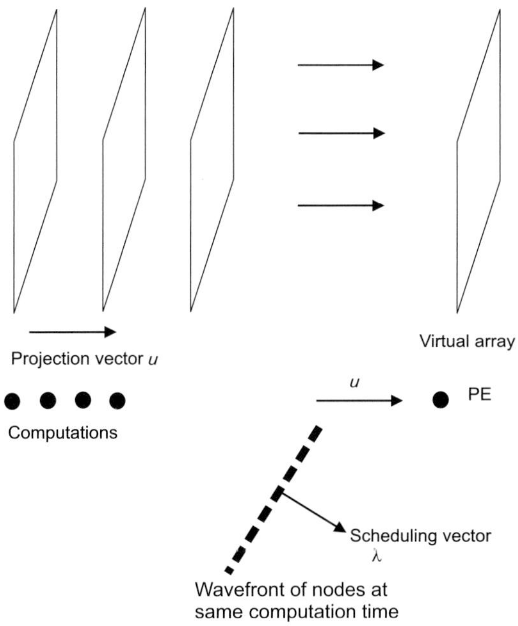

Figure 14.1 Projection from 3 to 2 dimensions

$$S_m(I) + h_m \leq S_m(J) \Rightarrow \lambda^T I + \gamma_m + h_m \leq \lambda^T I + \alpha\lambda^T u + \gamma_m \quad \Rightarrow \quad \alpha\lambda^T u \geq h_m,$$

$$S_m(I) \geq S_m(J) + h_m \Rightarrow \lambda^T I + \gamma_m \geq \lambda^T I + \alpha\lambda^T u + \gamma_m + h_m \quad \Rightarrow \quad -\alpha\lambda^T u \geq h_m,$$

for all $m = 1, \dots, V$ and $\alpha \in \mathbf{Z}$. In conclusion, we have that

$$|\lambda^T u| \geq \max_{m=1,\dots,V} h_m,$$

which shows the time **compatibility constraint** simplifies to

$$|\lambda^T u| > 0. \tag{14.8}$$

For the case of $h_m = 1$, (14.8) states that the λ and u vectors cannot be perpendicular. Observe that the quantity $c \triangleq |\lambda^T u|$ represents the number of time slots between successive calculations scheduled on the same processor.

Consider now the case in which the variable $X_m(I)$ depends on $X_n(I - D_{nm})$ as shown in Fig. 14.2. The calculation of $X_n(I - D_{nm})$ must be scheduled so as to be completed by the start of the calculation of $X_m(I)$. The calculation of $X_n(I - D_{nm})$ ends at time $\lambda^T I - \lambda^T D_{nm} + \gamma_n + \tau_n$, whereas the calculation of $X_m(I)$ begins at time $\lambda^T I + \gamma_m$. Hence, the scheduling vector λ and the additive γ constants must satisfy

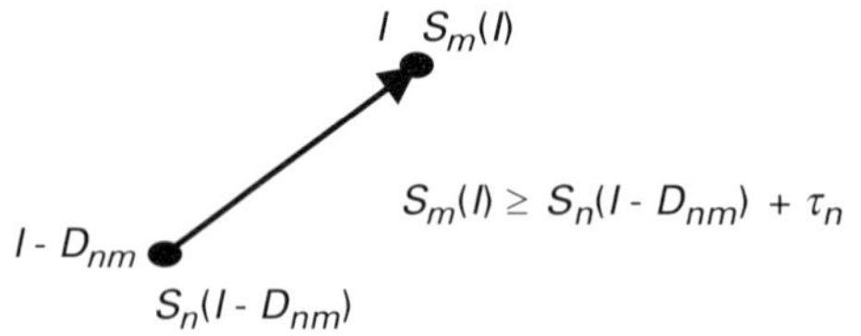

Figure 14.2 Dependence relation between node I and node J and associated scheduling functions

$$S_m(I) \geq S_n(I - D_{nm}) + \tau_n \Rightarrow \lambda^T I + \gamma_m \geq \lambda^T I - \lambda^T D_{nm} + \gamma_n + \tau_n$$
$$\Rightarrow \lambda^T D_{nm} + \gamma_m - \gamma_n \geq \tau_n = 1, \quad (14.9)$$

for all $1 \leq m \leq V$ and for all dependences D_{nm} (precedence constraint). If the γ constants are chosen to be all equal, the above **precedence constraint** becomes

$$\lambda^T D_{nm} \geq 1 \quad \forall n, m = 1, \ldots, V. \quad (14.10)$$

Suppose that the precedence and compatibility constraints are met (namely, $\lambda^T D_i > 0$, where D_i is any dependence vector, and $\lambda^T u \neq 0$). Since λ and u are coprime vectors, then it is possible to extend both vectors to two unimodular matrices. A matrix with integral entries is named *unimodular* when its determinant is equal to ± 1. This immediately implies that they admit integral inverses. The unimodular extension of coprime vectors is not unique. One way is to produce the Hermite form of the vectors. Let U and Λ be the matrices that have u and λ, respectively, as their first columns.

Now, it is possible to show that the columns of any S-dimensional unimodular matrix can constitute a basis for the space $\mathbf{Z}^S$. Since we have assumed that the algorithm at hand is a proper dense subset G of $\mathbf{Z}^S$, then it is easy to see that U represents a possible basis of G. Moreover, if we call $\sigma_1, \ldots, \sigma_S$ the columns of $\Sigma = U^{-T}$, by definition we have $\sigma_1^T u = 1$ and $\sigma_i^T u = 0$ for all $i = 2, \ldots, S$. Therefore, $\{\sigma_2, \ldots, \sigma_S\}$ will be a basis of the processor space of the resulting logic array.

Following analogous reasonings, it is possible to see that the first column of T (the inverse of Λ^T), t_1, represents the direction in which time increases by one step, that is, it is the vector defining the hyperplane of the points computed at the same time. The other columns of T (say, $t_2, \ldots, t_S$) are a basis of such a hyperplane.

If we denote by $\Sigma_+ = [\sigma_2, \ldots, \sigma_S]$ the matrix basis of the processor space, then the allocation function and the scheduling function have the form

$$A(I) = \Sigma_+^T I \quad (14.11)$$
$$S_m(i) = \lambda^T I + \gamma_m, \quad m = 1, \ldots, V. \quad (14.12)$$

See Fig. 14.3 for a pictorial representation of the four matrices U, Σ, Λ, and T.

With these elements, we can have the complete description of the final array: the processors are labeled by $A(I) = \Sigma_+^T I$ as I ranges over the index space; the

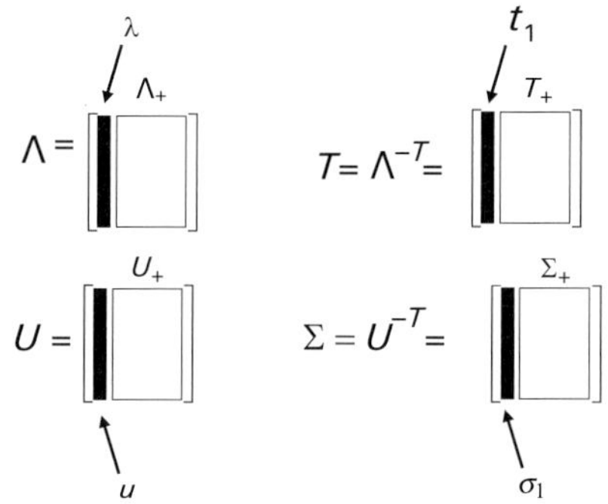

Figure 14.3 The four unimodular matrices Λ, T, U, Σ

dependences D_{nm} are mapped onto communication links $\Sigma_+^T D_{nm}$ and the delay registers on such links must be number equal to $\lambda^T D_{nm} + \gamma_m - \gamma_n - \tau_n = \lambda^T D_{nm} - 1$.

As we have mentioned before, $c = |\lambda^T u|$ represents the time that elapses between calculations at a particular processor. For this reason, $1/c$ is sometimes called *efficiency* of the processors, since the larger c, the more time the processors have to spend inactive. One common approach is to select the projection vector and the scheduling vector so as to achieve the highest efficiency (i.e., c as close to one as possible).

Example 14.2 Consider the systolic correlation problem as originally proposed by H. T. Kung, where the weights $\{w_1, \ldots, w_k\}$ and the data $\{x_1, \ldots, x_n\}$ are given. The correlation is given by

$$y_i = w_1 x_i + w_2 x_{i+1} + \cdots + w_k x_{i+k-1}, \quad 1 \le i \le n+1-k. \tag{14.13}$$

A recurrence equation formulation of (14.13) yields

$$y(i,j) = y(i,j-1) + w(i,j)x(i,j),\ y(i,0) = 0,\ y_i = y(i,k), \tag{14.14}$$
$$w(i,j) = w(i-1,j),\ w(0,j) = w_j, \tag{14.15}$$
$$x(i,j) = x(i+1,j-1),\ x(i,j) = x_{i+j-1}, \tag{14.16}$$

all with

$$1 \le i \le n+1-k, \quad 1 \le j \le k. \tag{14.17}$$

A graphical representation of (14.14)–(14.17) is shown in Fig. 14.4. Note that the directions of propagation of the variables in (14.14)–(14.16) is arbitrary and can be reversed without affecting the result of the computations.

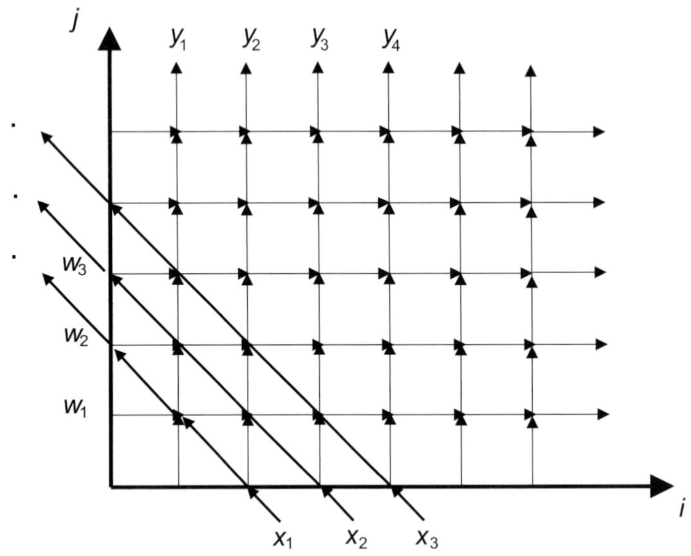

Figure 14.4 Two-dimensional dependence graph

A URE reformulation of (14.14)–(14.17) yields the index point

$$I = [i, j]^T, \tag{14.18}$$

where i and j are given by (14.17),

$$X_1(I) = y(i,j) = F_1(X_1(I - D_1), X_2(I - D_2), X_3(I - D_3)), \tag{14.19}$$

$$X_2(I) = w(i,j) = X_2(I - D_2), \tag{14.20}$$
$$X_3(I) = x(i,j) = X_3(I - D_3), \tag{14.21}$$

with displacement vectors

$$D_1 = [0,1]^T, \tag{14.22}$$
$$D_2 = [1,0]^T, \tag{14.23}$$
$$D_3 = [-1,1]^T. \tag{14.24}$$

In particular, consider the URE representation of the B1 design based on the choice

$$u = [1,0]^T, \tag{14.25}$$
$$\Sigma_+ = [0,1]^T, \tag{14.26}$$
$$\lambda = [1,1]^T, \tag{14.27}$$

$$\gamma_m = 0, \quad m = 1,2, \text{ and } 3, \tag{14.28}$$

and

$$\tau_n = h_n = 1, \quad n = 1,2, \text{ and } 3. \tag{14.29}$$

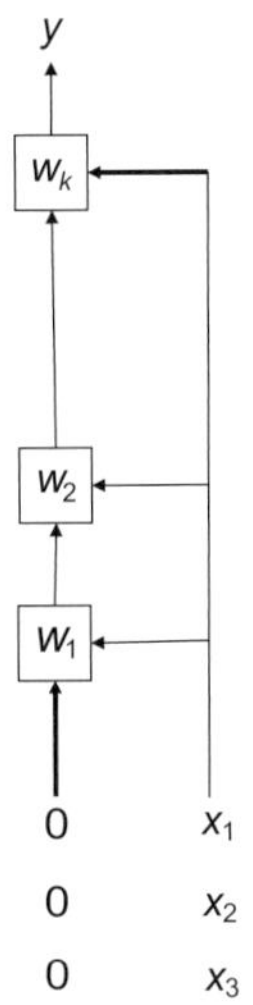

Figure 14.5 One-dimensional array for Design B1 with $\lambda = [1, 1]^T$

Then the two-dimensional graph of Fig. 14.4 is projected onto the one-dimensional graph of Fig. 14.5.

Specifically, for any index $I = [i, j]^T$, the processor allocation function yields,

$$A(I) = \Sigma_+^T I = j, \quad 1 \le j \le k, \tag{14.30}$$

which is a valid projection from two-dimensions to one-dimension. On the other hand, the index point for each input data x_ℓ (with variable name of X_3) is given by

$$I = [i, \ell - i + 1]^T, \quad \ell = 1, \ldots, n. \tag{14.31}$$

Then the scheduling function $S_3(I)$ is given by

$$S_3(I) = \lambda^T I = [1, 1][i, \ell - i + 1]^T = i + \ell - i + 1 = \ell + 1, \quad \ell = 1, \ldots, n. \tag{14.32}$$

This indicates that each x_ℓ (or variable $X_3(I)$ with I given by (14.18)) must be availabe at all the processors at time $\ell + 1$. Thus, there is no propagation and this means all the x_ℓ must be broadcast to all the processors. However, the systolic design based on (14.11) and (14.12) does not allow broadcasting. Indeed, the precedence constraint of (14.9) is *not* satisfied with D_3. That is

$$\lambda^T D_3 = [1, 1][-1, 1]^T = 0 \not\ge \tau_n = 1. \tag{14.33}$$

A better choice for λ would have been, for instance, $\lambda = [1,2]^T$, since in this case $\lambda^T D_i > 0$ for $i = 1,2,3$ and still $c = |\lambda^T u| = 1$, thus achieving maximum efficiency. The four unimodular matrices, in this case, would be as follows

$$\Lambda = \begin{bmatrix} 1 & 0 \\ 2 & 1 \end{bmatrix}, \quad T{=}\Lambda^{-T} = \begin{bmatrix} 1 & 2 \\ 0 & 1 \end{bmatrix},$$

$$U = \begin{bmatrix} 1 & 0 \\ 0 & 1 \end{bmatrix}, \quad \Sigma{=}U^{-T} = \begin{bmatrix} 1 & 0 \\ 0 & 1 \end{bmatrix}.$$

Example 14.3 Consider the matrix–vector multiplication problem in Subsection 13.1.3 using the dependence graph mapping method. Using the single assignment algorithm notation, we have

$$\begin{aligned}
&y(i,j) = y(i,j-1) + a(i,j)x(i,j),\ y(i,0) = 0,\ y(i,4) = y_i,\ 1 \le i,j \le 4,\\
&a(i,j) = a(i,j) = a_{ij},\ 1 \le i,j \le 4,\\
&x(i,j) = x(i-1,j),\ x(0,j) = x_j,\ 1 \le i,j \le 4.
\end{aligned}$$

By using the URE notation, we have

$$\begin{aligned}
&X_1(I) = y(i,j) = X_1(I - D_1) + X_2(I)X_3(I),\\
&X_2(I) = a(i,j) = X_2(I),\\
&X_3(I) = x(i,j) = X_3(I - D_3),\\
&I = \begin{bmatrix} i \\ j \end{bmatrix},\ 1 \le i,j \le 4,\\
&D_1 = \begin{bmatrix} 0 \\ 1 \end{bmatrix},\ D_2 = \emptyset,\ D_3 = \begin{bmatrix} 1 \\ 0 \end{bmatrix}.
\end{aligned}$$

We note, since there are no dependencies of the a_{ij} on other a_{kl}, we can set the D_2 vector to the null set.

In order to use the dependence graph mapping approach, we take

$$u = \lambda = \begin{bmatrix} 1 \\ 1 \end{bmatrix},\ \sigma_2 = \Sigma_+ = \begin{bmatrix} 1 \\ -1 \end{bmatrix},\ t_2 = \begin{bmatrix} 1 \\ -1 \end{bmatrix},$$

and

$$U = \begin{bmatrix} 1 & 0 \\ 1 & -1 \end{bmatrix},\ \Sigma = \begin{bmatrix} 1 & 1 \\ 0 & -1 \end{bmatrix},\ \Lambda = \begin{bmatrix} 1 & 0 \\ 1 & -1 \end{bmatrix},\ T = \begin{bmatrix} 1 & 1 \\ 0 & -1 \end{bmatrix}.$$

We first note, $|U| = -1$, $|\Sigma| = -1$, $|\Lambda| = -1$, and $|T| = 1$. Thus, all four of these matrices are unimodular. Furthermore, we note

$$\Sigma U^T = \begin{bmatrix} 1 & 1 \\ 0 & -1 \end{bmatrix}\begin{bmatrix} 1 & 1 \\ 0 & -1 \end{bmatrix} = I_2,\ T\Lambda^T = \begin{bmatrix} 1 & 1 \\ 0 & -1 \end{bmatrix}\begin{bmatrix} 1 & 1 \\ 0 & -1 \end{bmatrix} = I_2.$$

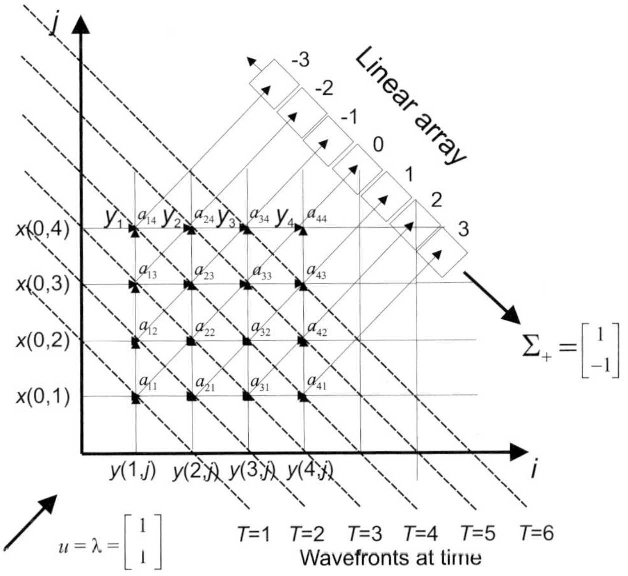

Figure 14.6 Dependence graph systolic array design of the matrix–vector multiplication problem

From the dependence graph for this problem in Fig. 14.6 we note that $x_j = x(0, j) = X_3(I)$, $I = [0, j]^T$, $j = 1, 2, 3, 4$, enter the j-axis from the left and moves to the right (consistent with $D_3 = [1, 0]^T$). The a_{ij} values do not propagate (consistent with $D_2 = \emptyset$). However, the computation of $X_2(I) = a_{ij}$ at the node $I = [i, j]^T$, is projected along the projection vector $u = [1, 1]^T$, which is at 45 degrees in the northeastern direction toward the linear array. In particular, since the allocation function is given by $A(I) = \Sigma_+^T I = [1, -1][i, j]^T = (i - j)$, this means the computations involving $\{a_{11}, a_{22}, a_{33}, a_{44}\}$ are computed by the linear array processor cell 0. In general, computations of all the $a_{ij}x(i, j)$ are performed by cell $(i - j)$, which are consistent with what we observe in Fig. 14.1. Furthermore, the $X_1(I) = y(i, j)$ starts on the i-axis with $y(i, 0)$ and move upward for $j = 1, 2, 3, 4$, (consistent with $D_1 = [0, -1]^T$). When $j = 4$, then $y(i, 4) = y_i$, $i = 1, 2, 3, 4$. However, after the completion of the formation of the y_i, they are projected to the processing cells of the linear array, and then output from cell -3. These results are also consistent with what we observe in Fig. 14.1. By using the scheduling function $S_m(I) = \lambda^T I + \gamma_m = (i + j) - 1 = T$, $m = 1, 2, 3$, $1 \leq i, j, \leq 4$, where $\gamma_m = -1$, the wavefront of processing at time T shown on diagonal lines at -45 degrees, moves upward in the $\lambda = [1, 1]^T$ northeastern direction. The hyperplanes of the wavefronts have the basis of $t_2 = [1, -1]^T$.

Since $c = |\lambda^T u| = |[1, 1][1, 1]^T| = 2$, the compatibility constraint of $|\lambda^T u| \geq 1$ is satisfied. With $c = 2$, this shows the efficiency of this SA is only $1/c = 0.50$, and the output y_i comes out at every other time instant. Furthermore, since $\lambda^T D_1 = [1, 1][0, 1]^T = 1$ and $\lambda^T D_3 = [1, 1][1, 0]^T = 1$, the precedence constraints of $\lambda^T D_m \geq 1, m = 1, 3$ are satisfied. Since $D_2 = \emptyset$, the precedence constraint for D_2 is not applicable.

14.5 Conclusion

Section 14.2 introduced the single assignment algorithm. Section 14.3 defined three different classes of shift-invariant algorithms. Section 14.4 introduced the scheduling and allocation functions of a formal systolic design procedure based on the use of the integral matrices. Various examples were given.

14.6 References

The materials in Sections 14.2 and 14.3 can be found in [3], [4], [5], and [6]. The materials in Section 14.4 can be found in [7] and [8]. Basic materials on integral matrices appeared in [9].

[1] H.T. Kung and C.E. Leisserson, "Systolic Arrays for VLSI," in *Introduction to VLSI Systems*, ed. C.A. Mead and L.A. Conway, Addison-Wesley, 1980, Section 8.3.

[2] P.R. Cappello and K. Steiglitz, "Unifying VLSI Array Designs with Geometric Transformations," *Proc. Inter. Conf. Parallel Processing*, 1983, pp. 448–457.

[3] D.I. Moldovan and J.A.B. Fortes, "Partitioning and Mapping of Algorithms into Fixed Size Systolic Arrays," *IEEE Trans. Computers*, 1986, pp. 1–12.

[4] P. Quinton, "Automatic Synthesis of Systolic Arrays from Uniform Recurrence Equations," *Proceedings of the IEEE 11th Inter. Symp. on Computer Arch.*, 1984, pp. 208–214.

[5] S.K. Rao,"Regular Iterative Algorithms and Their Implementations on Processor Arrays," PhD thesis, Stanford University, Stanford, California, 1985.

[6] S.Y. Kung, *VLSI Array Processors*, Prentice-Hall, 1988.

[7] A. Darte, "Regular Partitioning for Synthesizing Fixed-Size Systolic Arrays," *Technical Report LIP-IMAG 91-10, Laboratoire de lInformatique du Parallelisme, Ecole Superieure de Lyon*, 1991.

[8] P. Quinton and Y. Robert, *Algorithmes et Architectures Systoliques*, Masson, 1989.

[9] M. Newman. *Integral Matrices*, Academic Press, 1972.

14.7 Exercises

1. Consider a general FIR filter given by

$$y(n) = \sum_{i=0}^{N-1} a(i)\,x(n-i). \quad (1)$$

However, for implementational purposes, (1) may be implemented using the form of (2) or (3) after using the delay operator Z defined by $Zx(n) = x(n-1)$.

$$y(n) = \sum_{i=0}^{N-1} a(i)\,Z^i x(n), \quad (2)$$

$$y(n) = \sum_{i=0}^{N-1} Z^i \, (a(i)x(n)) . \quad (3)$$

Discuss the differences in the implementation of (2) versus that of the implementation of (3).

2. Using the systolic dependence graph mapping approach, formulate the B2 design to perform systolic correlation (i.e., find the allocation function $A(I)$ and the scheduling function $S(I)$ and the associated U; $+$; I. Show the dependence graph mapping for your systolic design.
3. Using the systolic dependence graph mapping approach, formulate both R1 and F designs to perform systolic correlation (i.e., find the allocation function $A(I)$ and the scheduling function $S(I)$ and the associated U; $+$; I. Show the dependence graph mapping for both of your systolic designs.
4. Read the paper, J.P. Charlier, M. Vanbegin, and P. Van Dooren, "Systolic Algorithms for Digital Signal Processing," *Philips J. Res.*, 1988, pp. 268–290, on some early applications of systolic algorithms to signal processing.

15 Systolic Array Processing for Least-Squares Estimation

The least-squares (LS) technique constitutes one of the most basic components of all modern signal processing algorithms dealing with linear algebraic and optimization of deterministic and random signals and systems. Specifically, some of the most computationally intensive parts of spectral analysis, beamformation, direction finding, adaptive arrays, image restoration, robotics, data compression, parameter estimation, and Kalman filtering all depend crucially on LS processing.

Regardless of any specific application, an LS estimation problem can be formulated as

$$Ax \approx y, \tag{15.1}$$

where the $m \times n$ data matrix A and the $m \times 1$ vector y are known, and the $n \times 1$ solution x is unknown. When $m = n$ and A is non-singular, then an exact solution for x exists. For most signal processing problems, $m > n$ and the optimum solution $\hat{x}$ of (15.1) in the LS sense is defined by

$$\| A\hat{x} - y \| = \min_{x} \| Ax - y \| . \tag{15.2}$$

A classical approach toward the LS solution is given by

$$\hat{x} = A^{+} y, \tag{15.3}$$

where A^{+} is the pseudo-inverse of A given by

$$A^{+} = (A^T A)^{-1} A^T . \tag{15.4}$$

This approach is not desirable from the complexity, the finite precision sensitivity, and processing architecture points of view. This is due to the need for a matrix inversion, the "squaring of the condition number" in the $A^T A$ operation, and the block nature of the operation in preventing a systolic sequential updating processing and architecture.

15.1 QR Decomposition

There are many ways to view the formulation of QR decomposition. Consider a real-valued $m \times n$ matrix A with $m \geq n$ and all the columns are linearly independent (i.e., rank $A = n$). Then from the QR decomposition theorem, we can find an $m \times m$ orthogonal matrix Q such that

$$Q^T A = R = \begin{bmatrix} R_1 \\ R_2 \end{bmatrix} \begin{matrix} n \times n \\ (m-n) \times n \end{matrix} . \tag{15.5}$$

The matrix R is an $m \times n$ matrix that can be expressed as an $n \times n$ upper triangular matrix R_1 (with positive-valued diagonal elements) and an all zero $(m-n) \times n$ matrix R_2.

Let $\alpha(i), i = 1, \ldots, m$ be the mth row of the matrix A and

$$y^T = [y_1, y_2, \ldots, y_m]$$

be a real-valued known $m \times 1 (m \gg n)$ observation vector. Then

$$Ax = \begin{bmatrix} \alpha(1) \\ \vdots \\ \alpha(m) \end{bmatrix} x \approx \begin{bmatrix} y_1 \\ \vdots \\ y_m \end{bmatrix} . \tag{15.6}$$

15.2 QR Decomposition to Solve the LS Estimation Problem

As was discussed in Section 15.1, QR decomposition provides a numerically stable approach to solve the LS estimation. From (15.2) and (15.5), the LS problem becomes

$$||Ax - y||^2 = ||Q^T(Ax - y)||^2 = ||Q^T Ax - Q^T y||^2 = ||Rx - f||^2, \tag{15.7}$$

where R is given by (15.5) and the $m \times 1$ matrix f is given by

$$f = Q^T y = \begin{matrix} \\ n \\ (m-n) \end{matrix} \begin{matrix} 1 \\ \begin{bmatrix} u \\ v \end{bmatrix} \end{matrix} . \tag{15.8}$$

In (15.7), we used the basic property that the norm of any vector is invariant with respect to any orthogonal transformation. Denote $e = Ax - y$ as the residual of the LS problem. From (15.5), (15.7) becomes

$$||e||^2 = ||Ax - y||^2 = ||[R_1 x, R_2 x]^T - [u, v]^T||^2 = ||R_1 x - u||^2 + ||v||^2. \tag{15.9}$$

Since R_1 is a non-singular upper triangular square matrix, the back substitution procedure of the traditional Gaussian elimination method can be used to solve for the exact solution of

$$R_1 x = u. \tag{15.10}$$

Let $\hat{x}$ denote the solution of (15.10). Thus, the LS problem reduces to

$$\begin{aligned} \min_x ||Ax - y||^2 &= ||A\hat{x} - y||^2 = ||R\hat{x} - f||^2 \\ &= ||R_1\hat{x} - u||^2 + ||v||^2 = ||v||^2. \end{aligned} \tag{15.11}$$

Equation (15.11) shows that the optimum LS solution is indeed given by the solution $\hat{x}$ of (15.10) and the minimum residual norm squared is given simply by $||v||^2$. As far as the LS problem is concerned (i.e., (15.7)–(15.11)), any of the QR decomposition techniques including G-S, M-G-S, Givens transformation, and Householder transformation are all equivalent under infinite precision computations to find the matrix R_1 and the vector v.

15.3 Systolic Least-Squares QR Processing Array

The complexity involved in the computation of the optimum residual $\hat{e}$ and the optimum LS solution vector $\hat{x}$ can become arbitrarily large as the number of samples (i.e., m) in the column vectors of A and y increases. In practice, we must limit m to some finite number greater than the number of columns n. We also periodically discard the existing set of old data and start collecting a new set of data vectors of length up to m. For each such set of data, we find the optimum LS solution weight vector $\hat{x}$ in a recursive manner. As the statistics of the signal changes over each such interval, these $\hat{x}$ vectors change adaptively with time. This observation motivates the development of a recursive implementation of QR decomposition for LS estimation problems.

15.3.1 LS Solution by QR Decomposition

Consider the use of QR decomposition to perform the optimum residual $\hat{e}$ and the associated LS solution $\hat{x}$ in (15.9) and (15.10). Denote the $m \times n$ matrix $A(m)$, the $m \times 1$ column vector $y(m)$, the $n \times 1$ solution weight column vector $x(m)$, and the $m \times 1$ residual column vector $e(m)$ as

$$A(m) = \begin{bmatrix} \alpha(1) \\ \vdots \\ \alpha(m) \end{bmatrix} = \begin{bmatrix} A(m-1) \\ \alpha(m) \end{bmatrix} \begin{matrix} (m-1) \times n \\ 1 \times n \end{matrix}, \tag{15.12}$$

$$y(m) = \begin{bmatrix} y_1 \\ \vdots \\ y_m \end{bmatrix} = \begin{bmatrix} y(m-1) \\ y_m \end{bmatrix} \begin{matrix} (m-1) \times 1 \\ 1 \times 1 \end{matrix}, \tag{15.13}$$

$$x(m) = \begin{bmatrix} x_1(m) \\ \vdots \\ x_n(m) \end{bmatrix}, \tag{15.14}$$

$$e(m) = A(m)x(m) - y(m) = \begin{bmatrix} e_1(m) \\ \vdots \\ e_m(m) \end{bmatrix}. \tag{15.15}$$

The QR decomposition of the $m \times n$ matrix $A(m)$, previously given by (15.5), now indexed by m is given by

$$Q^T(m)A(m) = \begin{bmatrix} R_1(m) \\ R_2(m) \end{bmatrix} \begin{matrix} n \times n \\ (m-n) \times n \end{matrix}, \tag{15.16}$$

where $Q^T(m)$ is an $m \times m$ orthogonal matrix, $R_1(m)$ is an $n \times n$ upper triangular matrix of the form

$$R_1(m) = \begin{bmatrix} r_{11} & r_{12} & \dots & r_{1n} \\ 0 & r_{22} & \dots & r_{2n} \\ \vdots & \vdots & & \\ 0 & 0 & \dots & r_{nn} \end{bmatrix}, \tag{15.17}$$

and $R_2(m)$ is an $(m-n) \times n$ matrix of all zeros. Since $Q^T(m)$ is an orthogonal matrix, then the transpose of $Q^T(m)$ is also an orthogonal matrix and $Q^T(m)Q(m) = Q(m)Q^T(m) = I_m$. Furthermore, denote

$$Q^T(m) = \begin{bmatrix} Q_1(m) \\ Q_2(m) \end{bmatrix} \begin{matrix} n \times m \\ (m-n) \times m \end{matrix}, \tag{15.18}$$

where $Q_1(m)$ is an $n \times m$ matrix and $Q_2(m)$ is an $(m-n) \times m$ matrix. A left multiplication of $A(m)$ in (15.16) and $y(m)$ in (15.13) by $Q^T(m)$ yields

$$Q^T(m)A(m) = \begin{bmatrix} R_1(m) \\ R_2(m) \end{bmatrix} = R_0(m), \tag{15.19}$$

$$Q^T(m)y(m) = \begin{bmatrix} Q_1(m) \\ Q_2(m) \end{bmatrix} y(m) = \begin{bmatrix} u(m) \\ v(m) \end{bmatrix} \begin{matrix} n \times 1 \\ (m-n) \times 1 \end{matrix}, \tag{15.20}$$

where

$$u(m) = Q_1(m)y(m), \tag{15.21}$$

$$v(m) = Q_2(m)y(m). \tag{15.22}$$

The square of the l_2 norm of the residual e is given by

$$\epsilon(m) = \| e(m) \|^2 = \| A(m)x(m) - y(m) \|^2 \tag{15.23}$$
$$= \| Q^T(m)(A(m)x(m) - y(m)) \|^2 \tag{15.24}$$
$$= \| \begin{bmatrix} R_1(m) \\ 0 \end{bmatrix} x(m) - \begin{bmatrix} u(m) \\ v(m) \end{bmatrix} \|^2 \tag{15.25}$$
$$= \| R_1(m)x(m) - u(m) \|^2 + \| v(m) \|^2 . \tag{15.26}$$

We note the expression (15.24) follows from (15.23) by using the orthogonal property of $Q^T(m)$, which preserves norm in the sense of

$$\| Q^T(m)e(m) \|^2 = \| e(m) \|^2 . \tag{15.27}$$

In order to minimize $\epsilon(m)$ wth respect to $x(m)$ in (15.23), we can set the first term in (15.26) to be zero. Then the optimum LS solution $\hat{x}(m)$ must satisfy

$$R_1(m)\hat{x}(m) = u(m) = \begin{bmatrix} u_1(m) \\ \vdots \\ u_n(m) \end{bmatrix}. \tag{15.28}$$

Since the columns of $A(m)$ are assumed to be linearly independent, then the $n \times n$ upper triangular matrix $R_1(m)$ is non-singular and the unique solution $\hat{x}(m)$ exists. Indeed, the upper triangular property of $R_1(m)$ shows (15.28) can be re-expressed as

$$r_{11}\hat{x}_1 + r_{12}\hat{x}_2 + \cdots + r_{1n}\hat{x}_n = u_1, \tag{15.29}$$

$$r_{22}\hat{x}_2 + \cdots + r_{2n}\hat{x}_n = u_2, \tag{15.30}$$

$$\vdots$$

$$r_{nn}\hat{x}_n = u_n, \tag{15.31}$$

where the dependencies of r_{ij}, $\hat{x}_j$, and u_j on m are omitted for simplicity of notation. The actual solution of $\hat{x} = [\hat{x}_1, \ldots, \hat{x}_n]^T$ by the back substitution method will be used explicitly in Section 15.3.3. We also note that

$$\min_x \epsilon(m) = \epsilon_{\min}(m) = \| v(m) \|^2 . \tag{15.32}$$

From (15.15), it is clear that the optimum residual $\hat{e}(m)$ is available after the optimum LS solution $\hat{x}(m)$ is available as seen from

$$\hat{e}(m) = A(m)\hat{x}(m) - y(m). \tag{15.33}$$

However, it is most interesting to note that it is not necessary to first obtain $\hat{x}(m)$ explicitly and then solve for $\hat{e}(m)$ as in (15.33). As we shall consider below, it is possible to use some properties of the orthogonal matrix $Q^T(m)$ in the QR decomposition of $A(m)$ and the vector $y(m)$, to obtain $\hat{e}(m)$ explicitly. By using (15.18), we can re-express (15.16) as

$$A(m) = Q(m)\begin{bmatrix} R_1(m) \\ 0 \end{bmatrix} = [Q_1^T(m), Q_2^T(m)]\begin{bmatrix} R_1(m) \\ 0 \end{bmatrix} = Q_1^T(m)R_1(m). \tag{15.34}$$

Then

$$\hat{e}(m) = A(m)\hat{x}(m) - y(m) \tag{15.35}$$

$$= Q_1^T(m)R_1(m)\hat{x}(m) - y(m) \tag{15.36}$$

$$= Q_1^T(m)u(m) - y(m) \tag{15.37}$$

$$= Q_1^T(m)Q_1(m)y(m) - y(m) \tag{15.38}$$

$$= [Q_1^T(m)Q_1(m) - I_m]y(m) \tag{15.39}$$

$$= -Q_2^T(m)Q_2(m)y(m), \tag{15.40}$$

since

$$I_m = Q(m)Q^T(m) = [Q_1^T(m), Q_2^T(m)]\begin{bmatrix} Q_1(m) \\ Q_2(m) \end{bmatrix} = Q_1^T(m)Q_1(m) + Q_2^T(m)Q_2(m).$$

Furthermore, from (15.20) we obtain

$$y(m) = Q(m)\begin{bmatrix} u(m) \\ v(m) \end{bmatrix} = [Q_1^T(m), Q_2^T(m)]\begin{bmatrix} u(m) \\ v(m) \end{bmatrix} \tag{15.41}$$

$$= Q_1^T(m)u(m) + Q_2^T(m)v(m) \tag{15.42}$$

$$= Q_1^T(m)Q_1(m)y(m) + Q_2^T(m)Q_2(m)y(m). \tag{15.43}$$

By using (15.43) in (15.39), then

$$\hat{e}(m) = \quad Q_2^T(m)Q_2(m)y(m) - -Q_2^T(m)v(m) \tag{15.44}$$

Thus, from either (15.39) or (15.44), we conclude $\hat{e}(m)$ can be obtained explicitly from the QR decomposition submatrix $Q_1(m)$ or $Q_2(m)$ (which depends only on $A(m)$) and $y(m)$ without needing to first explicitly evaluate $\hat{x}(m)$.

15.3.2 Recursive QR Decomposition

Now, we want to consider the recursive solution of the QR decomposition as given by (15.19). First, we assume the decomposition at step $(m-1)$ has been completed as given by

$$Q^T(m-1)A(m-1) = \begin{bmatrix} R_1(m-1) \\ 0 \end{bmatrix} \begin{matrix} n \times n \\ (m-n-1) \times n \end{matrix}. \tag{15.45}$$

Here $Q^T(m-1)$ is $(m-1) \times (m-1)$, $A(m-1)$ is $(m-1) \times n$, $R_1(m-1)$ is $n \times n$, and the 0 matrix of all zeros below $R_1(m-1)$ is a matrix of size $(m-n-1) \times n$. Next, we define a new $m \times m$ orthogonal transformation $T(m)$ as

$$T(m) = \begin{bmatrix} Q^T(m-1) & 0 \\ 0 & 1 \end{bmatrix}. \tag{15.46}$$

We note, $T(m)$ is an orthogonal matrix, since

$$T(m)T^T(m) = \begin{bmatrix} Q^T(m-1) & 0 \\ 0 & 1 \end{bmatrix}\begin{bmatrix} Q(m-1) & 0 \\ 0 & 1 \end{bmatrix} = \begin{bmatrix} I_{m-1} & 0 \\ 0 & 1 \end{bmatrix} = I_m.$$

By applying $T(m)$ on the new $m \times n$ data $A(m)$, which consist of the previously available $A(m-1)$ and the newly available row vector $\alpha^T(m)$ as represented in (15.12), we have

$$
\begin{aligned}
T(m)A(m) &= \begin{bmatrix} Q^T(m-1) & 0 \\ 0 & 1 \end{bmatrix} \begin{bmatrix} A(m-1) \\ \alpha^T(m) \end{bmatrix} = \begin{bmatrix} Q^T(m-1)A(m-1) \\ \alpha^T(m) \end{bmatrix} \\
&= \begin{bmatrix} R_1(m-1) \\ 0 \\ \alpha(m) \end{bmatrix} \begin{array}{c} n \times n \\ (m-n-1) \times n \\ 1 \times n \end{array} = R_3(m)\,(m \times n).
\end{aligned}
\tag{15.47}
$$

While $R_1(m-1)$ is an $n \times n$ upper triangular matrix, $R_3(m)$ of (15.47) does not have the same form as the desired $R_0(m)$ of (15.19).

Next, we want to transform $R_3(m)$ to $R_0(m)$ by an orthogonal transformation $G(m)$. While there are several possible approaches to perform this orthogonal transformation, we shall use the Givens transformation approach. Specifically, denote

$$
G(m) = G_n(m)G_{n-1}(m) \cdots G_1(m), \tag{15.48}
$$

where $G(m)$ as well as each $G_i(m), i = 1, \ldots, n$, are all $m \times m$ orthogonal matrices. Define

$$
G_i(m) = \begin{array}{c} \\ 1 \\ \\ i \\ \\ m \end{array}
\begin{array}{c}
\begin{array}{ccccc} 1 & & i & & m \end{array} \\
\left[\begin{array}{ccccc}
1 & & 0 & & 0 \\
 & 1 & & & \\
0 & & c_i(m) & & s_i(m) \\
 & & & 1 & \\
0 & & -s_i(m) & & c_i(m)
\end{array} \right]
\end{array}
, \; i = 1, \ldots, n, \tag{15.49}
$$

as an $m \times m$ identity matrix except the (i, i) and (m, m) elements are specified as $c_i(m)$, the (i, m) element as $s_i(m)$, and the (m, i) element as $-s_i(m)$. In (15.49), $c_i(m)$ and $s_i(m)$ are real-valued parameters depending on m and satisfy

$$
-1 \le c_i(m) \le 1, \qquad -1 \le s_i(m) \le 1, \qquad c_i(m)^2 + s_i(m)^2 = 1. \tag{15.50}
$$

In (15.50), we can consider $c_i(m) = \cos\theta_i(m)$ and $s_i(m) = \sin\theta_i(m)$, where $\theta_i(m)$ represents some rotation angle at that iteration. From (15.48) and (15.49), $G(m)$ can be re-expressed as

$$
G(m) = \begin{bmatrix} k(m) & 0 & d(m) \\ 0 & I_{m-n-1} & 0 \\ h^T(m) & 0 & \gamma(m) \end{bmatrix}, \tag{15.51}
$$

where $k(m)$ is $n \times n$, $d(m)$ and $h(m)$ are $n \times 1$, and $\gamma(m)$ is 1×1. In general $k(m), d(m)$, and $h(m)$ are quite involved functions of $c_i(m)$ and $s_i(m)$, but $\gamma(m)$ is given simply as

$$
\gamma(m) = \prod_{i=1}^{n} c_i(m). \tag{15.52}
$$

We shall make use of the form of (15.52) in the evaluation of the optimum residual in Section 15.3.3.

Now, apply $G(m)$ to (15.47) to obtain

$$
G(m)T(m)A(m) = G(m)R_3(m) = R_0(m). \tag{15.53}
$$

In order to follow the n orthogonal transformation operations of $G(m)$, first consider

$$
G_1(m)R_3(m) = \left[\begin{array}{cccccc} c_1(m) & 0 & \cdots & 0 & s_1(m) \\ 0 & 1 & \cdots & 0 & 0 \\ 0 & 0 & \ddots & 0 & 0 \\ 0 & 0 & \cdots & 1 & 0 \\ -s_1(m) & 0 & \cdots & 0 & c_1(m) \end{array}\right]
\times \left[\begin{array}{ccccc} x & x & \cdots & x & x \\ 0 & x & \cdots & x & x \\ & & \ddots & x & x \\ 0 & 0 & \cdots & 0 & x \\ 0 & 0 & 0 & 0 & 0 \\ x & x & x & x & x \end{array}\right]
\begin{array}{c} \uparrow \\ | \\ n \times n \\ \downarrow \\ (m-n-1) \times 1 \\ 1 \times n \end{array}
= \left[\begin{array}{ccccc} x & x & \cdots & x & x \\ 0 & x & \cdots & x & x \\ & & \ddots & x & x \\ 0 & 0 & \cdots & 0 & x \\ 0 & 0 & 0 & 0 & 0 \\ 0 & x & x & x & x \end{array}\right]
\begin{array}{c} \uparrow \\ | \\ n \times n \\ \downarrow \\ (m-n-1) \times 1 \\ 1 \times n \end{array} .
\tag{15.54}
$$

In (15.54), an x denotes some non-zero-valued element. The purpose of $G_1(m)$ operating on $R_3(m)$ is to obtain a zero at the $(m, 1)$ position without changing the $(m-n-2)\times n$ submatrix from the second to the $(m-1)$st rows of the r.h.s. of (15.54). In general, at the ith iteration, we have

$$
G_i(m) \left[\begin{array}{ccccc} x & x & \cdots & x & x \\ 0 & x & \cdots & x & x \\ & & \ddots & x & x \\ 0 & 0 & \cdots & 0 & x \\ 0 & 0 & 0 & 0 & 0 \\ 0 & \cdots & 0 & x & x \end{array}\right]_{\quad i-1 \quad i \quad \cdots}
= \left[\begin{array}{ccccc} x & x & \cdots & x & x \\ 0 & x & \cdots & x & x \\ & & \ddots & x & x \\ 0 & 0 & \cdots & 0 & x \\ 0 & 0 & 0 & 0 & 0 \\ 0 & \cdots & 0 & 0 & x \end{array}\right]_{\quad i-1 \quad i \quad \cdots}
\begin{array}{c} \uparrow \\ | \\ n \times n \\ \downarrow \\ (m-n-1) \times 1. \\ 1 \times n \end{array}
\tag{15.55}
$$

The zeroing operation in (15.55) can be explained by noting that the Givens matrix $G_i(m)$ operates as an $(m-2) \times (m-2)$ identity matrix on all the rows on the right of it except the ith and the mth rows. The crucial operations at the ith iteration on these two rows in (15.55) can be represented as

$$\begin{bmatrix} c & s \\ -s & c \end{bmatrix} \begin{bmatrix} 0 & \dots & 0 & r_i & r_{i+1} & \dots & r_n \\ 0 & \dots & 0 & a_i & a_{i+1} & \dots & a_n \end{bmatrix}$$
$$\hphantom{=}\quad\quad\quad\quad\quad\quad\quad\quad\quad\quad i$$

$$= \begin{bmatrix} 0 & \dots & 0 & r_i' & r_{i+1}' & \dots & r_n' \\ 0 & \dots & 0 & 0 & a_{i+1}' & \dots & a_n' \end{bmatrix}. \tag{15.56}$$
$$\quad\quad\quad\quad\quad i$$

For simplicity of notation, in (15.56) as well as in all the following equations in this section, we suppress the dependencies of i and m on c and s. Specifically, we want to force $a_i' = 0$ as given by

$$0 = a_i' = -sr_i + ca_i = \frac{-a_i r_i}{\sqrt{a_i^2 + r_i^2}} + \frac{r_i a_i}{\sqrt{a_i^2 + r_i^2}}, \tag{15.57}$$

in conjunction with $c^2 + s^2 = 1$ of (15.50), and

$$c^2 = \frac{r_i^2}{(a_i^2 + r_i^2)}, \tag{15.58}$$

$$s^2 = \frac{a_i^2}{(a_i^2 + r_i^2)}. \tag{15.59}$$

From (15.56), we also have

$$r_i' = cr_i + sa_i = \frac{r_i r_i}{\sqrt{a_i^2 + r_i^2}} + \frac{a_i a_i}{\sqrt{a_i^2 + r_i^2}}. \tag{15.60}$$

By using (15.58) and (15.59) in (15.60), we obtain

$$r_i' = \sqrt{(a_i^2 + r_i^2)}, \tag{15.61}$$

$$c = \frac{r_i}{r_i'}, \tag{15.62}$$

$$s = \frac{a_i}{r_i'}. \tag{15.63}$$

Indeed, the dependencies of $\{a_j', r_j'\}$ on $\{r_j, a_j\}$ satisfy

$$a_j' = -sr_j + ca_j,\ j = i, i+1, \dots, n \tag{15.64}$$

$$r_j' = cr_j + sa_j,\ j = i, i+1, \dots, n. \tag{15.65}$$

From the individual results of $G_1(m), G_2(m), \ldots, G_n(m)$, the overall results of (15.51) can be expressed as

$$Q^T(m)A(m) = G(m)R_3(m) = \begin{bmatrix} R_1(m) \\ 0 \end{bmatrix} = R_0(m), \tag{15.66}$$

with

$$Q^T(m) = G(m)T(m). \tag{15.67}$$

In (15.67), $Q^T(m)$ is indeed an orthogonal transformation due to the cascading of orthogonal transformations of the form of (15.49). However, the relevant operations involve only two rows at each iteration and are of the form given in (15.56).

15.3.3 Recursive Solution of the Optimum Residual

Now, consider the recursive solution of the last component $\hat{e}_m(m)$ of the optimum residual. From (15.44), we have

$$\hat{e}(m) = \begin{bmatrix} \hat{e}_1(m) \\ \vdots \\ \hat{e}_m(m) \end{bmatrix} = -Q_2^T(m)v(m) = -Q_2^T(m)\begin{bmatrix} v_1(m) \\ \vdots \\ v_m(m) \end{bmatrix}. \tag{15.68}$$

From (15.18), (15.46), (15.51), and (15.67), recursive time updating of $Q(m)$ can be expressed as

$$Q^T(m) = \begin{bmatrix} Q_1(m) \\ Q_2(m) \end{bmatrix} = G(m)T(m) \tag{15.69}$$

$$= \begin{bmatrix} k(m) & 0 & d(m) \\ 0 & I_{m-n-1} & 0 \\ h^T(m) & 0 & \gamma(m) \end{bmatrix} \begin{bmatrix} Q_1(m-1) & 0 \\ Q_2(m-1) & 0 \\ 0 & 1 \end{bmatrix} \tag{15.70}$$

$$= \begin{bmatrix} k(m)Q_1(m) & d(m) \\ \cdots & \\ Q_2(m-1) & 0 \\ h^T(m)Q_1(m-1) & \gamma(m) \end{bmatrix}. \tag{15.71}$$

Then

$$Q_2(m) = \begin{bmatrix} Q_2(m-1) & 0 \\ h^T(m)Q_1(m-1) & \gamma(m) \end{bmatrix}. \tag{15.72}$$

Thus, by using (15.72) in (15.68), we have

$$\begin{bmatrix} \hat{e}_1(m) \\ \vdots \\ \hat{e}_m(m) \end{bmatrix} = -\begin{bmatrix} Q_2^T(m-1) & Q_1^T(m-1)h(m) \\ 0 & \gamma(m) \end{bmatrix} \begin{bmatrix} v_1(m) \\ \vdots \\ v_m(m) \end{bmatrix}. \tag{15.73}$$

Then the last component of the optimum residual given by

$$\hat{e}_m(m) = -\gamma(m)v_m(m), \tag{15.74}$$

where $\gamma(m)$ is given by (15.52) consisting of all the products of the cosine parameters $c_i(m)$ in the Givens QR transformation and $v_m(m)$ is just the last component of $v(m)$, which is the result of $Q^T(m)$ operating on $y(m)$ in (15.20).

15.3.4 Recursive LS Solution by Back Substitution Method

As considered in Section 15.3.1, the LS solution $\hat{x}$ satisfies the triangular system of equations (15.31). After the QR operation on the extended matrix $[A(m), y(m)]$, all the $r_{ij}, j \geq i = 1, \ldots, n$, and $u_i, i = 1, \ldots, n$, in (15.31) are available. Thus, $\{\hat{x}_1, \ldots, \hat{x}_n\}$ can be obtained by using the back substitution method of

$$\hat{x}_i = \left(u_i - \sum_{j=i+1}^{n} \frac{r_{ij}\hat{x}_j}{r_{ij}} \right), \qquad i = n, n-1, \ldots, 1. \tag{15.75}$$

In (15.75), the summation operates only for those j indices less than or equal to n. It is interesting to note that the solution of $\hat{x}$ in (15.28) can be obtained as the multiplication of the inverse of the upper triangular matrix $R(m)$ with $u(m)$ (i.e., $\hat{x} = R(m)^{-1}u(m)$). Indeed, by writing out the explicit expression for the inverse of $R(m)$, we obtain the identical expression of (15.75) for the solution of $\hat{x}$. Thus, the back substitution method is identical to the upper triangular matrix inversion method.

Specifically, if $n = 1$, then (15.75) yields

$$\hat{x}_1 = \frac{u_1}{r_{11}}. \tag{15.76}$$

If $n = 2$, then (15.75) yields

$$\hat{x}_2 = \frac{u_2}{r_{22}}, \qquad \hat{x}_1 = \frac{u_1 - r_{12}\hat{x}_2}{r_{11}} = \frac{u_1}{r_{11}} - \frac{u_2 r_{12}}{r_{11} r_{22}}. \tag{15.77}$$

If $n = 3$, then

$$\hat{x}_3 = \frac{u_3}{r_{33}}, \tag{15.78}$$

$$\hat{x}_2 = \frac{u_2 - r_{23}\hat{x}_3}{r_{22}} = \frac{u_2}{r_{22}} - \frac{r_{23}u_3}{r_{22}r_{33}}, \tag{15.79}$$

$$\hat{x}_1 = \frac{u_1 - r_{12}\hat{x}_2 - r_{13}\hat{x}_3}{r_{11}} \tag{15.80}$$

$$= \frac{u_1}{r_{11}} - \frac{r_{12}u_2}{r_{11}r_{22}} + u_3 \left[\frac{-r_{13}}{r_{11}r_{33}} + \frac{r_{12}r_{23}}{r_{11}r_{22}r_{33}} \right]. \tag{15.81}$$

Simple Processing Cells for Givens QR Transformation

The recursive QR decomposition considered in Section 15.3.2 can be implemented in a two-dimensional processing systolic array based on the usage of four kinds of processing cells: boundary cells (Fig. 15.1(a)) for the generation of the sine and cosine parameters, s and c, needed in the Givens rotations in (15.54); internal cells (Fig. 15.1(b)) for the proper updating of the transformations in (15.62) and (15.63); a single output

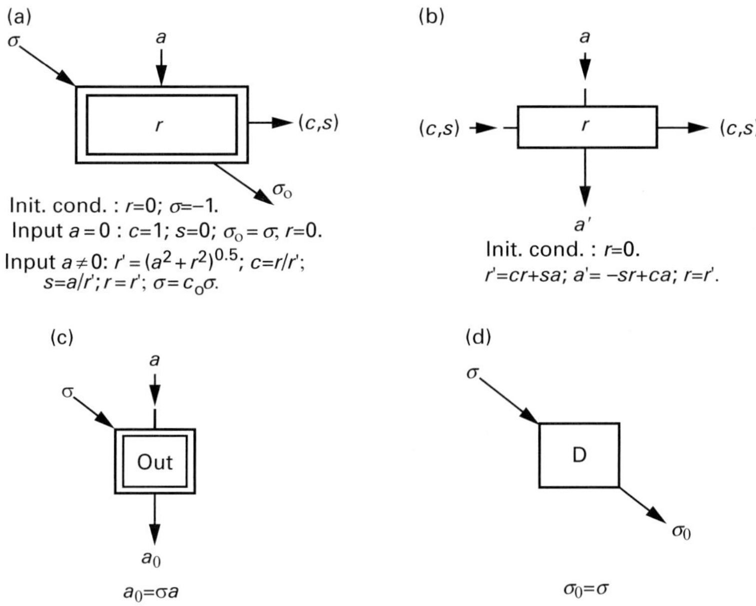

Figure 15.1 (a) Boundary cell; (b) Internal cell; (c) Output cell; (d) Delay cell

cell (Fig. 15.1(c)) needed in the generation of the last component of the residual $e_m(m)$ defined in (15.15); and also delay cells (Fig. 15.1(d)) for a unit time delay.

15.3.5 Systolic Array for QR Decomposition and LS Solution

A triangular systolic array capable of performing the recursive QR decomposition of Section 15.3.2 and the recursive LS solution of Section 15.3.4 by utilizing the basic processing cells is given in Fig. 15.2. In particular, the associated LS problem uses an augmented matrix $[A, y]$ consisting of n column data vectors A and the column vector y. The number of processing cells in the triangular array of Fig. 15.2, consists of:

- n boundary cells,
- $n(n+1)/2$ internal cells,
- n delay cells, and
- 1 output cell.

The input to the array in Fig. 15.2 uses the augmented matrix

$$[A, y] = \begin{bmatrix} a_{11} & a_{12} & \dots & a_{1n} & y_1 \\ a_{21} & a_{22} & \dots & a_{2n} & y_2 \\ & & \vdots & & \\ a_{m1} & a_{m2} & \dots & a_{mn} & y_m \end{bmatrix}. \quad (15.82)$$

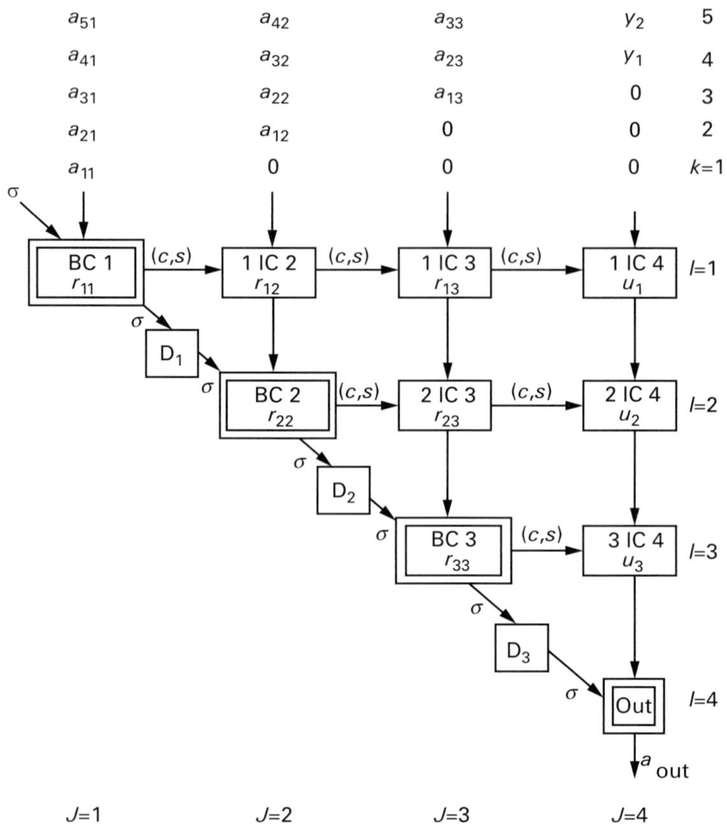

Figure 15.2 Systolic array implementation of $n=3$

skewed in a manner such that each successive column from left to right is delayed by a unit time as given by

$$
\begin{bmatrix}
a_{11} & 0 & & \cdots & 0 & 0 \\
a_{21} & a_{12} & 0 & \cdots & 0 & 0 \\
a_{31} & a_{22} & a_{13} & \cdots & 0 & 0 \\
 & & & \vdots & & \\
a_{n1} & a_{(n-1)2} & & & a_{1n} & 0 \\
a_{(n+1)1} & a_{n2} & & & a_{2n} & y_1 \\
 & & & \vdots & & \\
a_{m1} & a_{(m-1)2} & & & a_{(m-n+1)n} & y_{m-n} \\
 & & & \vdots & & \\
0 & & & & a_{mn} & y_{m-1} \\
0 & & & & 0 & y_m
\end{bmatrix}
\begin{matrix}
k=1 \\ 2 \\ 3 \\ \vdots \\ n \\ n+1 \\ \vdots \\ m \\ \vdots \\ m+n-1 \\ m+n
\end{matrix}
\qquad (15.83)
$$

From (15.83), we see that at time k, input data consist of the kth row of the matrix, and move down with increasing time. However, in Fig. 15.2, the relevant rows of data are shown as moving up with increasing k.

Let us now consider some of the detailed iterative operations of the QR decomposition for the augmented matrix $[A, y]$ for the systolic array in Fig. 15.2. At time $k = 1$, a_{11} enters BC 1 and results in $c = 0$, $s = 1$, and $r_{11} = a_{11}$. All other cells are inactive. At $k = 2$, a_{21} enters BC 1 with results $c = a_{11}/\sqrt{(a_{11}^2 + a_{21}^2)}$, $s = a_{21}/\sqrt{(a_{11}^2 + a_{21}^2)}$, $r_{11} = \sqrt{a_{11}^2 + a_{21}^2}$. This r_{11} corresponds to that of r_i' in (15.61), while the above c and s correspond to the c and s in (15.62) and (15.63). Indeed, the new a_i' of (15.57) is zero and does not need to be saved in the array. Still at $k = 2$, a_{12} enters 1 IC 2 and outputs $a' = 0$ and $r_{12} = a_{12}$. At $k = 3$, a_{13} enters BC 1, and the Givens rotation operation of (15.56) continues where the new r_i is given by the previously processed r_i' and a_i is now given by a_{13}. Meanwhile, a_{22} enters at 1 IC 2; it outputs $a' = -a_{21}a_{12}/\sqrt{(a_{11}^2 + a_{21}^2)} + a_{22}a_{21}/\sqrt{(a_{11}^2 + a_{21}^2)}$, which corresponds to that of a_{i+1}' of (15.56), and $r_{12} = a_{11}a_{12}/\sqrt{(a_{11}^2 + a_{21}^2)} + a_{21}a_{22}/\sqrt{(a_{11}^2 + a_{21}^2)}$, which corresponds to that of r_{i+1}' of (15.56). In general, the top (i.e., $I = 1$) row of the processing cells performs Givens rotation by using the 1st row to operate on the 2nd, 3rd, ..., mth rows (each row with $n + 1$ elements) such that the $\{21, 31, \ldots, m1\}$ locations in (15.82) are all zero. The next row ($I = 2$) of cells uses the 2nd row to operate on the 3rd, ..., mth rows (each row with n elements) such that locations at $\{32, 42, \ldots, m2\}$ are zero. Finally, at row $I = n$, by using the nth row to operate on the $(n + 1)$st, ..., mth rows, elements at locations $\{(n+1)n, (n+2)n, \ldots, mn\}$ are zero. We also note that the desired cosine values in $\gamma(m)$ of (15.52) are being accumulated by the C along the diagonal of the array. Delay cells $\{D_1, D_2, \ldots, D_n\}$ are used to provide the proper timing along the diagonal.

The cell BC 1 (at $I = J = 1$) terminates in the QR operation at time $k = m$, while the cell at $I = 1$ and $J = 2$ terminates at $k = m + 1$. In general, the processing cell at location (I, J) terminates at $k = I + J + m - 2$. In particular, the last operation in the QR decomposition on the augmented matrix is performed by the cell at $I = n$ and $J = n + 1$ at time $k = 2n + m - 1$. Then the last component of the optimum residual $e_m(m)$ exits the output cell at time $k = 2n + m$.

After the completion of the QR decomposition obtains the upper triangular system of equations in the form of (15.31), we can "freeze" the r_{IJ} values in the array to solve for the optimum LS solution x by the back substitution method. Specifically, we can append $[I_n, 0]$, where I_n is an $n \times n$ identity matrix and 0 is an $n \times 1$ vector of all zeros, to the bottom of the augmented matrix $[A, y]$ in (15.82). Of course, this matrix is skewed as in (15.83) when used as input to the array. In particular, immediately after the completion of the QR operation at BC 1, we can input the unit value at time $k = m + 1$. This is stage 1 of the back substitution method as shown in the upper left-hand side of Fig. 15.3. Due to skewing, a unit value appears at the $I = 1$ and $J = 2$ cells at stage 3. Finally, at stage $(2n - 1)$ which is time $k = m + 2n - 1$, the last unit value appears at the $I = n$ and $J = 1$ cell. For our example of $n = 3$, this happens at stage 5 at the upper left-hand side of Fig. 15.5. As shown in the upper left-hand side of Fig. 15.6, the desired LS solution

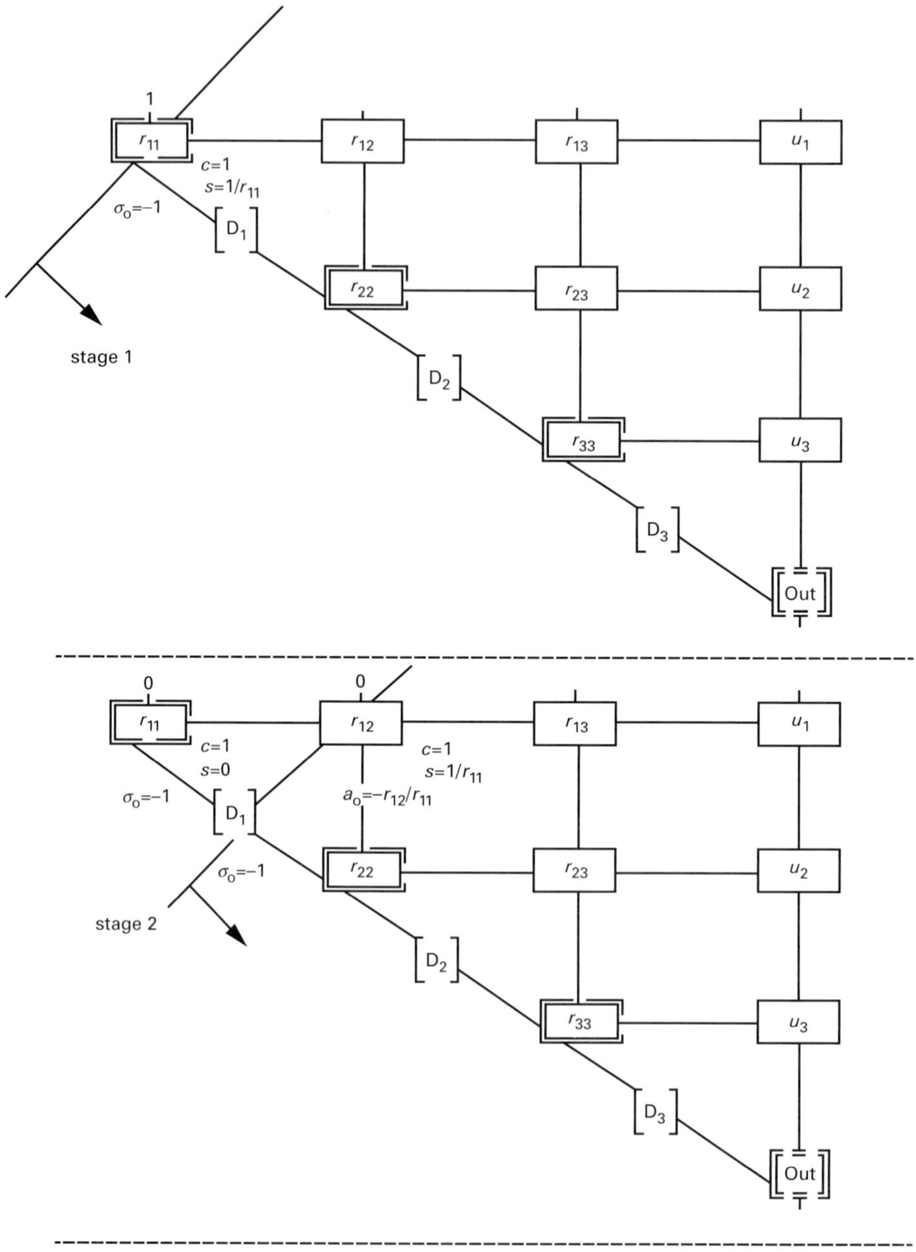

Figure 15.3 Propagation of back substitution wavefront for an upper triangularized systolic array for $n = 3$ at stages 1–2

$\hat{x}_1$ appears at stage $(2n + 1)$ (i.e., stage 7 for $n = 3$), which is time $k = 2n + m + 1$, while the last solution $\hat{x}_n$ appears at stage $3n$, (i.e., stage 9 for $n = 3$ in Fig. 15.7), which is time $k = 3n + m$. We note the values of $\{\hat{x}_1, \hat{x}_2, \hat{x}_3\}$ in Fig. 15.6 and Fig. 15.7) are identical to those obtained from those given in (15.76)–(15.81) and are indeed the back substitution method solution of the LS problem.

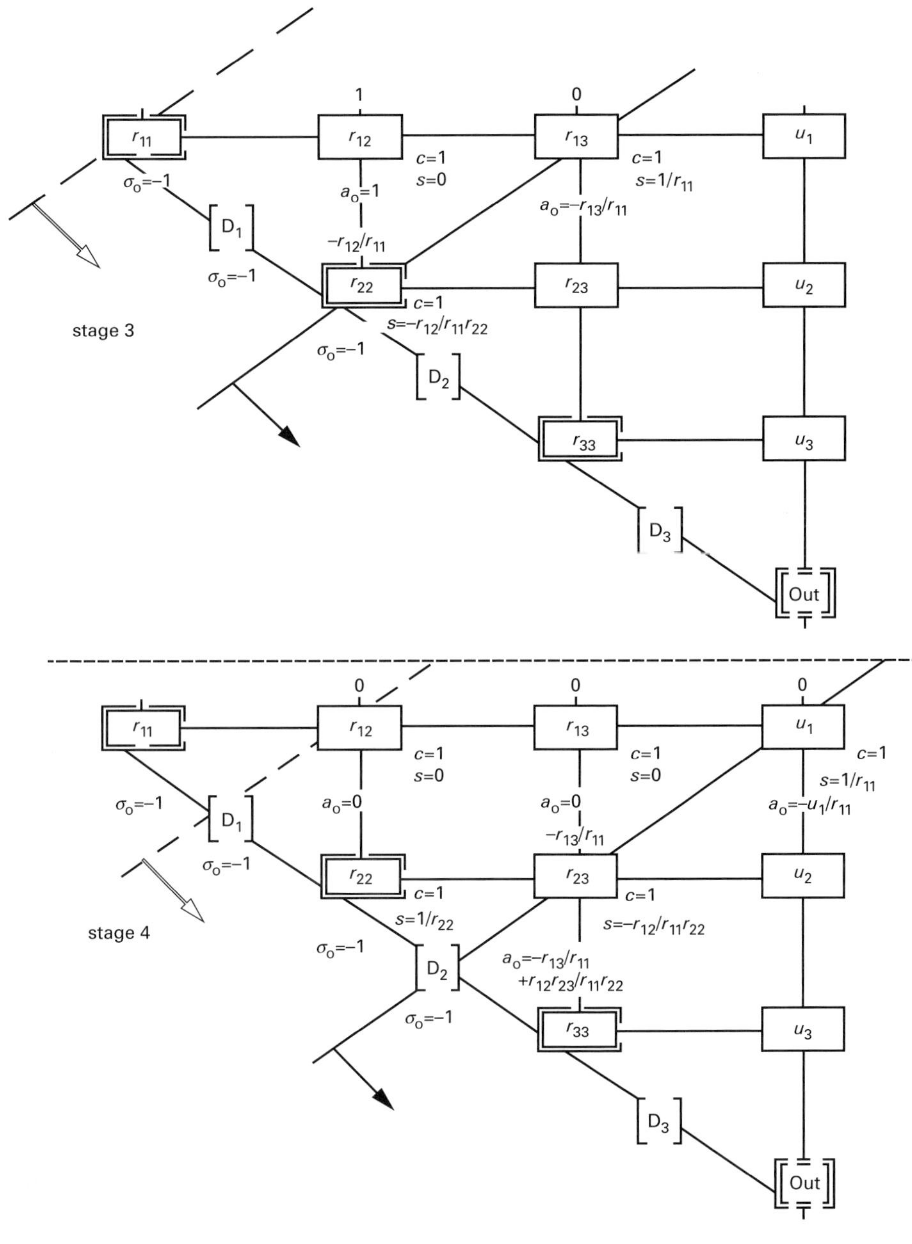

Figure 15.4 Propagation of back substitution wavefront for an upper triangularized systolic array for $n = 3$ at stages 3–4

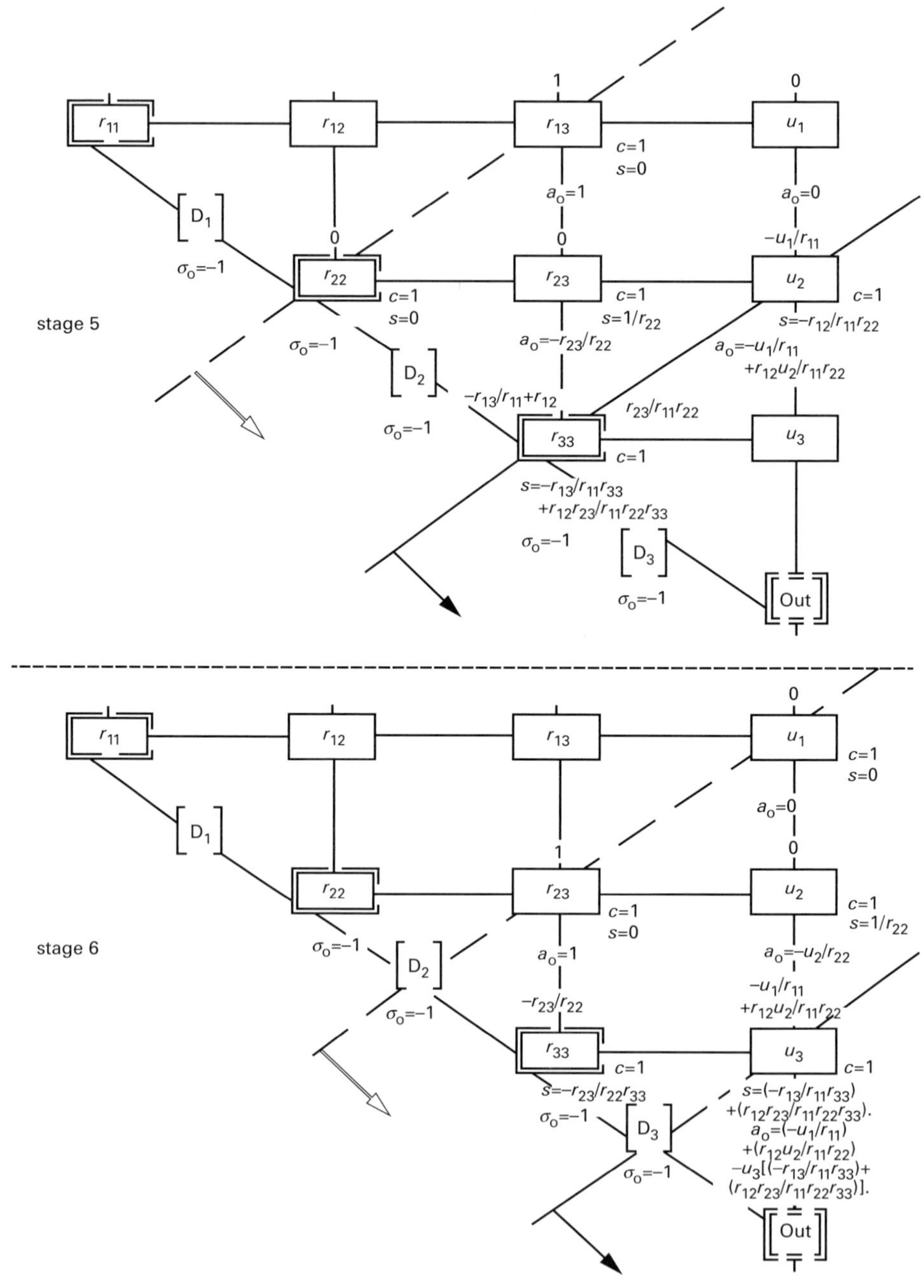

Figure 15.5 Propagation of back substitution wavefront for an upper triangularized systolic array for $n = 3$ at stages 5–6

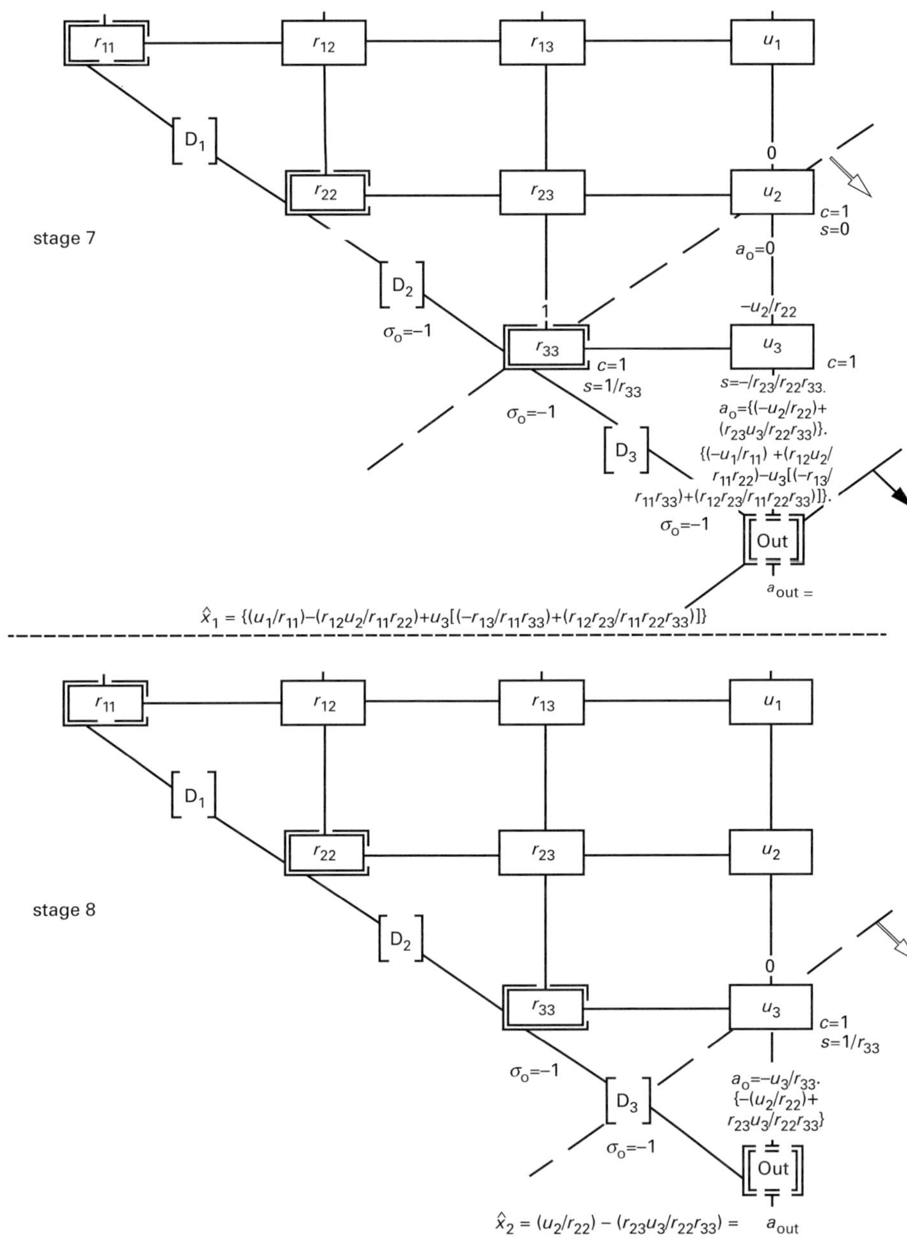

Figure 15.6 Propagation of back substitution wavefront for an upper triangularized systolic array for $n = 3$ at stages 7–8

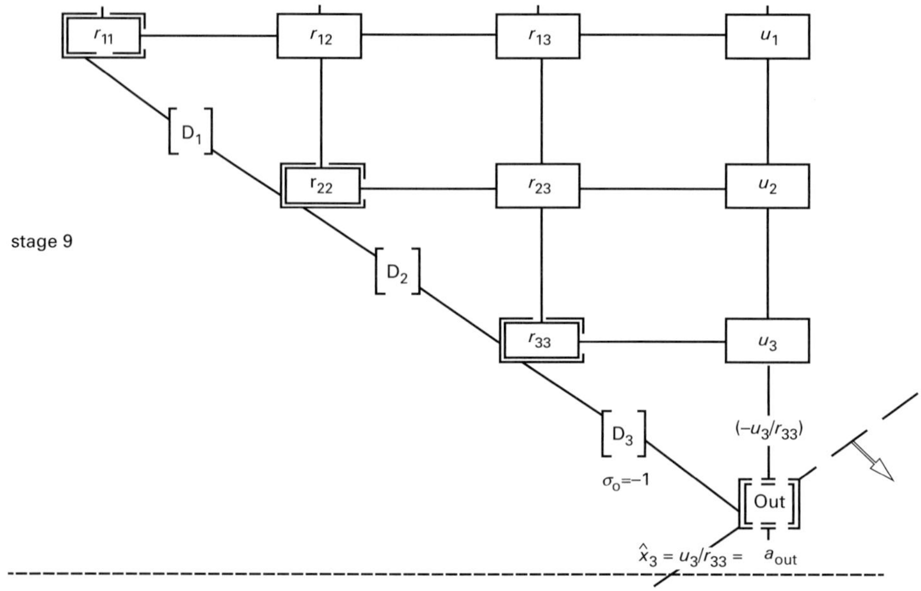

Figure 15.7 Propagation of back substitution wavefront for an upper triangularized systolic array for $n = 3$ at stage 9

Next, we consider an explicit $m = 4$ and $n = 3$ example to illustrate the detailed workings of the array. Let

$$A = \begin{bmatrix} 9 & 11 & 5 \\ 15 & 27 & 28 \\ 2 & 16 & 10 \\ 30 & 8 & 3 \end{bmatrix} \quad \text{and} \quad y = [10, 1, 6, 4]^T. \tag{15.84}$$

The optimum LS solution $\hat{x} = [-0.0468599, 1.19702, -1.10468]^T$ and the optimum residual $\hat{e} = [-2.77787, -0.31423, 2.01188, 0.856352]^T$. For each time instant indexed by k, the internal matrix-valued parameters of the cosines c, sines s, register values R as well as the input values AS of all the cells are given. The array computations in a serial computer are performed in the order of $I = 4, J = 4; I = 3, J = 4; I = 3, J = 3; I = 2, J = 4; I = 2, J = 3; I = 2, J = 2; I = 1, J = 4; I = 1, J = 3; I = 1, J = 2; I = 1, J = 1$. Of course, in a true array computer (i.e., systolic array or equivalent systems), all these computations are performed at the same time index k. As expected, the last component of $e_4(4)$ appears at the output at time $k = 10$ while $\hat{x}_1$ appears at $k = 11$, $\hat{x}_2$ at $k = 12$, and $\hat{x}_3$ at $k = 13$. The output of the systolic array for these 13 time instants are given in Fig. 15.8, Fig. 15.9, and Fig. 15.10.

```
>> run
Conventional least-squares block solution approach

A =
     9    11     5
    15    27    28
     2    16    10
    30     8     3

Y =
    10
     1
     6
     4

X(opt)      is the conventional LS block solution of AX ˜ Y given by X(opt) = A\Y
   -0.0469
    1.1970
   -1.1047

E = A*X(opt) -Y is the Residual of the LS solution =
   -2.7779
   -0.3142
    2.0119
    0.8564

----------------.
----------------.

 LS-QR Decomposition solution based on triangular systolic array approach

----------
k =
1

ans =

     0

----------
k =
2

ans =

     0

----------
k =
3

ans =

     0

----------
k =
4

ans =

     0
```

Figure 15.8 Systolic array output for $k = 1$ to 4

```
----------
k =
5

ans =

     0

----------
k =
6

ans =

     0

----------
k =
7

ans =

     0

----------
k =
8

ans =

     0

----------
k =
9

ans =

     0

----------
k =
10

ans =

    0.8564

----------
k =
11

ans =

   -0.0469

----------
k =
12

ans =

    1.1970

----------
```

Figure 15.9 Systolic array output for $k = 5$ to 12

```
k =
13

ans =

    -1.1047

>> quit

  2235 flop(s).
```

Figure 15.10 Systolic array output for $k = 13$

15.4 Conclusion

Chapter 15 used the Givens transformation QR decomposition to address the systolic array approach to solve the recursive least-squares linear system of equations problem. The practical use of a systolic array for adaptive nulling in radar using this problem was treated by [3].

15.5 References

Basic aspects of QRD in Section 15.1 can be found in [1]. The use of Givens transformation for the systolic least-squares problem was first given by [2]. Other forms of systolic QRD algorithms and architectures have also been used, including in [4], [5], and [6].

[1] G.H. Golub and C.F. Van Loan, *Matrix Computations*, 3rd edn., J. Hopkins University Press, 2016.

[2] J.G. McWhirter, "Recursive Least-Squares Minimization Using a Systolic Array," *Proc. SPIE*, 1983, pp. 105–112.

[3] C.M. Rader, "VLSI Systolic Arrays for Adaptive Nulling," *IEEE Signal Processing Magazine*, 1996, pp. 29–49.

[4] F. Ling and J.G. Proakis, "A Generalized Multichannel Least-Squares Lattice Algorithm Based on Sequential Processing Stages," *IEEE Trans. Acoustics Speech and Signal Processing*, 1984, pp. 381–389.

[5] K.J.R. Liu, S.F. Hsieh, and K. Yao, "Recursive LS Filtering Using Block Householder Transformations," *IEEE Trans. Acoustics, Speech and Signal Processing*, 1990, pp.1631–1634.

[6] S.Z. Kalson and K. Yao, "A Class of Least-Squares Filtering and Identification Algorithms with Systolic Array Architectures," *IEEE Trans. on Information Theory*, 1991, pp. 43–52.

15.6 Exercises

1. Read and understand the first proposed use of a triangular systolic array to solve the recursive least-squares solution of an over-determined linear system of

equations in J. McWhirter, "Recursive Least-Squares Minimization Using a Systolic Array," *Proc. SPIE,* 1983.

2. While a triangular systolic array offers a high degree of parallelism with a computation time linear in n (the number of columns in the array), the hardware complexity for large n may be objectionable. One possible solution is to use a one-dimensional linear array by folding and projecting the triangular array. Specifically, consider a systematic approach on the operation of starting with a 5×5 systolic triangular array and ending up with a 3×1 linear systolic array. Show a block diagram of the folded linear array.
3. In some applications, the LS problem involves the covariance matrix of the data rather than the data matrix. In other words, the problem is to solve $Rw = D$, where $R = X^H X$ is now the covariance matrix. Show the block diagram of a conventional approach to this problem, which may square the condition number of matrix X. Are there alternative approaches to the solution of this problem without first forming the R matrix?
4. J. McWhirter has shown that the optimum residual in the least-squares solution of the over-determined linear system of equations can be obtained efficiently using a triangular systolic array. Describe the inputs to this array in obtaining the least-squares solution after the completion of the Givens transformation in the QR operation of this array.
5. Relate the Hermit normal form and the Smith normal form over integral matrices in *Integral Matrices,* Academic Press, 1972, to similar properties on the real-line useful in signal and systolic array processing.

16 Systolic Algorithms and Architectures for Communication and Radar Systems

16.1 Introduction

In this chapter, we will deal with the systolic Kalman filter, the systolic MIMO receiver, and the systolic algorithm for SVD updating.

16.2 Kalman Filter

Kalman filtering (KF) [1] is a natural extension of the classical Wiener filtering. It has had a profound influence in the theoretical and practical aspects of estimation and filtering. KF is almost universally used for tracking and estimation and guidance of aircraft, satellite, GPS, and missiles, as well as in system estimation and identification problems. KF is not one unique method but is a generic name for a class of optimum linear state estimators based on the unbiased minimum-variance estimation criterion for multiple inputs and multiple outputs with possibly non-stationary and time-varying system characteristics. In the following discussion, we first introduce the basic concepts of a KF (in the covariance updating form), then consider the algorithm of the square-root information filter form of a KF that can have a systolic implementation.

The KF model consists of a discrete-time linear dynamical system equation and a measurement equation. A linear discrete-time dynamical system with an $N \times 1$ dimensional state vector $x(K+1)$, at time $K+1$ is given by

$$x(K+1) = F(K)x(K) + w(K), \tag{16.1}$$

where $x(K)$ is the state vector at time K, $F(K)$ is an $N \times N$ system coefficient matrix, and $w(K)$ is an $N \times 1$ zero-mean system noise vector with a covariance matrix denoted by $R_W(K)$. The input to the KF is the $M \times 1$ measurement (also called observation) vector $y(K)$ modeled by

$$y(K) = C(K)x(K) + v(K), \tag{16.2}$$

where $v(K)$ is an $M \times 1$ zero-mean measurement noise vector with an $M \times M$ positive-definite covariance matrix $R_V(K)$, and $C(K)$ is an $M \times M$ observation model matrix that maps the true state space into the observed space. The requirement of the positive-definite condition on $R_V(K)$ is to guarantee the Cholesky (square-root) factorization of $R_V(K)$ needed in certain KF algorithms. In general, $M \leq N$. It is also assumed that

$w(K)$ is uncorrelated to $v(K)$ and both the system noise and the measurement noise vectors are white sequences as a function of time.

The KF provides a recursive linear estimation of $x(K)$ under the minimum variance criterion based on the observed data $y(K)$. Let $\hat{x}(K)$ denote the optimum filter state estimate of $x(K)$ given measurements up to and including $y(K)$, while $x_+(K)$ denotes the optimum predicted state estimate of $x(K)$ given measurements up to and including $y(K-1)$. The $N \times N$ optimum estimation error covariance matrix is given by

$$P(K) = E\{(x(K) - \hat{x}(K))(x(K) - \hat{x}(K))\}, \tag{16.3}$$

while the minimum estimation error covariance is given by

$$J(K) = \mathrm{tr}\{P(K)\} = E\{(x(K) - \hat{x}(K))(x(K) - \hat{x}(K))\}. \tag{16.4}$$

The $N \times N$ optimum prediction error covariance matrix is given by

$$P_+(K) = E\{(x(K) - x_+(K))(x(K) - x_+(K))\}. \tag{16.5}$$

The original KF recursively updates the optimum error covariance and the optimum state estimate vector by using two sets of update equations. Thus, the KF is often called the covariance KF. The time update equations for $K = 1, 2, \ldots,$ are given by

$$x_+(K) = F(K-1)\hat{x}(K-1) \tag{16.6}$$

and

$$P_+(K) = F(K-1)P(K-1)F(K-1) + R_W(K-1). \tag{16.7}$$

The $N \times N$ Kalman gain matrix $K_0(K)$ is given by

$$K_0(K) = P_+(K)C(K)[C(K)P_+(K)C(K) + R_V(K)]^{-1}. \tag{16.8}$$

The measurement update equations are given by

$$\hat{x}(K) = x_+(K) + K_0(K)(y(K) - C(K)x_+(K)) \tag{16.9}$$

and

$$P(K) = P_+(K) - K_0(K)C(K)P_+(K). \tag{16.10}$$

The first equation shows the update relationship of $\hat{x}(K)$ to the predicted state estimate $x_+(K)$ and when the latest observed value $y(K)$ is available. The second equation shows the update relationship between $P(K)$ and $P_+(K)$. Both equations depend on the Kalman gain $K_0(K)$ which depends on the measurement coefficient matrix $C(K)$ and the statistical property of the measurement noise, covariance matrix $R_V(K)$. Furthermore, $K_0(K)$ involves an $M \times mM$ matrix inversion.

16.2.1 Systolic Kalman Filter

While the original KF was formulated as an error covariance matrix updating problem, various alternative formulations as the square-root covariance matrix, information matrix, and square-root information matrix are also possible [2]. In particular, Paige and Saunders [3] proposed a modified square-root information filter (SRIF) for Kalman filtering which transforms the problem into a least-squares computation at each filter iteration or recursion. Triangular systolic array implementation of the measurement updating equations based on the square-root covariance approach was considered by Jover and Kailath [4]. Chen and Yao [5], and [6] presented a systolic array for Kalman filtering based jointly on time and measurement updating equations, utilizing the SRIF approach which solved the LS computation by the QRD method. This LS formulation was made possible by the use of a whitening operation on the system equations. Denote the Cholesky factorization of the inverses of the noise covariance matrices as $R_W(K+1)^{-1} = W(K+1)$ and $R_V(K)^{-1} = V(K)V(K)$, where the matrices $W(K+1)$ and $V(K)$ are upper triangular. Multiplying the dynamical system equation and the observation equation by the corresponding whitening operators, that is, $W(K+1)$ and $V(K)$, results in the noise processes being white, for which the minimum variance solution can be obtained by an LS computation. At each Kalman filter recursion, the QR triangularization of a $(2N+M) \times (2N+1)$ matrix is given by

$$T \begin{bmatrix} R(K) & 0 & b(K) \\ \tilde{F}(K) & W(K+1) & 0 \\ \tilde{C}(K) & & \tilde{y}(K) \end{bmatrix} = \begin{bmatrix} R_{K,K} & Z(K+1) & a(K) \\ 0 & R(K+1) & b(K+1) \\ 0 & 0 & r_K \end{bmatrix},$$

where T is the orthogonal transformation matrix for the triangularization. $R_{K,K}$ and $R(K+1)$ are $N \times N$ upper triangular matrices. The matrices $\tilde{F}(K)$ and $\tilde{C}(K)$, that is, the coefficient matrices after whitening, are defined by $\tilde{F}(K) = -W(K+1)F(K)$ and $\tilde{C}(K) = V(K)C(K)$. Similarly, the observation vector after whitening is $\tilde{y}(K) = V(K)y(K)$. The optimal predictor of the state vector, $x(K+1)$, given the measurements $y(1), \ldots, y(K)$ is $\hat{x}(K+1) = R(K+1)^{-1}b(K+1)$. The QR triangularization can also be set up to obtain the optimal filtered state vector in a fashion similar to that for the predictor.

Kung and Hwang [9] observed that by performing the nullification of the elements of the $\tilde{C}$ and $\tilde{F}$ matrices in a particular order, the W matrix remains triangular. Specifically, the matrix $\tilde{C}$ is first zeroed, and then the zeroing of the matrix $\tilde{F}$ starts in the lower left component and proceeds up each column before starting at the bottom of the next column. The design of the proposed array was based partly on the use of dependence graphs and partly on heuristic observations. The Lincoln and Yao [7] paper used the dependence graph mapping methodology of Rao [8] to derive, in a completely systematic manner, a systolic array Kalman filter. In addition to providing an alternative approach in deriving the Kung–Hwang array, a new array is derived which has a greater speed-up factor and processor utilization. Block diagrams of the Kung–Hwang array and the Lincoln–Yao are given in Fig. 16.1. The detailed operations of these two arrays can be found in their papers. The array proposed by

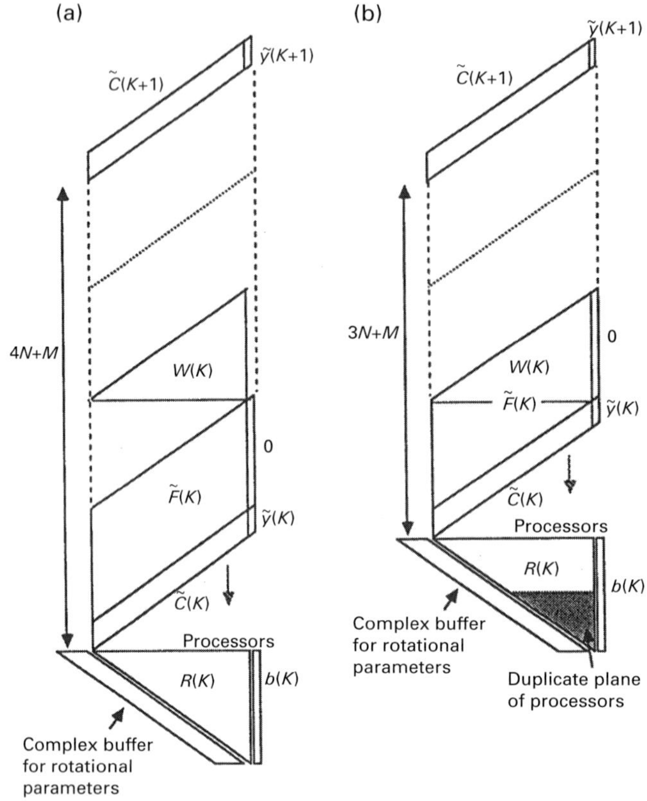

Figure 16.1 (a) Kung–Hwang systolic Kalman filter and (b) Lincoln–Yao systolic Kalman filter

Gaston and Irwin [10] is also derived using dependence mapping techniques. Comparisons of the advantages and the unique features of the Chen–Yao array, the Kung–Hwang array, the Gaston–Irwin array, and the Lincoln–Yao array are summarized in Table 16.1. Kung and Hwang [9] observed that by performing the nullification of the elements of the $\tilde{C}$ and $\tilde{F}$ matrices in a particular order, the W matrix remains triangular. Specifically, the matrix $\tilde{C}$ is the first zero, and then the zeroing of the matrix $\tilde{F}$ starts in the lower left component and proceeds up each column before starting at the bottom of the next column. The design of their proposed array was based partly on the use of dependence graphs and partly on heuristic observations. In addition to providing an alternative approach in deriving the Kung–Hwang array, a new array is derived which has a greater speed-up factor and processor utilization. Block diagrams of the Kung–Hwang array and the Lincoln–Yao array are given in Fig. 16.1. The detailed operations of these two arrays can be found in their papers. The array proposed by Gaston and Irwin [10] is also derived using dependence mapping techniques. In Table 16.1, comparisons of the

Table 16.1 Comparisons of systolic array Kalman filters

Array	# Processors	Recursion	Avg. PE utilization		Particular utilization
		Steps	$N \gg M$	$N = M \gg 1$	$N \gg M$
Chen–Yao	$(N/2)(3N + 5)$	$2N + M$	1/3	1/3	Not comparable
Kung–Hwang	$(N/2)(N + 3)$	$4N + M$	1/2	3/5	1/2
Gaston–Irwin (PEs for stage two only)	$N(N + 1)$	$3N + M$	1/3	3/8	1/3 to 2/3 0 to 1/3
Lincoln–Yao (PEs for stage one or two only)	$(N/8)(5N + 13)$	$3N + M$	8/15	3/5	2/3 1/3

advantages and unique features of the Chen–Yao array, the Kung–Hwang array, the Gaston–Irwin array and the Lincoln–Yao array, are summarized. Column 1 lists the four arrays of interest. Column 2 shows the number of processors utilized in these arrays. Column 3 provides the number of recursion steps. Column 4 shows the average PE utilization defined as the ratio of processing power used to the available processing power for the two cases of $N \gg M$ and $N = M \gg 1$. This ratio gives the percentage of the time that an average processor is performing algorithm computations. Column 5 shows the processor utilization of a particular array, which can be determined using the dependence graph mapping specification. The allocation function specifies the index space points assigned to a particular processor. The ratio of the number of these indices to the recursion steps gives the particular processor utilization for the given processor. Detailed discussions of the results presented in Table 16.1 are given in [7].

16.3 MIMO Receiver

In recent years, multi-input and multi-output (MIMO) technology has been shown to be a great improvement in spectral efficiency and achieve higher channel capacity in fading communication environments [11]. However, the system benefits from MIMO at the expense of dramatically increasing computation complexity at the detector. One approach is to develop complexity efficient algorithms for MIMO detection and also to apply systolic array technology on the detection algorithms in order to increase the throughput rate.

We consider a MIMO system with m transmit and n receive antennas in a rich-scattering flat-fading channel. Let $\mathbf{x}$ be the transmitted M-QAM signal vector, $\mathbf{y}$ the received signal vector, and $\mathbf{H}$ the channel matrix where the entries are uncorrelated, zero-mean, unit-variance complex Gaussian fading gains. The baseband model for this MIMO system is

$$\mathbf{y} = \mathbf{H}\mathbf{x} + \mathbf{n}. \tag{16.11}$$

where $\mathbf{n}$ is the white Gaussian noise vector. Additionally, we assume the channel matrix entries are fixed during each frame interval, and the receiver has perfect knowledge of the realization of $\mathbf{H}$.

In maximum likelihood detection, we seek the solution of

$$\hat{\mathbf{x}}_{ML} = \arg \min_{\mathbf{x} \in S^m} \|\mathbf{y} - \mathbf{H}\mathbf{x}\|, \tag{16.12}$$

where $\in S^m$ is in the signal set. By exhaustive search, the computation complexity of ML is $O(m^n)$. Zero-forcing detection and minimum mean-square-error (MMSE) detection are viable sub-optimum detection schemes.

16.3.1 Systolic MIMO Receiver

A complex lattice L contains a set of complex basis $\mathbf{h}_1, \mathbf{h}_2, \ldots, \mathbf{h}_m \in C^n$, and any set of points in the lattice can be expressed as a linear combination of these basis vectors. The purpose of lattice reduction is to select good basis vectors such that these vectors are reasonably orthogonal to each other and as short as possible. The Lenstra–Lenstra–Lovász (LLL) algorithm is a widely used polynomial time lattice reduction algorithm [15]. The objective of the lattice reduction algorithm is to derive a better-conditioned matrix $\tilde{\mathbf{H}}$ along with a unimodular matrix $\mathbf{T}$ from the original channel matrix $\mathbf{H}$ under a given criterion such that $\tilde{\mathbf{H}} = \mathbf{H}\mathbf{T}$ [12]. Linear lattice-reduction-aided detection (LRAD) combines lattice reduction algorithm with the linear detection, such as ZF and MMSE. Consider ZF first, and the estimated signal $\hat{\mathbf{x}}$ can be written as

$$\hat{\mathbf{x}} = \tilde{\mathbf{H}}^{\dagger}\mathbf{y} = \tilde{\mathbf{H}}^{\dagger}\left((\mathbf{H}\mathbf{T})(\mathbf{T}^{-1}\mathbf{x}) + \mathbf{n}\right) = \mathbf{T}^{-1}\mathbf{x} + \tilde{\mathbf{H}}^{\dagger}\mathbf{n}. \tag{16.13}$$

Let $\hat{\mathbf{x}}_q$ be a version of the $\hat{\mathbf{x}}$ quantized element. From (16.12), it is clear that $\hat{\mathbf{x}}_q$ is an estimate of $\mathbf{T}^{-\mathbf{1}}\mathbf{x}$, rather than of $\mathbf{x}$. Hence, the last step is to transform $\hat{\mathbf{x}}_q$ back into an estimate of $\mathbf{x}$. (16.12) also applies to MMSE detection if the extended system model is considered. There are three steps for an LRAD detector. There is a QRD, followed by a modified LLL algorithm, and then a linear detector, N. Wang [12] has shown an efficient systolic algorithm and associated architecture for the implementation of this LRAD MIMO detector. Details of this problem are omitted here but can be found in [12].

16.4 Systolic Algorithm for SVD Updating

In Chapter 15, we have seen that a high throughput clutter filter can be implemented using systolic algorithms and architectures for a QRD approach to the solution of the over-determined linear system of equations. In high performance radar and aerospace applications, real-time SVD is known to be useful for parameter estimation and rank determination in source classification. Now, we introduce a systolic algorithm for SVD updating.

We consider the updating algorithm for singular value decomposition (SVD) based on Jacobi rotations [14]. Ideally, an infinite number of rotations are required for the exact computation of the SVD. In practice, only a finite number of *sweeps* are used. Two questions arise: Is it possible to input new data for update *before* the SVD has converged to a (quasi) diagonal form? If so, is it possible to update after only a fraction of a sweep has been performed? We show that it is indeed possible to update after only a finite number of rotations. Since high update rates pose demanding hardware requirements, what would justify a fast update rate (after fractions of a sweep) is a higher tracking capability. In environments in which data are almost stationary, a high update rate is logically not needed. We analytically show that the SVD of the full data matrix and the SVD of the matrix obtained by selecting a subset of equally spaced rows are actually very closely related. The question remains whether for highly non-stationary input data there is any advantage in increasing the update rate so that only a fraction of a sweep is performed between consecutive updates. An approximate analysis seems to imply that the tracking error is dependent on the non-stationarity of the input data and not on the number of sweeps (or fractions thereof) executed between updates. Simulations of a direction of arrival (DOA) estimation problem also show that the tracking performance does not seem to depend on the update rate. Indeed, slower updating algorithms may sometimes provide better estimates. Details on this problem are omitted here but can be found in [14].

16.5 Conclusion

Section 16.2 introduced the Kalman filter. Subsection 16.2.1 discussed many versions of systolic Kalman filter. Section 16.3 considered the MIMO receiver problem and Subsection 16.3.1 treated the systolic version of that problem. Section 16.4 introduced the SVD updating problem.

16.6 References

The Kalman filter problem was proposed in [1] and extensive theoretical and practical publications have appeared on this problem. Some earlier versions of parallel and systolic Kalman filtering algorithms have appeared in [2], [3], [4], [5], and [6]. Systolic algorithms and architectures for Kalman filtering included: [7], [8], [9], and [10]. An introduction to MIMO receiver was treated in [11]. The LLL algorithm was introduced in [15] and a systolic architecture for a lattice-reduced MIMO receiver was given by [12]. An earlier paper of Jacobi relevant to the Jacobi SVD problem appeared in [13]. A Jacobi SVD updating problem was addressed in [14].

[1] R.E. Kalman, "A New Approach to Linear Filtering and Prediction Problems," *J. Basic Engineering*, March 1960, vol. 82, pp. 35–45.
[2] P.G. Kaminski, "Discrete Square Root Filtering: A Survey of Current Techniques," *IEEE Tr. Autom. Control*, 1971, vol. 16, pp. 727–735.
[3] C.C. Paige and M.A. Saunders, "Least Squares Estimation of Discrete Linear Dynamic Systems Using Orthogonal Transformation," *SIAM J. Numerical Analysis*, 1977, vol. 4, pp. 180–193.
[4] J.M. Jover and T. Kailath, "A Parallel Architecture for Kalman Filter Measurement Update and Parameter Estimation," *Automatica*, 1986, vol. 22, pp. 43–57.
[5] M.J. Chen and K. Yao, "On Realizations of Least-Squares Estimation and Kalman Filtering by Systolic Array," *Proc. 1st Inter. Workshop on Systolic Arrays*, ed. W. Moore, Adam Hilger, 1986, pp. 161–170.
[6] M.J. Chen and K. Yao, "Systolic Kalman Filtering Based on QR Decomposition," *Proc. SPIE*, 1987, pp. 25–32.
[7] R.A. Lincoln and K. Yao, "Efficient Systolic Kalman Filtering Design by Dependence Graph Mapping," *Proceedings of the VLSI Signal Processing III*, 1988, pp. 396–407.
[8] S.K. Rao, "Regular Iterative Algorithms and Their Implementations on Processor Arrays," Ph.D. thesis, Stanford University, Oct. 1985.
[9] S.Y. Kung and J.N. Hwang, "Systolic Array Designs for Kalman Filtering," *IEEE Trans. Signal Processing*, 1991, pp. 171–182.
[10] F.M.F. Gaston and G.W. Irwin, "A Systolic Square Root Information Kalman Filter," *Proc. of International Conference on Systolic Arrays*, 1988, pp. 643–656.
[11] E. Biglieri, R. Calderbank, A. Constantinides, A. Goldsmith, A. Paulraj, and H. V. Poor, *MIMO Wireless Communications*, Cambridge University Press, 2007, Ch. 1.
[12] N.C. Wang, E. Biglieri, and K. Yao, "Systolic Arrays for Lattice-Reduction-Aided MIMO Detection," *Journal of Communications and Networks*, 2011, pp. 481–493.
[13] C.G.J. Jacobi, "Über ein leichtes Verfahren die in der Theorie der Säcularstörungen vorkommende Gleichungen numerisch aufzulösen," *J. Reine Angew. Math.*, 1846, 30, pp. 51–94.
[14] F. Lorenzelli and K. Yao, "Effect of Update Rate on Tracking Error for Jacobi SVD Arrays," *UCLA Report No. Eng-93-49*, Oct. 1993, pp. 1–16.
[15] A.K. Lenstra, H.W. Lenstra, and L. Lovász, "Factoring Polynomials with Rational Coefficients," *Math. Ann*, 1982, pp. 515–534.

16.7 Exercises

1. While the least-squares solutions of a linear system of equations are needed in many scientific and engineering problems, Charles M. Rader of the MIT Lincoln Lab was the first to describe in some detail the use of a triangular systolic array for solving the side-lobe canceling radar problem. Read the paper, C.M. Rader, "Wafer-Scale Integration of a Large Systolic Array for Adaptive Nulling," *IEEE Signal Processing Magazine*, 1996, pp. 29–49.
2. Compare the systolic structures and throughput rates of the various systolic Kalman filters by Chen–Yao, Kung–Hwang, Gaston–Irwin, and Lincoln–Yao in the reference section of Chapter 16.

3. In this problem, all the vectors $\mathbf{a}$, $\mathbf{a_i}$, $\mathbf{s}$, $\mathbf{I}$, $\mathbf{y}$ are all m-dimensional column vectors. The main sensor $\mathbf{y} = \mathbf{s} + \mathbf{I}$, where $\mathbf{s}$ is the signal and $\mathbf{I}$ is some unwanted interference.

 a. Consider $n = 1$ sensor and $\mathbf{a}_1 = \alpha_1 \mathbf{I}$. Denote $\mathbf{e} = \mathbf{y} - \mathbf{a}_1 x$. Find the optimum $\hat{x}_1$ that minimizes $||\mathbf{e}||^2$. Find this $||\mathbf{e}||^2$.

 b. Suppose there are K interferers $\mathbf{I}_1, \ldots, \mathbf{I}_K$. Let $\mathbf{y} = \mathbf{s} + \mathbf{I}_1 + \cdots + \mathbf{I}_K$ and $\mathbf{a_i} = \sum_{k=1}^{K} \alpha_{ki} \mathbf{I}_k$, $i = 1, \ldots, n$. What is the minimum number of sensors, n, that is needed to be used to cancel all the interference effects in $\mathbf{y}$? Find the optimum $\hat{\mathbf{x}} = \begin{bmatrix} \hat{x}_1 \\ \vdots \\ \hat{x}_n \end{bmatrix}$ and the $\min_{\mathbf{x}} ||\mathbf{e}||^2$.

4. Consider the least-squares solution of $\mathbf{Ax} = \mathbf{b}$, where $\mathbf{A}$ is an $m \times n$ matrix, $\mathbf{b}$ is an $m \times 1$ vector, and $\mathbf{x}$ is an $n \times 1$ vector. The optimum residual and the least-squares solution $\hat{\mathbf{x}}$ can be obtained using an $n \times n$ triangular systolic array, where $\mathbf{b}$ is fed into the right-most column of the triangular array. Now, let $\mathbf{b}$ be an $m \times p$ matrix. Show the optimum residual, and the $n \times p$ least-squares solution of $\hat{\mathbf{x}}$ also can be obtained by appending an $n \times (p - 1)$ matrix to the right-most column of the triangular array.

Index